Agricultural Resilience

Perspectives from Ecology and Economics

Agriculture as a social–ecological system embraces many disciplines. This book breaks through the silos of individual disciplines to bring ecologists and economists together to consider agriculture through the lens of resilience. It explores the economic, environmental and social uncertainties that influence the behaviour of agricultural producers and their subsequent farming approach, highlighting the importance of adaptability, innovation and capital reserves in enabling agriculture to persist under climate change and market volatility. The resilience concept and its relation to complexity theory are explained and the characteristics that foster resilience in agricultural systems, including the role of biodiversity and ecosystem services, are explored. The book discusses modelling tools, metrics and approaches for assessing agricultural resilience, highlighting areas where interdisciplinary thinking can enhance the development of resilience. It is suitable for those researching sustainable agriculture or those engaged in agricultural policy decisions and analysis, as well as students of ecology, agriculture, social science and economics.

SARAH M. GARDNER works at the interface of ecology and environmental economics. She has worked with agricultural policymakers, land managers and farmers for over 20 years as adviser, researcher and lecturer. Her current work at GardnerLoboAssociates involves the design of data management systems for the livestock sector.

STEPHEN J. RAMSDEN is Associate Professor in Farm Business Management at the University of Nottingham, UK. He has taught Agriculture for 25 years and worked on a wide range of research, with a central focus on farmers as decision-makers.

ROSEMARY S. HAILS is Director of Nature & Science at the National Trust, UK, and formerly Director of Biodiversity and Ecosystem Science at the Centre for Ecology & Hydrology. She leads the coordination of the Valuing Nature Programme (funded across the UK Research Councils), is Chair of the Advisory Committee on Releases to the Environment (ACRE) and a member of the Natural Environment Research Council (NERC) Science Board.

Ecological Reviews

SERIES EDITOR Philip Warren *University of Sheffield, UK*
SERIES EDITORIAL BOARD
Andrea Baier *British Ecological Society, UK*
Mark Bradford *Yale University, USA*
David Burslem *University of Aberdeen, UK*
Alan Gray *CEH Wallingford, UK*
Kate Harrison *British Ecological Society, UK*
Sue Hartley *University of York, UK*
Mark Hunter *University of Michigan, USA*
Ivette Perfecto *University of Michigan, USA*
Heikki Setala *University of Helsinki, Finland*

Ecological Reviews publishes books at the cutting edge of modern ecology, providing a forum for volumes that discuss topics that are focal points of current activity and likely long-term importance to the progress of the field. The series is an invaluable source of ideas and inspiration for ecologists at all levels from graduate students to more-established researchers and professionals. The series has been developed jointly by the British Ecological Society and Cambridge University Press and encompasses the Society's Symposia as appropriate.

Biotic Interactions in the Tropics: Their Role in the Maintenance of Species Diversity
Edited by David F. R. P. Burslem, Michelle A. Pinard and Sue E. Hartley

Biological Diversity and Function in Soils
Edited by Richard Bardgett, Michael Usher and David Hopkins

Island Colonization: The Origin and Development of Island Communities
By Ian Thornton
Edited by Tim New

Scaling Biodiversity
Edited by David Storch, Pablo Margnet and James Brown

Body Size: The Structure and Function of Aquatic Ecosystems
Edited by Alan G. Hildrew, David G. Raffaelli and Ronni Edmonds-Brown

Speciation and Patterns of Diversity
Edited by Roger Butlin, Jon Bridle and Dolph Schluter

Ecology of Industrial Pollution
Edited by Lesley C. Batty and Kevin B. Hallberg

Ecosystem Ecology: A New Synthesis
Edited by David G. Raffaelli and Christopher L. J. Frid

Urban Ecology
Edited by Kevin J. Gaston

The Ecology of Plant Secondary Metabolites: From Genes to Global Processes
Edited by Glenn R. Iason, Marcel Dicke and Susan E. Hartley

Birds and Habitat: Relationships in Changing Landscapes
Edited by Robert J. Fuller

Trait-Mediated Indirect Interactions: Ecological and Evolutionary Perspectives
Edited by Takayuki Ohgushi, Oswald Schmitz and Robert D. Holt

Forests and Global Change
Edited by David A. Coomes, David F. R. P. Burslem and William D. Simonson

Trophic Ecology: Bottom-Up and Top-Down Interactions across Aquatic and Terrestrial Systems
Edited by Torrance C. Hanley and Kimberly J. La Pierre

Conflicts in Conservation: Navigating Towards Solutions
Edited by Stephen M. Redpath, R. J Gutiérrez, Kevin A. Wood and Juliette C. Young

Peatland Restoration and Ecosystem Services
Edited by Aletta Bonn, Tim Allott, Martin Evans, Hans Joosten and Rob Stoneman

Rewilding
Edited by Nathalie Pettorelli, Sarah M. Durant and Johan T. du Toit

Grasslands and Climate Change
Edited by David J. Gibson and Jonathan A. Newman

Agricultural Resilience

Perspectives from Ecology and Economics

Edited by

SARAH M. GARDNER
GardnerLoboAssociates

STEPHEN J. RAMSDEN
University of Nottingham

ROSEMARY S. HAILS
Centre for Ecology & Hydrology

CAMBRIDGE
UNIVERSITY PRESS

University Printing House, Cambridge CB2 8BS, United Kingdom

One Liberty Plaza, 20th Floor, New York, NY 10006, USA

477 Williamstown Road, Port Melbourne, VIC 3207, Australia

314–321, 3rd Floor, Plot 3, Splendor Forum, Jasola District Centre,
New Delhi – 110025, India

79 Anson Road, #06–04/06, Singapore 079906

Cambridge University Press is part of the University of Cambridge.

It furthers the University's mission by disseminating knowledge in the pursuit of education, learning, and research at the highest international levels of excellence.

www.cambridge.org
Information on this title: www.cambridge.org/9781107067622
DOI: 10.1017/9781107705555

© British Ecological Society 2019

This publication is in copyright. Subject to statutory exception
and to the provisions of relevant collective licensing agreements,
no reproduction of any part may take place without the written
permission of Cambridge University Press.

First published 2019

Printed in the United Kingdom by TJ International Ltd, Padstow Cornwall

A catalogue record for this publication is available from the British Library.

ISBN 978-1-107-06762-2 Hardback
ISBN 978-1-107-66587-3 Paperback

Cambridge University Press has no responsibility for the persistence or accuracy of URLs for external or third-party internet websites referred to in this publication and does not guarantee that any content on such websites is, or will remain, accurate or appropriate.

Contents

List of Contributors	page ix
Preface and acknowledgements	xiii
1. Introducing resilience *Sarah M. Gardner and Stephen J. Ramsden*	1
PART I: BIODIVERSITY, ECOSYSTEM SERVICES AND RESILIENCE IN AGRICULTURAL SYSTEMS	11
2. Complexity and resilience in agriculture *Sarah M. Gardner*	13
3. Biodiversity and agriculture *David Tilman*	39
4. Determining the value of ecosystem services in agriculture *Rosemary S. Hails, Rebecca Chaplin-Kramer, Elena Bennett, Brian Robinson, Gretchen Daily, Kate Brauman and Paul West*	60
5. Resilience in agricultural systems *Stephen J. Ramsden and James Gibbons*	90
6. Building resilience into agricultural pollination using wild pollinators *Neal M. Williams, Rufus Isaacs, Eric Lonsdorf, Rachael Winfree and Taylor H. Ricketts*	109
7. Conflicts and challenges to enhancing the resilience of small-scale farmers in developing economies *Richard Ewbank*	135
8. Modern biotechnology and sustainable intensification: chances and limitations *Rolf Meyer*	159
9. Pastoralism, conservation and resilience: causes and consequences of pastoralist household decision-making *Katherine Homewood, Marcus Rowcliffe, Jan de Leeuw, Mohamed Y. Said and Aidan Keane*	180

PART II: INTEGRATING BIODIVERSITY AND BUILDING RESILIENCE INTO AGRICULTURAL SYSTEMS — 209

10. Delivering sustainability in agricultural systems: some implications for institutional analysis — 211
    Ian Hodge
11. The resilience of Australian agricultural landscapes characterised by land-sparing versus land-sharing — 232
    David J. Abson, Kate Sherren and Joern Fischer
12. Ecological–economic modelling for designing cost-effective incentives to conserve farmland biodiversity — 253
    Martin Drechsler and Frank Wätzold
13. Viability analysis as an approach for assessing the resilience of agroecosystems — 273
    Sophie Martin
14. Integrating economics and Resilience Thinking: the context of natural resource management in Australia — 295
    Michael Harris, Graham R. Marshall and David J. Pannell
15. Integrating biodiversity and ecosystem services into European agricultural policy: a challenge for the Common Agricultural Policy — 315
    Allan Buckwell
16. Using environmental metrics to promote sustainability and resilience in agriculture — 340
    Jonathan R. B. Fisher and Peter Kareiva
17. Conclusions on agricultural resilience — 362
    Sarah M. Gardner, Stephen J. Ramsden and Rosemary S. Hails

Index — 380

Colour plates can be found between pages 208 and 209.

Contributors

DAVID J. ABSON
Future of Ecosystem Services
(FuturES) Research Center
Leuphana University
Lüneburg
Germany
abson@uni-leuphana.de

ELENA BENNETT
School of Environment and
Department of Natural
Resource Sciences
McGill University
Ste. Anne de Bellevue
Québec
Canada
elena.bennett@mcgill.ca

KATE BRAUMAN
Institute on the
Environment
University of Minnesota
Saint Paul, Minnesota
USA
kbrauman@umn.edu

ALLAN BUCKWELL
Institute for European
Environmental Policy
London
UK
Abuckwell@ieep.eu

REBECCA CHAPLIN-KRAMER
Natural Capital Project
Woods Institute for the Environment
Stanford University
Stanford, California
USA
bchaplin@stanford.edu

GRETCHEN DAILY
Natural Capital Project
Stanford University
Stanford, California
USA
gdaily@stanford.edu

MARTIN DRECHSLER
Department of Ecological
Modelling
Helmholtz Centre for
Environmental Research–UFZ
Leipzig
Germany
martin.drechsler@ufz.de

RICHARD EWBANK
Programme Performance &
Learning (PPL)
Christian Aid London
UK
REwbank@christianaid.org

JOERN FISCHER
Leuphana University Lüneburg

Germany
joern.fischer@uni.leuphana.de

JONATHAN R. B. FISHER
The Nature Conservancy (Central Science)
Arlington, Virginia
USA
jon_fisher@tnc.org

SARAH M. GARDNER
GardnerLoboAssociates
Windsor
UK
sarah.gardnerlobo@gmail.com

JAMES GIBBONS
School of Environment, Natural Resources and Geography
Bangor University
Bangor, Gwynedd
UK
j.gibbons@bangor.ac.uk

ROSEMARY S. HAILS
The National Trust
Swindon
UK
Rosie.Hails@nationaltrust.org.uk

MICHAEL HARRIS
Environmental Protection Authority
Sydney South, New South Wales
Australia
michael.harris@epa.nsw.gov.au

IAN HODGE
Department of Land Economy
University of Cambridge
Cambridge
UK
idh3@cam.ac.uk

KATHERINE HOMEWOOD
Department of Anthropology
University College London
UK
k.homewood@ucl.ac.uk

RUFUS ISAACS
Michigan State University
East Lansing, Michigan
USA
isaacsr@msu.edu

PETER KAREIVA
Institute of the Environment and Sustainability
University of California,
Los Angeles, California
USA
pkareiva@ioes.ucla.edu

AIDAN KEANE
School of Geosciences
University of Edinburgh
Edinburgh
UK
aidan.keane@ed.ac.uk

JAN DE LEEUW
ISRIC World Soil Information
Wageningen
The Netherlands
jan.deleeuw@wur.nl

ERIC LONSDORF
Department of Biology
Franklin & Marshall College
Lancaster, Pennsylvania
USA
eric.lonsdorf@fandm.edu

GRAHAM R. MARSHALL
The Institute for Rural Futures
University of New England
Armidale, New South Wales
Australia
gmarshal@une.edu.au

SOPHIE MARTIN
Institut national de recherche en sciences et technologies pour l'environnement et l'agriculture
Aubière
France
sophie.martin@irstea.fr

ROLF MEYER
Institute for Technology Assessment and Systems Analysis
Karlsruhe Institute of Technology
Karlsruhe
Germany
rolf.meyer@kit.edu

DAVID J. PANNELL
Centre for Environmental Economics and Policy
The University of Western Australia
Crawley
Australia
david.pannell@uwa.edu.au

STEPHEN J. RAMSDEN
School of Biosciences
The University of Nottingham
Sutton Bonington
UK
stephen.ramsden@nottingham.ac.uk

TAYLOR H. RICKETTS
Gund Institute for Ecological Economics
University of Vermont Burlington, Vermont
USA
Taylor.Ricketts@uvm.edur

BRIAN ROBINSON
Department of Geography
McGill University
Montréal, Québec
Canada
brian.e.robinson@mcgill.ca

MARCUS ROWCLIFFE
Institute of Zoology
Zoological Society of London
London
UK
marcus.rowcliffe@ioz.ac.uk

MOHAMED Y. SAID
Institute for Climate Change and Adaptation (ICCA)
University of Nairobi
Nairobi
Kenya
msaid362@gmail.com

KATE SHERREN
School for Resource and Environmental Studies
Dalhousie University
Halifax, Nova Scotia
Canada
kate.sherren@dal.ca

DAVID TILMAN
Department of Ecology, Evolution and Behavior
University of Minnesota
St. Paul, Minnesota
USA
tilman@umn.edu

FRANK WÄTZOLD
Brandenburg University of Technology Cottbus
Germany
waetzold@tu-cottbus.de

PAUL WEST
Institute on the Environment
University of Minnesota
Saint Paul, Minnesota
USA
pcwest@umn.edu

NEAL M. WILLIAMS
Department of Entomology
University of California
Davis, California
USA
nmwilliams@ucdavis.edu

RACHAEL WINFREE
Department of Ecology, Evolution
and Natural Resources
Rutgers University
New Brunswick, New Jersey
USA
rwinfree@rutgers.edu

Preface and acknowledgements

Agriculture is facing a time of significant change and uncertainty. In many countries, increased productivity, driven by technical developments during the twentieth century, has led to agricultural systems that are productive, but often very specialised, while reforms to agricultural policy have reduced the amount of protection provided by the state. Climate change is already increasing the probability of adverse weather-related shocks to farm systems and the likelihood of extreme events and their severity are predicted to increase. Concerns have been raised that modern agricultural systems are therefore vulnerable to increased levels of uncertainty across a range of environments – natural, market and political. Moreover, farmers are increasingly expected to produce food and other farm products with a lower environmental footprint. Faced with these demands, within increasingly uncertain external environments, how equipped are current agricultural systems to adapt and maintain production for a growing population? Do modern 'efficient' farm systems have sufficient buffering capacity to withstand future uncertainty and shocks? This book considers the adaptability of agriculture through the lens of *resilience*. It explores the economic, environmental and social uncertainties that influence the behaviour of agricultural producers and their subsequent choice of farming approach and considers the factors that foster resilience in agricultural systems, especially the role of biodiversity and ecosystem services.

The book arose from discussions and ideas that emerged during three interdisciplinary meetings organised by the Natural Capital Initiative, the British Ecological Society and the UK Agricultural Economics Society in 2012 and 2015. These meetings were designed to promote dialogue between ecologists and economists on the sustainable management of biodiversity and ecosystems services within agriculture and several of the themes developed in this book were highlighted at these meetings. Agriculture is a social–ecological system that embraces many disciplines. To reflect this, we have deliberately invited a diverse range of authors, including those with non-ecological backgrounds. As editors we were interested in determining whether the resilience concept might offer new insights on how agricultural systems should develop in the future, on the role that biodiversity and ecosystems services might play in enhancing the adaptive capacity of agriculture and on approaches that place greater emphasis on natural capital assets, such as soil, within agricultural production systems. In presenting a cross-disciplinary view of resilience, our aim was to consider the broad range of factors that influence decision-making in agriculture, to expose readers to the approaches and thinking adopted by different disciplines, e.g. to tackle questions of risk and sustainability in agriculture; and to identify areas where agriculture and the environment might benefit from future interdisciplinary collaboration.

The book is divided into two parts. After a brief review of the resilience concept, Part I sets out the context within which agriculture operates and its relationship with the global food system, before considering how this context influences the capacity of agricultural producers to adapt to change. The main chapters in this section focus on the contribution of biodiversity and ecosystem services to fostering resilience in agricultural systems and examine approaches for valuing and assessing this contribution. Part II considers different approaches for integrating biodiversity and ecosystem services into agricultural systems and sets out some of the challenges facing producers and policy-makers for building resilience in present-day agricultural systems. The book concludes by drawing together the lessons learned from the different economic and ecological approaches set out in each chapter and highlighting areas where research and policy development are needed to build agricultural resilience for the twenty-first century.

In examining the studies presented in this volume, we are reminded that resilience is an emergent property of complex adaptive systems which arises from the interactions and behaviour of many, diverse and independent, individual agents. It is the self-organising behaviour of these agents that facilitates adaptation and the emergence of resilience in systems. Within agriculture, the self-organising agents are the agricultural producers that build agricultural systems and the biological species that 'build' the agroecosystems and ecosystem services on which agriculture is founded. The dependence of agriculture on natural capital and ecosystem services highlights the importance of assigning 'values' to these assets to ensure that they are accounted for in agricultural production systems. Development of valuation approaches for natural capital is a flourishing area of dialogue between ecologists and economists. A more challenging area for dialogue focuses on exploring ways of balancing efficiency and resilience in agricultural production systems. These properties, while apparently promoting diametrically opposed system requirements – efficiency encourages uniformity among agents while resilience requires diversity – are both needed for growth and adaptive capacity in agricultural systems. Differences in economic and ecological approaches to this problem can lead to divergent thinking and the implementation of trade-offs between agricultural production and ecosystem services, where in fact convergence and integration of objectives for agriculture and natural capital assets is needed – focusing, for example, on the multifunctional nature of agriculture, the value of the different goods arising from it and the maintenance of reserves of all types of capital. This is a critical area for discussion, argument and innovative thinking between ecologists and economists as the uncertainties surrounding agriculture and food production increase through the twenty-first century. Our outlook from this book is cautiously optimistic. Integrating resilience into agricultural systems is possible but requires policies and production paths that promote autonomy and heterogeneity and opportunities for self-organisation among agricultural producers and that incentivise the development of reserves of all types of capital. When viewed as complex adaptive systems, agriculture and the agroecosystems that underpin it possess many pathways to resilience. The key to these pathways is the multi-dimensional ecological, economic and social context within which agriculture operates. This context varies globally, both within and between nations. It is this variation that is the source of future adaptive ability for agriculture.

In presenting this volume, we are greatly indebted to all the authors of the chapters and to all the people who reviewed them. We especially thank Jenny van der Meijden at Cambridge University Press who has supported and guided us through the publishing process, the Editorial Board of Ecological Reviews and many friends and family who have supported us in the production of this volume. We hope that together we have been able to present a better understanding of the parameters that enhance and constrain resilience in agriculture and its co-existence with nature.

CHAPTER ONE

Introducing resilience

SARAH M. GARDNER
GardnerLoboAssociates
STEPHEN J. RAMSDEN
University of Nottingham

The power of the unexpected
It is a common experience that the best laid plans can be thwarted by unexpected events: events that alter the rate or direction of an expected path of development. Such events may originate from seemingly unrelated sources. Shutting down oil rigs on the US Gulf Coast in the wake of Hurricane Katrina in 2005 drives a spike in Mexican corn prices, which leads in turn to food riots (Zolli & Healy, 2012); a highly 'improbable' event such as the explosion at the Chernobyl nuclear plant changes lamb markets in Wales for over 20 years. Sometimes disruption arises from an apparently insignificant change in a system's internal processes; for example, a change in feed quality triggering financial losses on specialised livestock farms. In highly connected systems, unexpected events can have a range of negative consequences. In the case of agriculture, these consequences can be severe if certain limits or 'thresholds' are crossed.

Modern agriculture is part of a complex, global system of food production that is driven primarily by economic, environmental and socio-cultural forces largely outside of the control of farmers. These external forces have been responsible for most of the events that have triggered recent concerns over food security and the sustainability of current agricultural production systems. Volatility in commodity prices, extreme weather, unexpected policy or political change (e.g. the exit of the UK from the European Union), social unrest (e.g. concerns over immigration) all heighten uncertainty over the viability and future direction of agricultural production, practice and management. Such events can stimulate innovation in production methods, new technologies and products, but can also result in production failure, environmental degradation and land abandonment.

How then can we manage the impacts of the unexpected on agriculture and the wider food production system? Which systems are vulnerable and what can be done to buffer systems against negative outcomes? Such questions are the focus for studies of resilience, a topic that touches many areas of human life and which considers the ability of systems to persist in the face of uncertainty and change. In this book, we examine the concept of resilience with respect to the current and future functioning of agricultural systems.

The evolution of the resilience concept

The term 'resilience' is used in many disciplines including engineering, ecology, economics, psychology, international development and social sciences. It refers to the property of a system or a system state. The resilience concept has developed significantly since the mid-1990s (Folke, 2006), although its definition and interpretation varies between disciplines (Martin-Breen & Anderies, 2011).

Most frequently, a resilient system is described as one that 'returns to its reference state (or dynamic) after a temporary disturbance' (Grimm & Wissel, 1997). This definition, referred to as 'engineering' resilience (Holling & Gunderson, 2002), implies that a resilient system maintains a particular set of characteristics and conditions which together constitute the stable state for that system. If disturbed, the system may be moved away from this stable reference state, but will reliably return to it over time. An example of such a system is a pendulum on a clock, which fluctuates around a single stable equilibrium point. For 'engineering resilience' the time taken for the system to return to its stable reference state is the measure of the system's resilience (Folke, 2006).

The concept of the 'reference state' has prompted researchers to focus on equilibrium-based analyses to identify the conditions under which a system is stable (May, 1974). These analyses use differential or integral equations to determine the rate of system change in response to changing input conditions (Grimm & Calabrese, 2011). Such equilibrium-centred approaches are frequently used to identify the limits within which natural resources – fish, timber, grazing land – can be harvested without risk of significant system change or population collapse.

For ecologists, the concept of a reference state is challenging because ecological systems typically exist in a number of states and orientate around multiple equilibria (Hastings, 2013): for example, lands such as the semi-arid Chaco of western Argentina which consists of a mosaic of grassland, thorny scrub and open hardwood forest. In unmanaged semi-arid Chaco, the balance between these habitats is maintained by periodic wildfires which burn the accumulated biomass of litter and scrub and trigger the regeneration of grassland (Bucher, 1987; Cabido et al., 1994). This succession of grass, scrub and woodland can be viewed as a set of states through which the semi-arid Chaco system transits, the external agent – fire – disrupting the cycle but rejuvenating it by resetting it to an earlier state. Thus, within a specific area, the temporal and spatial pattern of the vegetation will change following a fire, but the overall composition and the relationships that determine system function can continue. There is no single equilibrium state for this ecosystem because the system is constantly changing.

Holling (1973) was the first to recognise the emergence of multi-stable states in ecological systems and their implications for ecosystem dynamics and natural resource management. He viewed ecosystems as complex adaptive systems with dynamics driven by non-linear processes (e.g. spatial heterogeneity) and feedback loops that allow the system to self-organise (Levin, 1998). Holling recognised that high variability was an essential attribute for maintaining system existence and that 'surprise and inherent unpredictability was the inevitable consequence for (the dynamics of) ecological systems' (Holling, 2003, reported in Folke, 2006). He further postulated that for ecosystems, persistence of relationships and entities and a capacity to absorb the effects of disturbance events were the defining features of resilience (Holling, 1973; Holling et al., 1995). This led to 'ecological resilience' being defined as 'the capacity of a system to absorb disturbance and reorganise while undergoing change so as to retain essentially the same function, structure, identity and feedbacks' (Walker et al., 2004). This definition suggests that after a disturbance, resilience should be viewed as a measure of the probability of system failure/extinction rather than as a measure of return time to a particular system state (Folke, 2006; Grimm & Calabrese, 2011). Crucially, in ecological systems, it is the identification of the range of conditions under which system relationships and functions can persist that is important, not the set of conditions needed for a system to remain within a particular or optimal state (see Martin, Chapter 13, for further discussion).

Ecological systems are dynamic evolving systems that have the capacity to adapt and change in response to fluctuations in their external environment. Ecological resilience recognises that a system can exist as a set of states; this set includes all the possible combinations of variables (initial conditions) from which the system can develop its characteristic form, function and relationships. The probability of a system developing its characteristic form is dependent on its initial condition, the stochasticity of the endogenous and exogenous variables driving change, and the frequency, extent and sequence of disturbance events during development (Walker et al., 2004).

Ecological resilience has been defined by some as a system's ability to buffer change; it is the loss of this buffering ability that leads to system change (Grimm & Calabrese, 2011). For example, the introduction of cattle grazing into Argentine semi-arid Chaco has been sufficient to tip significant areas of the system from a characteristic mosaic of grass, scrub and open woodland to a landscape dominated by thorny scrub (Cabido et al., 1994; Zak & Cabido, 2002). This scrubland system is resilient to grazing animals but has a considerably reduced biological diversity of plants, invertebrates and birds

(Gardner et al., 1995), a reminder that resilience is a property of both desirable and less-desirable states. System change is often triggered by what have been called 'slow variables', where change may accumulate gradually over time without any apparent change in system function until a single event abruptly pushes the system into another state (a so-called tipping point). The classic example is the switch from clear to turbid water in lakes (Folke et al., 2004). Focusing on the rate of change (fast and slow) and behaviour (instantaneous and cumulative response) of key system variables is important for understanding how systems self-organise and for identifying the conditions and drivers that may tip a system from one equilibrium state to another (O'Riordan & Lenton, 2013).

In considering the complex adaptive nature of ecological systems as multi-state self-organising systems, Holling and Gunderson (2002) proposed the metaphor of the Adaptive Cycle as a generic mechanism for understanding changes in system resilience (Grimm & Calabrese, 2011). The Adaptive Cycle suggests that a system can pass through different phases of development, each of which is characterised by a unique combination of 'potential (capacity) for change and degree of connectedness between internal controlling variables and processes' (Holling & Gunderson, 2002). Four phases are proposed:

(1) a growth and exploitative r phase: characterised by rapid growth, expansion and colonisation of new areas; in ecological systems typified by the life-history strategies of small, fast-breeding species that colonise new areas via scramble competition (Begon & Mortimer, 1986);
(2) a conservation/maturation phase: characterised by the accumulation/acquisition of resources, consolidation of position often to achieve dominance or maintain competitiveness; typified in ecological systems by slow-growing species adapted for persistence and contest competition to maintain position, lock-up resources and displace competitors;
(3) a release phase: characterised by a disturbance/disruption and sudden release of resources; this phase may be triggered by natural disasters or disruptive agents such as pests, disease or human intervention; and
(4) a reorganisation: characterised by the emergence of pioneers, innovation, new approaches and restructuring to enable system renewal and adaptation to changed circumstances or the emergence of new directions for system development and transformation.

The metaphor of the Adaptive Cycle is important, as it reminds us that natural systems are dynamic, and that the nature of and capacity for resilience differs between different phases of system development. This variation in adaptive capacity suggests that there are many ways of developing resilience and that stand-alone, prescriptive, quick-fix solutions are unlikely to foster system resilience.

The concept of disturbance as a trigger for reorganisation highlights a further interpretation of resilience as a measure of a system's capacity to learn, innovate, adapt and transform (Folke, 2006). Faced with global challenges of climate change, human population growth, migration and economic globalisation, the ease with which ecological and human systems can adapt and transform themselves to different circumstances will be critical for their survival. In considering the nature of system change we need to remember that change can cascade across different scales and rapidly undermine the resilience of systems operating at levels below and above the origin of the disturbance factor. The interaction of conservation and pastoralism in the chapter by Homewood et al. (Chapter 9) is a good example of this.

Resilience is thus the capacity of a system, be it an individual, a farm, a forest, a city or an economy, to deal with change, to continue to develop and retain function. It is also about the capacity to use shocks and disturbances – such as a financial crisis or climate change-induced weather extremes – to encourage system renewal and innovation. Resilience thinking embraces learning, diversity and above all the belief that humans and nature are strongly coupled to the point that they should be conceived as one social-ecological system (Folke, 2006).

Why is resilience important now?

> In our fast-moving global economy ... what matters most is not how successful a [system] is at present, but how resilient it will be facing future challenges.
>
> (Swanstrom, 2008)

It is pertinent to ask why this book is focusing on resilience rather than on the more widely discussed topics of sustainable agriculture and, more recently, sustainable intensification. Resilience recognises that change is inevitable and that systems such as human society and ecological systems are designed for change. They have change mechanisms built into them in the form of genetics, behaviour, social organisation, networks and learning. These systems are dynamic and are driven by feedback mechanisms that enable them to self-organise and adapt.

Resilience thinking has at its heart the requirement to maintain system function (Holling, 1973; Walker et al., 2004). In this respect, the outcome of resilience can be aligned with sustainability or sustainable development. In its most frequent formulation, the latter has been defined as 'development that meets the needs of the present without compromising the ability of future generations to meet their own needs' (Bruntland Commission, 1987). In economic terms, Neumayer (2010) has argued that sustainable development is that which 'does not decrease the capacity to provide non-declining per capita utility for infinity'. Resilience, by focusing on the adaptive process, recognises that there are many potential approaches to maintaining system function. Not all of these approaches will be desirable or sustainable, and it is the role of the resource manager and policymaker to determine those that are. By providing an environment that facilitates the emergence of a variety of system approaches, however, resilience thinking can enable system function delivery to be maintained in the face of change through continuous system development, innovation or transformation (Folke, 2006, see also Harris et al., Chapter 14). Thus, the resilience concept offers important insights and tools in the quest for sustainability.

Sustainability is interpreted differently by economists and ecologists, a difference that has been explored fully by Neumayer (2010) using the concepts of 'Weak' and 'Strong' Sustainability. Neumayer characterises Weak Sustainability as the substitutability paradigm. Advocates of this paradigm assume that natural capital (natural resources) is either abundant or substitutable, and that a reduction in the stock of natural capital can be compensated for by increased availability of manufactured and/or human capital (knowledge, creativity and skills). The implication is that 'a rise in consumption can compensate future generations for a decline in the stock of renewable resources or a rise in pollution stock' (Neumayer, 2010). Strong Sustainability, on the other hand, considers natural capital to be non-substitutable by other forms of capital and requires losses in the stock of natural capital stock to be compensated for by adequate 'shadow projects' (Barbier et al., 1990); for example, a reduction in natural energy resources due to coal-mining could be compensated by investment in renewable energy projects. Building on this definition of Strong Sustainability, certain forms of natural capital can be considered critical such that their use should be constrained within the regenerative capacity of the existing stock (Daly, 1992); for example, balancing the rate of topsoil erosion against the rate of soil formation. In this way, the stock (and therefore its function) is maintained. One of the challenges for this 'balancing' approach is exactly the one raised in our discussion of resilience above: the occurrence of external 'shocks', e.g. flood, financial collapse, war, etc., which can unexpectedly 'tip' a system from one state to another.

Using the concept of the Adaptive Cycle, resilience thinking offers a number of insights and approaches that may help in the quest for sustainability. First, it reminds us that systems change, that change is rarely linear and that systems can occupy several different states. It also suggests that system collapse is to be expected and that change offers opportunities for system renewal, the recombination of structures and processes and the discovery of new development pathways delivering the same service but in different operating environments. During the process of change, systems may embrace new approaches and technologies (e.g. the use of batteries, computers and electricity to drive machines replacing petrol engines and mechanical control systems) or old 'technologies' can re-emerge as the true cost of their replacement is revealed (e.g. use of wild pollinators to pollinate crops: see Williams et al., Chapter 6). The Cycle highlights the fragility of large, highly connected systems, suggesting that there is a limit on the size and level of complexity that a system may sustain. An expectation of system collapse highlights a need to nurture learning, to encourage a diversity of approaches and to maintain capital reserves to draw on in the event of unexpected 'shocks'.

The nature of agriculture and resilience in agriculture

This book focuses on agriculture, a livelihood that has endured for millennia and one that is critical for human health, well-being and survival. While agriculture is most readily identified with food production, the practice of agriculture has consequences well beyond those of the supply of food. The establishment of farmed land was the catalyst for human settlement, the organisation of human societies and trade (Mazoyer & Roudart, 2006). Today, agriculture continues to supply resources to industries such as energy, textiles and clothing, medicine and health, leisure and tourism, as well as managing resources for industries such as water, waste management, heritage and nature conservation. In many cultures, the ownership of land and livestock are assets symbolic of status, influence and wealth (see Ewbank, Chapter 7, and Homewood, Chapter 9). Agriculture is also responsible for many of the negative externalities experienced by human society today: water pollution, land degradation, loss of biodiversity and enhanced greenhouse gas emissions, particularly from ruminant livestock and irrigated rice production systems. Such externalities are also the product of increasing societal demand for year-round supplies of high-quality, low-cost, global food products and of the search for more efficient ways of producing these goods (see Gardner, Chapter 2).

The success of agriculture is in large part still dependent on the natural resources available to it. These dictate the type of livelihood (e.g. pastoralist, cropping, mixed farming, intensive or extensive agriculture) and assets that can be established. Uncertainty surrounding the supply of natural resources such as pollinators, water and nutrients to agriculture coupled with volatility in food markets, environmental change and changing policy frameworks have increased the vulnerability of many producers to unexpected events. In this book, we consider the ability of agriculture to respond to unexpected change, focusing particularly on the role of ecosystem services and biodiversity in enabling agriculture to maintain, adapt and transform its outputs and production approaches in the face of environmental and economic uncertainty.

We first review the context within which agriculture operates, as part of the global food system, before exploring how this context influences the capacity of agricultural producers to adapt to change (Gardner, Chapter 2, and Hodge, Chapter 10). We then consider ways in which ecosystem services and biodiversity can enhance the resilience of agriculture, exploring their role in:

- buffering farms against environmental uncertainty (e.g. through soil conservation, limiting nutrient and water loss – Tilman, Chapter 3, Hails et al., Chapter 4, and Abson et al., Chapter 11);
- enhancing crop productivity and reducing production costs (e.g. through enhancing nutrient supply, soil quality, pollination and pest control – Tilman, Chapter 3 and Williams et al., Chapter 6),
- managing agricultural externalities (e.g. water pollution, greenhouse gas emissions, salinity – Hails et al., Chapter 4, and Harris et al., Chapter 14);
- enhancing the 'value' of agriculture through, for example, diversifying farm income streams and the provision of Public Goods (e.g. carbon management, wildlife conservation, aesthetic and cultural landscapes – Hails et al., Chapter 4, Drechsler & Wätzold, Chapter 12 and Buckwell, Chapter 15).

These roles are explored in the context of agricultural systems in both the developed (Gardner, Chapter 2; Ramsden & Gibbons, Chapter 5; Abson et al., Chapter 11; Harris et al., Chapter 14) and developing world (Ewbank, Chapter 7; Homewood et al., Chapter 9), in relation to biotechnology as a potential contributor to agricultural sustainability (Meyer, Chapter 8) and with respect to the management of agriculture and natural resources, particularly the relationship between efficiency, resilience and sustainability (Gardner, Chapter 2; Ramsden & Gibbons, Chapter 5; Hodge, Chapter 10; Harris et al., Chapter 14).

Finally, we consider how the contributions of biodiversity and ecosystem services to agriculture might be evaluated (Hails et al., Chapter 4; Hodge, Chapter 10; Harris et al., Chapter 14) and integrated into policies that encourage producers to build and retain resilience in their agricultural systems for future generations (Drechsler & Wätzold, Chapter 12; Martin, Chapter 13; Buckwell, Chapter 15; Fisher & Kareiva, Chapter 16).

The book is divided into two parts:

Part I considers the resilience of agriculture and agroecosystems and how biodiversity and ecosystem services might contribute to this resilience;

Part II examines approaches for integrating biodiversity and ecosystem services into agricultural systems and sets out some of the challenges facing producers and policymakers in fostering resilience in agricultural systems.

The concluding chapter draws together the lessons learned from the different economic and ecological approaches set out in the volume, highlights areas where the two disciplines can work together and indicates areas where research and policy development are needed to build agricultural resilience for the twenty-first century.

References

Barbier, E.B., Pearce, D.W. & Markandya, A. (1990). Environmental sustainability and cost–benefit analysis. *Environment and Planning A*, **22**(9), 1259-1266.

Begon, M. & Mortimer, M. (1986). *Population Ecology: a unified study of animals and plants*. 2nd edition. Oxford: Blackwell Scientific Publications.

Bruntland Commission (1987). *Our Common Future*. Oxford: Oxford University Press.

Bucher, E.H. (1987). Herbivory in arid and semi-arid regions of Argentina. *Revista Chilena de Historia Natural*, **60**, 265-273.

Cabido, M., Manzur, A., Carranza, L. & González Albarracín, C. (1994). La vegetación y el medio físico del Chaco Arido en la provincia de Córdoba, Argentina Central. *Phytocoenologia*, **24**, 423-460.

Daly, H.E. (1992). *Steady-state Economics*. 2nd edition with new essays. London: Earthscan.

Folke, C. (2006). Resilience: the emergence of a perspective for social–ecological systems analyses. *Global Environmental Change*, **16**, 253-267.

Folke, C., Carpenter, S., Walker, B. et al. (2004). Regime shifts, resilience and biodiversity in ecosystem management. *Annual Review of Ecology, Evolution and Systematics*, **35**, 557-581.

Gardner, S.M., Cabido, M.R., Valladares, G.R. & Diaz, S. (1995). The influence of habitat structure on arthropod diversity in Argentine semi-arid Chaco forest. *Journal of Vegetation Science*, **6**(3), 349-356.

Grimm, V. & Calabrese, J.M. (2011). What is resilience? A short introduction. In: *Viability and Resilience of Complex Systems: concepts, methods and case studies from ecology and society*, edited by G. Deffuant & N. Gilbert. Berlin: Springer-Verlag, pp. 3-13.

Grimm, V. & Wissel, C. (1997). Babel, or the ecological stability discussions: an inventory and analysis of terminology and a guide for avoiding confusion. *Oecologia*, **109**, 323-334

Hastings, A. (2013). Multiple stable states and regime shifts in ecological systems. *Mathematics TODAY*, February 2013, 37-39.

Holling, C.S. (1973). Resilience and stability of ecological systems. *Annual Review of Ecology and Systematics*, **4**, 1-23.

Holling, C.S. & Gunderson, L.H. (2002). Resilience and adaptive cycles. In: *Panarchy: understanding transformations in human and natural systems*, edited by L.H. Gunderson & C.S. Holling. Washington, DC: Island Press, pp. 25-62.

Holling, C.S., Schindler, D.W., Walker, B.H. & Roughgarden, J. (1995). Biodiversity in the functioning of ecosystems: an ecological primer and synthesis. In: *Biodiversity Loss: ecological and economic issues*, edited by C. Perrings, K.-G. Mäler, C. Folke, C.S. Holling & B.-O. Jansson. Cambridge: Cambridge University Press, pp. 44-83.

Levin, S.A. (1998). Ecosystems and the biosphere as complex adaptive systems. *Ecosystems*, **1**, 431-436.

Martin-Breen, P. & Anderies, J.M. (2011). *Resilience: a literature review*. Brighton: Institute of Development Studies, Bellagio Initiative, University of Sussex. http://opendocs.ids.ac.uk/opendocs/handle/123456789/3692

May, R.M. (1974). *Stability and complexity in model ecosystems*. Princeton, NJ: Princeton University Press.

Mazoyer, M. & Roudart, L. (2006). *A History of World Agriculture: from the Neolithic age to the current crisis*. London: Earthscan.

Neumayer, E. (2010). *Weak versus Strong Sustainability: exploring the limits of two opposing paradigms*. 3rd edition. Cheltenham: Edward Elgar Publishing Limited.

O'Riordan, T. & Lenton, T. (2013). *Addressing Tipping Points for a Precarious Future*. Oxford: Oxford University Press for the British Academy.

Swanstrom, T. (2008). *Regional Resilience: a critical examination of the ecological framework*. Working Paper No. 2008, 07. Berkeley, CA: Institute of Urban and Regional Development, University of California.

Walker, B., Holling, C.S., Carpenter, S.R. & Kinzig, A. (2004). Resilience, adaptability and transformability in social–ecological systems. *Ecology and Society*, **9**(2), 5.

Zak, M.R. & Cabido, M., (2002). Spatial patterns of the Chaco vegetation of central Argentina: integration of remote sensing and phytosociology. *Applied Vegetation Science*, **5**, 213–226.

Zolli, A. & Healy, A.M. (2012). *Resilience: why things bounce back*. London: Headline Publishing Group.

PART I

Biodiversity, ecosystem services and resilience in agricultural systems

CHAPTER TWO

Complexity and resilience in agriculture

SARAH M. GARDNER
GardnerLoboAssociates

Introduction

By many measures, today's agriculture is very successful. From constrained resources we now produce more food for more people than ever before, largely as a result of technological developments – genetic, chemical, mechanical – that have raised the productivity of agricultural ecosystems. Current crop varieties and livestock breeds convert more fertiliser and feed into more usable biological yield than older varieties and breeds. An examination of the development of agriculture at a global or national level reveals a history of change, adaptation and transformation in production practices that has enabled producers to deal with hostile environments, pest and disease challenges and the demands of an increasingly urbanised population (Mazoyer & Roudart, 2006). This facility to change over time and the ability to produce food in very different socio-economic and natural environments may suggest that agriculture is one of the most resilient of human activities.

Agriculture has developed over thousands of years, its development marking the transition of humans from hunter-gatherers to settled agriculturalists (Mazoyer & Roudart, 2006); from foraging for what nature provides, to managing nature to provide what humans need or desire. By comparison, the technological developments that have enabled today's agriculture to emerge during the twentieth century have developed relatively rapidly and have spread to many parts of the globe. Despite these technological developments, agricultural systems still depend on ecosystem services provided by ecological systems that have evolved over millions of years (Levin, 1999). It is not clear, however, whether today's agricultural systems, which humans have designed, are sufficiently sensitive to ensure the continuing persistence and renewal of these ecosystem services (see Hails et al., Chapter 4). This concern is amplified when we consider the future demands and uncertainties that will be placed on agriculture, most notably from a growing global food market and climate change.

Agriculture is the world's most extensive land management practice, extending over 37.5 per cent of the global land mass (World Bank, 2017).

The ecological systems that agriculture replaces are diverse and complex. Within these ecological systems, large-scale patterns emerge from the interactions of many individual agents operating over small spatial scales and over short timescales (Levin, 2009). These large-scale patterns give rise to fundamental ecosystem services and cycles – the carbon and water cycles, for example, that regulate the functioning of the planet. Replacing ecological systems with 'agricultural systems' manipulated by humans carries the risk that these ecosystem services will fail, or more precisely as argued by Hails et al. (Chapter 4), that services are used at a rate greater than their regenerative capacity.

Agriculture is a social–ecological system, founded on complex adaptive ecological systems (Folke, 2006; Biggs et al., 2015). Resilience is a property that emerges from this complex adaptive nature (Holland, 2014). This chapter considers the extent to which agriculture retains sufficient elements of the nature and behaviour of complex adaptive systems to enable agricultural production to be resilient. The chapter starts with a brief review of the nature and behaviour of complex adaptive systems before exploring whether agriculture, positioned within a global food system, can behave as a complex adaptive system. In particular, the structure of the food supply chain and the responses of producers (farmers) to environmental and economic uncertainty are considered, together with their potential effects on producer resilience to unexpected change. The challenge of balancing efficiency and resilience in agricultural production is also examined with examples of producers who have transitioned to an arguably more resilient yet profitable form of production. The chapter concludes with some suggestions on the characteristics and approaches that might be fostered among producers and within the wider food industry to help build the future resilience of agricultural production.

Resilience in complex adaptive systems

Resilience is a property that characterises adaptive capacity and emerges from system behaviour (Holland, 1998), but is not one that can be quantified and optimised (Levin et al., 2013) in the sense, for example, of the optimal farm plans that are an output from the bio-economic models discussed by Ramsden and Gibbons in Chapter 5. Resilience focuses on a system's ability to absorb or adapt to change and persist (Holling, 1973). In considering the characteristics that foster system resilience, Biggs et al. (2015) set out seven principles: (i) maintain diversity and redundancy within a system; (ii) manage the amount of connectedness within the system; (iii) identify and manage slow (e.g. loss of soil organic matter) and fast variables (e.g. spread of forest fire) that act on the system; (iv) consider the nature of complex adaptive systems; (v) encourage learning among the system's agents; (vi) ensure participation from all disciplines and players; and (vii) encourage the development of polycentric governance. Many of these principles reflect the characteristics and emergent behaviour of complex adaptive systems, making the retention of these characteristics desirable for social–ecological systems such as agriculture (Folke, 2006; Darnhofer et al., 2010).

Characteristics of complex adaptive systems

Complex adaptive systems are not static. They are dynamic and self-organising, changing in response to feedback from higher levels of system organisation which influence the dynamics of local interactions between agents at lower[1] levels of the system (Levin, 1999, 2005). The ability to self-organise arises from the presence of numerous autonomous agents in the system set that interact freely with each other. The outcome of these free interactions is the formation of higher-level patterns, behaviours or properties that persist over time (Holland, 1992). An example is the arrowhead pattern observed in a flock of flying geese. This pattern arises from the tendency of each bird in the flock to position itself to take advantage of a reduction in wind resistance from the bird in front. Although individual birds change position, as lead birds drop back when tired, for example, the overall arrowhead pattern persists because it confers an aerodynamic and energy-saving advantage both to individuals and the flock (Weimerskirch et al., 2001; Portugal et al., 2014). This example illustrates how the behaviour of autonomous individuals and the interactions between them generates a self-organising pattern that persists under changing conditions.

There is no agreed definition of complex adaptive systems, but one proposed by Ladyman et al. (2013) captures the necessary features. They define a complex adaptive system as 'an ensemble of many elements which are interacting in a disordered way, resulting in robust organisation and memory'. Complex adaptive systems tend to exist as hierarchies of complexity (Holland, 2014), within which agents operating autonomously at one level tend to ignore the effects of their actions at other levels (Pigou, 1920). Agents

---

[1] In complex adaptive systems, use of the terms higher and lower agents are not indicative of levels of superiority. Instead, they reflect the dependence of an agent on other agents in the system. In an ecological food chain, for example, the fundamental building blocks are resources such as oxygen, nitrogen, carbon dioxide, water and minerals. In this chain, primary producers such as algae, bacteria and plants are the lowest-level agents because they utilise some or all of these building blocks to grow and provide resources for other organisms. Herbivores, which depend on primary producers for food, occur at the next level and predators, which feed on herbivores and can utilise neither the building blocks or primary producers directly, are seen as higher-level agents. Thus, higher agents depend on the agents 'below' them for their survival, while 'lower' agents might be considered as the 'independent' agents, as they build resources for themselves and other organisms. In this sense, 'higher' might be considered as 'increasingly vulnerable' agents because they are dependent on the presence of 'independent' agents for their survival.

act selfishly but are able to learn from each other and macroscopic properties, such as resilience and the flight pattern described above, arise from the free interactions between individual agents and the aggregate behaviour of individuals acting collectively together (Levin et al., 2013). In a complex situation, however, individual agents act with incomplete knowledge and cannot know all the choices or all the consequences of the choices available to them. Thus, while agents may act in their own best interest, their information is incomplete (the 'bounded' rationality of Arrow, as reported in Holland, 2014, p. 24) and their choices may appear disordered. Consequently, in a complex adaptive system, the interactions and behaviour of the numerous, autonomous and heterogeneous agents present may appear to be disordered.

Within complex adaptive systems, macroscopic properties will only persist if the feedback from these properties confers benefits to the underlying agents. In the goose flight example above, each bird receives an energy-saving benefit and adjusts its position to maintain this. If individual agents receive no benefits from a macroscopic property or benefits emerge slowly due to a delay in feedback, then agents' actions will change and the property is lost. Maintaining tight feedback is important for the persistence of a system's macroscopic properties, and feedback that is slow, is not recognised or is interrupted by an external agent can disrupt system function (Levin, 1999).

Behaviour of complex adaptive systems

In examining the behaviour of complex adaptive systems, Levin et al. (2013) suggest that consideration of features such as non-linear feedback, varying timescales, individual and spatial heterogeneity and strategic interactions is essential for understanding the behaviour of these systems.

Nonlinear feedback arises when a change triggers a disproportionate effect. For example: a heavy rainfall event tips lake water from clear to turbid (Folke et al., 2004), a change in stocking rate shifts moorland vegetation from dwarf shrub to grass (Gardner et al., 2009), the movement of 16 diseased sheep through the livestock markets of England and Wales precipitates a national disease epidemic (Comptroller and Auditor General, 2002). In each case, the mechanism triggering change is different: run-off of polluted water exceeding the rate of phosphorus assimilation in a lake, a change in grazing pressure releasing an unpalatable grass from plant competition, connectivity between markets facilitating the rapid spread of disease. Such effects are non-linear and make the future direction of system behaviour hard to predict.

The challenge of managing non-linear effects is increased by variable response rates. In social–ecological systems, economic responses (e.g. human capacity to catch fish – a fast variable) often change more rapidly than ecological ones (e.g. recovery of fish populations – a slow variable) and the impact of these differences is exacerbated when the relevant agents occur at different organisational (higher or lower) levels of the system. This hierarchical separation of fast and slow variables combined with the tendency for economic agents to heavily discount the future can lead to system collapse (Crépin, 2007). To facilitate system persistence, it is important to understand how fast and slow variables interact.

Heterogeneity among the underlying agents is essential for enabling complex adaptive systems to self-organise, respond to and persist within new environments and under changing conditions. Within ecological systems, heterogeneity is the resource on which natural selection acts to generate a diversity of ecological forms and functions (Levin, 1999). It is this diversity that facilitates system adaptation, either through the occurrence of several forms delivering the same or similar functions – 'functional redundancy' – or through the occurrence of several forms delivering different functions – 'functional diversity' (Cabell & Oeloefse, 2012). Functional redundancy can help buffer a system against shocks by providing insurance against loss of system function arising from variation in current conditions. Functional diversity can help a system adapt by providing alternative forms, pathways or functions to enable the system to meet new challenges. Levin et al. (2013) state that 'maintaining the capacity to absorb change is costly', as retaining features that enhance performance under current conditions occurs at the expense of variation that facilitates adaptation to new environmental challenges. Thus, a trade-off exists between functional redundancy, which helps maintain current function but reduces system efficiency, and functional diversity, which helps a system adapt or transform but at a risk of reducing current function.

Agents within complex adaptive systems are able to learn or adapt and their interactions may be copied by other agents resulting in the formation of complementary groups, collective behaviour and/or cooperation between agents. Factors such as spatio-temporal heterogeneity in the distribution of resources can also trigger such responses. Cooperation between agents affects system organisation and structure, such that some agents become tightly connected to each other and maintain fewer connections with other parts of the system, while others are excluded (Levin et al., 2013). This modular structure allows groups to maintain a degree of independence within the system with opportunities to innovate and self-organise. Modularity contributes to system heterogeneity and diversity and increases the chance of system survival in the face of shocks. The modular structure provides the building blocks for system renewal and reorganisation (Lengnick, 2015) but slows down the spread of properties through the system – an advantage when such properties are harmful, but disadvantageous when they are beneficial. Thus, in complex adaptive systems, trade-offs also occur between modularity and connectivity with respect to system change and adaptation.

As noted earlier, the ability to self-organise is critical to the persistence of complex adaptive systems. Factors that constrain the self-organising behaviour of the autonomous agents that underpin the system will reduce system resilience and increase the probability of system collapse (Ladyman et al., 2013). It is important to note that resilience can be as much a property of undesirable systems as desirable ones (Holling, 1973) and in the context of agriculture, society must ultimately decide on the system values (environmental, social, economic) that are desirable for agriculture (see discussions in Hails et al., Chapter 4 and Hodge, Chapter 10). In the following section we consider agriculture as a complex adaptive system before exploring the challenges this poses for producers working within the wider food system.

Agriculture as a complex adaptive system

In today's global food system, agriculture is just one link in a multi-layered chain of production, processing, distribution, retail, consumption and waste management (Lang & Ingram, 2013). Within this chain, large-scale structures, patterns and behaviours can be observed at several levels. Are these structures that emerge from the behaviour of lower-level agents or are they structures that have been developed at a higher organisational level? The question is important for resilience because structures that emerge from the 'bottom-up' can lead to resilience, but those developing from 'top-down' tend to constrain the behaviour of lower-level agents and can lead to fragility.

At the agroecosystem level, the production of agricultural goods depends on natural capital[2] and ecosystem services that arise from many autonomous biological species interacting with each other and with the physical environment within which they exist (Levin, 1999). This heterogeneous set of species and the interactions between them represents a wide variety of resources and hence opportunities for agricultural development. Those resources occur, however, in variable quantities, creating uncertainty for agricultural production.

---

[2] The capital types discussed in this chapter are based on the five types of sustainable capital described by Forum for the Future: www.forumforthefuture.org/the-five-capitals. They define the capital types as follows: *Natural capital* – any stock or flow of energy and material that produces goods and services. It includes natural resources – renewable and non-renewable materials; sinks – that absorb, neutralise or recycle wastes and ecosystem services such as climate regulation. *Human capital* includes people's health, knowledge, skills and motivation; all of which are needed for productive work. *Social capital* includes the institutions that encourage social cohesion, e.g. families, communities, businesses, trade unions, schools, and voluntary organisations. *Manufactured capital* includes the material goods or fixed assets that contribute to the production process rather than being the output itself, e.g. tools, machines and buildings. *Financial capital* provides the means by which other types of capital are owned and traded. Of itself, it has no real value but is representative of natural, human, social or manufactured capital, e.g. shares, bonds or banknotes.

At the production level, the agricultural system can be considered as multiple sets of autonomous humans (e.g. producers, advisers, contractors, farm suppliers) who interact and build the various farming approaches that occur globally today. These interactions generate a variety of forms, practices and products and are important for maintaining the human and social capital on which present-day and future agricultural systems are built.

At the market level, producers interact with any one or more of wholesalers, food processors, retailers or consumers, depending on the strategy they use to market their produce. Such interactions provide opportunities for the innovation of new products and the development of new market outlets. The rise of organic farming and various farm certification schemes, e.g. LEAF and the 'Red Tractor' mark in the UK, are evidence of producers' responses to consumer concern over farming practices and animal welfare.

Agricultural resilience within a multi-layered food system

Considering agricultural production as a layer within a multi-layered complex food system and the nature of the structures within each layer provides some insights on how the resilience of agriculture might build up and break down. Within complex adaptive systems, heterogeneity and autonomy among agents is necessary for enabling resilience. Factors that constrain these characteristics and the ability of agents to interact freely also constrain a system's potential to develop resilience (Ladyman et al., 2013). In hierarchical systems, a lack of heterogeneity at one level adversely affects the heterogeneity of agents in the levels surrounding it (Sengupta, 2006). Because agricultural production is nested within a wider food system, we consider the effects of the structure and behaviour of that system on producer resilience.

The food supply chain

A feature of today's food systems is the concentration of food production within a relatively small number of centralised processing units. Goods from many producers pass to a limited number of processing companies to supply numerous consumers. Macfayden et al. (2016) illustrate this structure with respect to wholesale traders in the Dutch food supply system where five traders supply 16.5 million consumers. Using data from Hoogervorst et al. (2012), the authors infer that each individual trader serves the equivalent of 13,000 producers, 1300 manufacturers, 300 distributors and five supermarket chains – the latter retailing to 3.3 million consumers via 880 outlets. A similar picture can be seen in Table 2.1 for the European Union (EU) as a whole. In 2014, 289,000 food and drink companies served 508.4 million people. Together these companies added £212 billion to EU Gross Value Added (GVA), but over 50 percent of this GVA was accounted for by just 4000 companies

Table 2.1 *Numbers of farm holdings × 10^3 (Eurostat, 2013), food and drink companies × 10^3 (FoodDrinkEurope, 2016) and population size (Eurostat 2015 rounded to nearest 100,000) of the top six food-producing countries in the EU and for the EU as a whole.*

Country	France	Germany	Italy	Spain	UK	Netherlands	EU
Farm holdings	472.2	285	1010	965	185.2	67.5	10800
Food and drink companies	62.2	5.8	54.9	28.3	6.4	5.6	289 (4*)
Population size	66,400	81,200	60,800	46,400	64,700	16,900	508,200

* Just 4000 companies accounted for over 50 percent of the £212 Bn generated by the food and drink industry to EU GVA in 2014.

(FoodDrinkEurope, 2016). These 4000 companies thus have the potential to exert considerable influence on both the flow of goods through the EU food supply chain and on the 10.8 million EU farm holdings that supply them (Table 2.1). With a significant proportion of goods being funnelled through a small number of companies, the choice of buyers available to agricultural producers in the EU is constrained. The scale of purchasing by a limited number of companies also enables them to impose conditions on the types of goods accepted and on how they are produced (Lang et al., 2009). In this way, the structure of the food supply chain and the purchasing scale of some companies act to constrain producer autonomy. Retention of autonomy among lower-level agents is essential for the functioning of complex adaptive systems and for the emergence of resilience (Holland, 2014).

Using the data in Table 2.1, it is possible to infer that the number of farm holdings and consumers served by an individual food and drinks company varies considerably among the top six food-producing countries in the EU (Table 2.2). Thus, in Germany and the UK, the number of farm holdings that potentially supply each company is six and three times greater than in France, respectively, while the number of consumers served by each company in these two countries is approximately 10 times larger than in France (Table 2.2). There are 10 times more food and drink manufacturers in France compared to Germany and the UK (Table 2.1), reflecting policies that fiercely foster French food and farmers (Lang et al., 2009). Following the volatility in commodity prices in 2008–09, a report by Ernst & Young (2008) concluded that France was more resilient to external food shocks than the UK (Lang et al., 2009).

The influence of different sectors (producers, processors, retailers, etc.) within the food chain is evident from an examination of the GVA for each sector. Figures for the UK food system in 2015, for example, report producer GVA at £8.7bn, food and drink manufacturing at £28.2bn, wholesalers at £10.5bn and retailers at £29.5bn. UK household expenditure on food and

Table 2.2 *The number of farm holdings that supply and the number of consumers served by each individual food and drinks company, calculated pro rata using the data in Table 2.1.*

Country	France	Germany	Italy	Spain	UK	Netherlands	EU
Farm holdings per company	8	49	18	34	29	12	37 (2700*)
Consumers per company	1067	14,000	1107	1640	10,109	3017	1759 (127,100*)

* Figures calculated on the basis of the 4000 companies that account for over 50 percent of EU GVA for food and drink (FoodDrinkEurope, 2016).

drink for the same period was £115.3bn (Defra, 2016). Lang et al. (2009), reporting on the position of food manufacturers and retailers globally, noted that in 2005, 10 food companies captured 24 percent of the global market in packaged foods while the top 10 retailers had an estimated 24 percent of the $3.5 trillion world market. The figures reflect the shift that has occurred over the last century from a producer-driven to a buyer-driven food supply chain. Lang et al. (2009) argue that this shift is continuing to move increasingly towards the retailer – as the agent closest to the consumer. As noted earlier, the shift is accompanied by a loss of influence and autonomy for producers who are positioned at the bottom of the supply chain, remote from the consumer. As the ultimate purchaser of the producer's goods, interaction with consumers is important for producer innovation, experimentation and adaptation. A long supply chain where feedback is poor, delayed or constrained undermines the functioning of a complex adaptive system and is detrimental to its resilience.

Scale effects

Within a multi-layered food system, agents working at different levels (agro-ecosystem, production, market) operate at different spatial and temporal scales. The majority of producers work locally over a scale of hectares, or hundreds of hectares. They balance farm husbandry issues which build over weeks or years, e.g. herd nutrition, soil structure and health, against market issues which can change in days, e.g. commodity prices. Food processors and retailers, on the other hand, work at national and global scales with access to many producers (often via an intermediary) and to technologies (e.g. the barcode) that enable them to work to 'just-in-time' schedules (Lang et al., 2009).

This disparity in spatial and temporal scales between producers and processors/retailers influences the decision-making of producers and their approach to farm management. Global market prices change much faster and fluctuate more widely than ecosystem services such as soil fertility, and short-term uncertainty in the marketplace can come to dominate producers' decision-making on cropping, to the long-term detriment of their natural capital. This capital can then fall into a slow long-term decline and a consequent loss of resilience within the agroecosystem. This loss can be exacerbated by government programmes or market instruments that help manage the risks of crop losses and encourage producers to increase production and reduce unit costs via economies of scale, often at the expense of a gradual degradation in natural capital. Decline in natural capital can then become prevalent over a wide area. This was the case in Montana, where intensive cereal production and year-on-year cropping to combat the financial pressures of drought, rising fertiliser costs and falling commodity prices during the 1980s led to increasing soil erosion and farm foreclosures reminiscent of the Depression (Carlisle, 2015).

What can be done to address this disparity between market and agroecosystem response time? Processors absorbed in the task of maintaining their share of a competitive market have limited resources to consider the impact of market fluctuations on local producers, some of whom may be far removed geographically from processors. The task of developing resilience and the facility to withstand shocks to the agricultural production system is one for producers. Only they are on hand to monitor the condition of their resources – natural and others. This monitoring task can be aided by indicators such as those of Cabell and Oelofse (2012) or Fisher and Kareiva (Chapter 16), but in a complex adaptive system building resources to facilitate adaptation and resilience has to start at the lowest organisational level by enabling autonomy and heterogeneity among producers (Lewin, 1992; Holland, 2014).

Connectivity effects

Global transport has enabled the development of a highly connected food supply chain. This in turn has opened up markets for producers, as evidenced by the range of countries represented by the ingredients present in many goods (Macfayden et al., 2016) and the diversity of goods on many supermarket shelves. Connectivity facilitates rapid switching to alternative suppliers after a system shock. US retailers, for example, turned to Morocco and Spain for supplies after the Florida citrus crop was hit by Hurricane Irma in 2017 (CNBC, 2017). Such flexibility enables retailers to maintain supplies and prices to consumers but can undermine producer resilience, especially where the network is concentrated through a few processors/retailers (Rotz & Fraser, 2015).

The benefits of a highly connected supply chain for consumers include continuity of food supply with little seasonality, access to a large range of diverse goods and abundant, relatively low-cost food – UK household spending on food averaged 9 percent in the 2000s compared to 25 percent in 1950s (Lang et al., 2009). Proximity to a global market has been an important driver of lower food prices, although not the only one (see Mazoyer & Roudart, 2006).

Lower prices have, in turn, incentivised concentration and consolidation in all elements of the food sector – retail, processing and production – with a subsequent loss of heterogeneity among agents at all levels of the food system (Dobson et al., 2003). Consolidation has also increased uniformity of production practice among producers and led to a loss of autonomy (Rotz & Fraser, 2015). Loss of autonomy and heterogeneity among producers reduces their ability to adapt to changing conditions and in turn reduces their resilience and that of the food system overall.

Connectivity has its advantages in facilitating rapid flow of information, uptake of technologies, exchange of goods, etc. The nature of the network – the number of nodes and pathways – is, however, important and in a complex adaptive system concentration of flow on one or two highly connected nodes exposes the system to vulnerabilities. For example, the droughts and wildfires that destroyed almost a quarter of Russia's wheat crop in 2010 and the halting of wheat exports that followed drove wheat prices to a historic high (Kogan & Guo, 2014), and have been seen as an important contributor to the events that led to the Arab Spring (Johnstone & Mazo, 2013). Here, too much dependence on a single source had a disproportionately large effect when it was disrupted. Connectivity around a single, large, centralised unit is much greater, tighter and more direct than when several smaller units are organised as independently functioning modules (Lengnick, 2015). Collapse of a centralised unit within a simplified network cascades down to affect many more dependent units, while the loss of a single unit within a modular network can often be absorbed by other units and goods transferred via other pathways.

A network that facilitates the flow of goods also facilitates the flow of disturbances. The highly integrated nature of US hog production has been seen as exacerbating the rapid spread of porcine epidemic diarrhoea virus through the US pig population in 2013–14 (Becton, 2014). Similarly, the rapid movement of animals between UK livestock markets facilitated the spread of foot and mouth disease in a short period of time and triggered an epidemic that lasted 32 weeks and cost the tourist industry an estimated £5 billion (Comptroller and Auditor General, 2002). The inclusion of modules into a network slows down the flow of disturbances and goods with associated benefits and dis-benefits. There is a trade-off between connectivity and modularity in complex adaptive systems. Modularity is, however, seen as the most efficient way of building adaptability because the control mechanism underpinning each module tends to differ between modules, thus maintaining heterogeneity and autonomy at the system level (Levin, 1999). For producers, a more modular food system offers more outlets for their goods and builds in a degree of independence and self-organisation among producers that is needed to retain resilience within an increasingly centralised food supply system (Lengnick, 2015).

Standards and controls

Within the modern food system, little is left to chance. Supply chains are efficient and products standardised in order for processors and retailers to remain competitive and maintain profit and market share. Health and supply risk are a major concern and specifications about these are written into producer contracts (Lang et al., 2009). The latter may specify the volume, size, variety or type of good accepted, time period for delivery, and compliance conditions concerning their quality, production, storage and delivery, including compliance with national and international food standards.

Where manufacturers or retailers are large and few, their operations and practices can penalise small producers. For example, practices such as minimum volumes or bulk discounts for the supply or purchase of goods are non-linear functions that favour larger producers. Similarly, operations such as processing, transportation of goods and servicing arrangements tend to favour large over small producers (Carlisle, 2015). The cumulative effect of these approaches is to encourage large size and uniformity in production approach and to diminish heterogeneity and autonomy among producers. A focus on volume and standardisation limits opportunities for producers to introduce and trial new products. Rotz and Fraser (2015) offer an example of how meat safety regulation in Canada unwittingly penalised small producers. Regulation, introduced following outbreaks of bovine spongiform encephalopathy (BSE) and *E. coli* 157:H7, required all meat slaughter to be undertaken at centralised, publicly licensed facilities. This placed a significant transport requirement on small, rural, isolated producers supplying local markets or direct to customers. While acknowledging that in concentrated industrial production systems with widespread markets the risk of outbreak was high, small-scale producers argued that this was not the case for their isolated local systems. A study established to examine the risk associated with different scales and methods of production concurred, and the policy was amended to take account of the variation in risk associated with different production approaches and their contributions to buffering systemic risk (Miewald et al., 2013).

Resilience and agricultural production

Today's producers compete within a global market for prices that may not reflect production costs, and supply consumers who are increasingly isolated from the source of production. Producers work with agroecosystems and species to obtain the goods required by the food system. As argued in this chapter, agroecosystems are complex adaptive systems: as such, they offer many paths for producers to develop. The path taken will be determined by a producer's starting resources and opportunities: capital assets, market opportunities, future succession planning, among others. In turn, the chosen path will strongly influence future production resilience; the extent to which producers can adapt or transform their production approach to meet changing conditions. In this section we consider how producer decision-making with respect to environmental and economic uncertainty can influence production resilience.

Managing uncertainty

Agricultural producers are familiar with the concepts of variation and risk. Environmental uncertainty is driven by both small-scale variation such as within-field fertility or water availability and large-scale effects of weather and disease which affect the choice of crops and their rate of growth. Environmental variation presents a challenge for consistent agricultural productivity, but is important for maintaining the diversity of the underlying agroecosystem and ecosystem services on which agriculture depends (see Tilman, Chapter 3 and Hails et al., Chapter 4).

Producers manage environmental variation either by working with and building on it or by reducing and controlling it. Those working with environmental variation look to couple agricultural practice with changes in their natural capital (see Abson et al., Chapter 11). For example, producers manage their soils by adopting practices such as crop rotation, cover crops, no- or reduced tillage and manuring to build fertility, structure, organic matter content and soil biodiversity. Such practices help to support the structure and function of the soil system, building both soil health and the associated services that flow from it. In this way, the practices provide a balanced feedback that helps maintain the integrity and resilience of the soil system and mitigate the effects of environmental variation on agricultural productivity (Lengnick, 2015).

Producers seeking to control environmental variation adopt 'tools' such as fertilisers, pesticides and veterinary treatments, among others. Such tools help producers 'de-couple' their production husbandry from the environment by promoting sufficient uniformity in field variables such as soil fertility and pest and disease incidence, to minimise short-run production uncertainty and increase yield. These tools tend to focus on individual elements of the system and their continuous use can lead to system imbalance. Fertiliser application, for example, boosts levels of soil nutrients to sustain crop yields, but contributes little to other elements of soil structure and function. In the absence of other soil management practices, continuous fertiliser use reduces the integrity of the soil and diminishes its resilience.

Targeting individual elements within agroecosystems can impose strong evolutionary pressures on these elements. The repeated exposure of insect pests to the same insecticide and the consequent development of pesticide resistance is a well-known example. Similarly, management that focuses on reducing environmental variation leads to the selection of crop or livestock types best suited to current conditions. At the same time, those types that perform less well or are better suited to a different condition are squeezed out, thus reducing the diversity present and the capacity of the production system to adapt to changing environmental conditions. Market pressures, production and food processing methods have a similar selective effect (Levin et al., 2013). Requirements such as mechanisation, long storage life and reduced susceptibility to damage can act to lower the diversity of types by promoting those types with characteristics most suited to the food supply chain rather than a variable growing environment. The introduction of mechanised harvesters into Californian tomato production, for example, restricted growers to one particular variety of tomato which could be handled easily by the harvester and which in turn facilitated the scaling-up of tomato production (Friedland et al., 1981). Declining diversity in crop types (both the range of varieties within any one crop and the range of crops) is well documented among commercial crops (see Rotz & Fraser, 2015 for many examples in the USA and Canada) and is a factor in limiting a producer's ability to adapt to environmental change, to transition to different forms of production and to build agroecosystem and production resilience (Lengnick, 2015).

Balancing resilience and efficiency in production

Building diversity into production systems enhances the resilience of the underlying agroecosystem, but can also increase the opportunity costs and management costs for producers, reducing short-run productivity and profitability. In an uncertain environment, producers must reconcile a system's current productivity and functionality with future adaptability and persistence by considering whether today's management approach and mix of agricultural enterprises – the 'product portfolio' – is appropriate for tomorrow's production environment. Producers have always had to deal with local environmental variation, but greater uncertainty driven by climate change, rapidly changing technology, change in global markets, food policy or regulation suggests that a focus on productivity and production efficiency alone will be insufficient to enable adaptation to a changing production environment.

Using insights from information theory, Ulanowicz et al. (2009) highlight that both efficiency and resilience are needed to enable systems to undergo change, self-organise and adapt. Using a network capacity model, they show that efficiency and resilience are complementary: efficiency providing the 'driving force' to maintain system growth over time and resilience providing the reserve of possible pathways that allow systems to adapt following disturbance. Too much efficiency results in systems being too tightly constrained; too much resilience leads to stagnation. Ulanowicz et al. (2009) also show that both efficiency and resilience relate to levels of diversity (number of different agents) and connectivity (number of pathways), but in opposite directions. Highly efficient systems are characterised by minimal diversity and a dominant end-to-end pathway; highly resilient systems by multiple pathways and high diversity. The authors conclude that both resilience and efficiency are necessary for system sustainability and that the balance of these two properties within a system can be calculated from its configuration of diversity and connectivity (Goerner et al., 2009). In addition, the modelling work highlights the necessary role of maintaining reserve capacities in order to sustain ecosystems (Ulanowicz et al., 2009).

Efficiency is a key characteristic of today's intensive agriculture. For producers supplying processors or retailers operating within the global food supply chain, there is little opportunity to influence market price (Lang et al., 2009). In this situation, efficiency becomes synonymous with cost management, because the main route by which these producers can achieve profitability is by reducing the unit costs of production – cost per litre of milk, per kilogramme of meat, per tonne of grain. Working as a price-taker within a competitive market where profitability is achieved primarily by cost management offers little scope for investing in production adaptability. In these circumstances, growth is most usually achieved by scaling-up production, specialising in a narrow range of products and using inputs and other tools to maintain the conditions required for maximising productivity. However, in an increasingly uncertain environment – both climatic and market – the capacity to control variability becomes ever more difficult and costly. Moreover, as noted earlier, constraining a complex adaptive system to operate within a certain range of conditions is likely to render that system vulnerable to disturbance. As the work of Ulanowicz et al. (2009) suggests, under conditions of change and uncertainty, a focus on both growth and adaptability is needed. Currently efficiency in agriculture is focused on the production of system outputs per unit inputs, but for the future this focus needs to broaden to consider system adaptability as well (see Harris et al., Chapter 14, for further discussion). In the event of a shock (environmental or market), how easily might producers adapt their production systems? In the following section we consider examples of actions taken by some producers to address this question.

Building resilience
The nature of pathways
Complexity theory reminds us that adaptability to change and shocks comes from maintaining multiple pathways for development. The variety of development paths that arise within complex adaptive systems is made possible by the self-organising behaviour of the autonomous agents present at the lower levels of the system. The complex nature of agricultural systems at both agroecosystem and producer level thus offers a wide range of potential production paths to the producer. Ulanowicz et al. (2009) have shown that the configuration of a system's pathways influences the development of system resilience. The latter reflects a system's ability to adapt and requires flexibility and change. System change is triggered by feedback; the faster the feedback, the sooner change can be initiated. The structure of the production pathway influences how quickly

feedback is received by and sent from producers and thus determines the speed with which they can adapt their production system.

In a market situation, the two-way exchange of information is important to both customer and producer. The simplest pathway – direct selling to a customer – is the most useful for a producer for obtaining feedback concerning a product and production method. This pathway is not available to producers supplying today's food chains which typically pass through several intermediates (buyers, processors, wholesalers, retailers) before reaching the customer. Moreover, because there are relatively few processors and retailers, each handling a high volume of goods, the provision of feedback from customers to individual producers on their particular goods is impractical. In today's food chain, customer feedback can, however, be used by the processor or retailer to develop product standards which can then be passed back to the producer as a future production requirement (Lang et al., 2009). Thus, information exchange on this pathway can become a one-way flow from customer to retailer/processor to producer rather than the two-way exchange observed via direct selling. This change diminishes the role of producers in the food chain, undermining their autonomy and limiting opportunities for innovation and experimentation with new products or approaches.

Developing more resilient pathways
The ease with which a producer can switch from one production pathway to another will depend on a producer's current production pathway and the capital assets and skills available to him or her. In today's food supply chain, producers need to produce ever larger volumes of output in order to maintain their income (Winter & Lobley, 2016). This requirement reduces producer autonomy and limits opportunities for investigating new production pathways unless these are supported by their product buyer. This is particularly the case where producers are contracted to supply a specified volume of goods. For producers with limited capital resources and supplying single or a few commodities under contract to a wholesaler, there is little incentive to change, as this risks disrupting a key source of income (Rotz & Fraser, 2015). In these cases, investigating a different production pathway is difficult.

Box 2.1 gives examples of producers who have adopted production pathways that have enabled them to thrive independently while still maintaining, in some cases, a supply path with the mainstream food industry. As might be expected for complex adaptive systems, the transition has involved different drivers and different resources in each case. In these examples, key drivers of change have been extreme weather events (Rachel's Dairy, Timeless Seeds), increasingly variable weather (Tonnemaker and Almar Farms) and the financial squeeze of high input costs and falling commodity prices (Happy Cow Creamery, Tonnemaker and Almar Farms). To effect change, these producers have drawn on different resources and adopted quite different pathways. Thus, Rachel's Dairy revived redundant human and manufactured capital to convert milk that would have been wasted to butter and cream. The local success of these products encouraged further investment and the development of a diverse range of organic dairy products. Today, Rachel's is a well-known business supplying mainstream retailers. In contrast, Happy Cow Creamery emerged from a shift away from mainstream commercial milk production. Faced with mounting debt, the producer increased the use of previously under-utilised natural capital to switch from feed- to pasture-based production, and subsequently established an independent dairy.

> **Box 2.1. Examples of producers developing new production pathways**
>
> **Rachel's Dairy, Wales, UK (www.rachelsorganic.co.uk)**
> Established as the first certified organic dairy farm in the UK supplying premium organic milk. During freak snowstorms in 1982, which prevented tankers from reaching the farm to collect milk, the owners turned to making butter and cream by hand and selling them as emergency supplies to local stores. The success and popularity of these products encouraged the owners to expand and diversify their range, particularly into yoghurt. Increasing demand led to the construction of a purpose-built dairy in 1992 in Aberystwyth, which is still in operation today.
>
> **Tonnemaker Hill farm, Washington, USA (Lengnick, 2015)**
> Operating as a commercial fruit farm in 1981 supplying apples, cherries and pears for the wholesale commodity markets. Falling wholesale prices during the 1980s prompted the Tonnemaker brothers to diversify into higher-value direct markets and in 1997 to transition to certified organic production. Year-on-year variability in winter kill and spring frosts prompted further diversification into short-season annual vegetable crops which give flexibility in planting. Production of the vegetable crops can be easily scaled up or scaled back during the year, thus offering a complementary pathway to variable fruit production and an additional source of income.
>
> **Happy Cow Creamery, South Carolina, USA (Lengnick, 2015)**
> Established as one of the top industrial milk producers in South Carolina, rising costs and falling prices during the 1980s left Tom Trantham producing a lot of milk but barely making a profit. The refusal of an operating loan in 1987 left him with no apparent options for continuing his dairy business. One morning in April 1987, his cows broke through a fence to graze the neighbouring field of ryegrass, clover, fescue mix which had been abandoned because Tom could no longer afford to purchase seed and

> **Box 2.1. (cont.)**
>
> fertiliser to plant corn. The increased milk yield that evening gave him pause for thought. After researching annual forage crops and intensive grazing practices, he successfully transitioned his herd of 90 cows from feed- to pasture-based production. The switch increased both herd health and milk quality, while successive plantings of seasonal annual crops gave him flexibility, year-round provision of high-quality feed and very high-quality soils thoughout the farm. Tom has not used any chemicals or fertilisers for 27 years. Tom completed his transition from commodity dairyman to specialist milk retailer with the opening of Happy Cow Creamery in 2002, selling whole milk, buttermilk and chocolate milk direct to wholesale markets and to an on-farm store.
>
> **Almar Farm and Orchards, Michigan, USA (Lengnick, 2015)**
> A long-established fruit farm was growing apples and other fruits conventionally until the mid 1980s, when Jim Koan switched to Integrated Pest Management (IPM) in order to cut costs and reduce environmental impacts. Success with IPM encouraged Jim to transition to organic production. With increasing variability in weather, Jim switched from growing a diverse range of fruit types to focus on a diverse mix of 30 apple varieties. Some of the apples are marketed directly, the rest are processed on-farm into hard cider and other apple products. Jim points out that the processed products help buffer his business against weather variability and extremes. He has also integrated pigs into his orchard production system. The pigs clean up the fallen apples, which helps to prevent pest and disease outbreaks, manure the orchards and manage the weeds. The pigs also consume the apple waste left over from the cider press and their meat provides another source of income for the farm.
>
> **Timeless Seeds, Montana, USA (www.timelessfood.com; Carlisle, 2015)**
> Established in 1987 when four farmers in Montana set out to transition from conventional commodity wheat-growing to pulse crops, particularly lentils, and rotational organic farming in order to save their soils. Falling wheat prices, increased fertiliser cost, soil erosion and rising debt prompted the transition. Abandoned buildings, cooperative working and 'crowd-funding' from a dozen friends and family enabled the development of an independent processing and packaging plant. Premium-quality organic lentils, other pulses and speciality heirloom grains are marketed to the natural food industry and restaurants. Timeless remain a small, farmer-owned company in rural Montana with lentils and grains sold worldwide.

> **Box 2.1. (cont.)**
>
> Ferme de l'Or Blanc, Cheniers, France, independent cheesemakers (http://academie-mons.com/cheesemaking.shtml#tab-ivan; www.weeklytimesnow.com.au/agribusiness/farm-magazine/ivan-and-julie-larcher-are-choosing-smallscale-production-on-their-ferme-de-lor-blanc/news-story/84a00edb4390a327a0bf2e51af51e444)
>
> Starting in 2013, the Larchers deliberately chose to develop this farm as a small-scale, high-quality specialist dairy. Building on their skills and experience as dairy technicians and artisan cheesemakers and with access to loans from the French government, Ivan and Julie make cheese, butter, yoghurt and rice pudding on-farm with milk from their small herd of approximately 20 Jersey cows. The animals are pasture-fed to enhance the cheese-making. Seventy percent of production is sold at farmers' markets and the remainder at local supermarkets. Working in a culture where support for local products is strong encourages producers to innovate and specialise but, as Ivan says, also requires knowledge, investment and commitment.

All the examples in Box 2.1, except for the French cheesemakers (Ferme de l'Or Blanc), transitioned their businesses within the context of a highly competitive food market and apparently without the support of grants or subsidies. In four cases, producers reduced inputs and enhanced on-farm resources (e.g. soil management, pasture diversification, waste recycling) to help cut costs; this would seem to be a frequent response of producers facing economic hardship (Darnhofer, 2010; Parsonson-Ensor & Saunders, 2011; Winter & Lobley, 2016). Several producers changed their businesses either by increasing the range of crops or mix of varieties grown (Tonnemaker and Almar Farms, Box 2.1), shifting production paths (e.g. from feed- to pasture-based grazing: Happy Cow Creamery; wheat to lentils: Timeless Seeds; fruit to pigs and apples: Almar Farm) and/or increasing the range of products offered, often by developing an element of value-added processing (e.g. dairy products and cider, see Box 2.1). Capital assets were critical in facilitating change with human capital (e.g. skills in crop and grazing management, processing and marketing), social capital (e.g. cooperation with other farmers, crowd-funding), natural capital (e.g. building soils, diversifying crops and pastures) and manufactured capital (e.g. utilising unused buildings or equipment for product processing) – all playing as important a role as financial capital in the transition. All of the producers engaged (and most continued) in direct selling to customers to access better prices and to obtain customer feedback, albeit over a potentially smaller market. The availability of the internet also offered opportunities in this respect, enabling Timeless Seeds, for example, to sell to distributors across North America.

The shift towards more resilient production pathways by these producers has delivered a number of production benefits. By building and utilising natural resources and reducing their dependence on external inputs, these producers have undertaken actions to help enhance the resilience of their agroecosystems, reduce their costs, increase their autonomy and reduce their risk. Diversifying product range has helped producers cope with variable weather and commodity prices and helped ensure continuity of revenue. Thus, Almar Farm's cider, with its higher market value and longer storage, provides a buffer during a poor apple year, while Tonnemaker Farm can scale up its seasonal vegetable production during a similar shortfall. Collaboration and cooperation between local producers has been important for overcoming challenges such as the processing of small volumes of 'novel' products and sharing knowledge and experience in implementing new approaches (Timeless Seeds reported in Carlisle, 2015). As highlighted by Abson et al. (Chapter 11), opportunities to access advice can be limited when producers choose to work outside the 'mainstream'. Direct marketing helped producers in Box 2.1 to attain better prices for their products and also enabled them to inform customers about their products. Thus, Almar Farm sells apple varieties not found in supermarkets but better suited to the farm's growing conditions – helping the resilience of both the agroecosystem and production system (Lengnick, 2015).

Transitioning towards more autonomous and resilient production systems is not without problems. Ivan Larcher of Ferme de l'Or Blanc (Box 2.1) emphasises the need for 'commitment, investment and knowledge' in order to shift from milk production into specialist cheese-making. Each of the producers in Box 2.1 drew on all forms of capital assets to diversify and establish their businesses. In her study of 25 US producers, Lengnick (2015) highlighted the importance that producers placed on maintaining high levels of all types of capital to achieve production resilience. Conversely, Winter and Lobley (2016) noted that the benefits of a shortened supply chain had bypassed many small farmers in the UK because the investment of time and capital required was beyond their means. The authors also noted that a UK policy focus on consumers and environmental issues had overlooked the challenges facing small farmers and suggested that greater emphasis might be put on the development of local food movements; an approach and culture that seems to have helped support small farmers in France. The role of local and regional food hubs as a mechanism for supporting more resilient forms of food production is explored by Lengnick (2015) and Ackerman-Leist (2013).

Different types of resilience
As the examples in Box 2.1 illustrate, there is no 'standardised' path or approach for building resilience into a system; each situation is unique and

will involve different trade-offs. The form that resilience takes varies between systems and is subject to some debate (see Gardner & Ramsden, Chapter 1). Darnhofer (2014), reviewing the literature on resilience, states that systems can exhibit resilience by absorbing and/or adapting to change – a so-called 'bounce back' response – or by adjusting and moving forward to a renewed form – a 'bounce forward' response – following or in anticipation of changing circumstances. She sets out three capabilities – buffering, adaptive and transformative capabilities, that equip farming systems with the full range of adjustments needed for them to respond to change. All three capabilities need to be present and integrated to produce a resilient farm system.

Lengnick (2015), exploring these capabilities with respect to the resilience of different US farms coping with climate change, identified different actions and assets associated with each capability. Thus, for buffering, actions tended to use manufactured and financial assets rather than natural, human or social capital. Manufactured assets were used to protect system components, for example wind machines to reduce frost damage in fruit crops, and financial tools such as insurance, to aid recovery when manufactured protection failed. Such actions tended to be directed towards high-value crops, e.g. orchards. Buffering enables a farm to absorb a shock while avoiding any fundamental change in structure or function. Too much system buffering can, however, lead to mal-adaptation (Bennett et al., 2014), as these actions work to maintain the current system state, which may be inappropriate under conditions of continuing uncertainty. On the other hand, the retention of redundant elements that can maintain current function within the agroecosystem can provide insurance for services that are key for production – e.g. planting of wildflower strips to enhance populations of wild pollinators and natural enemies can enhance crop pollination and pest control and can help in transitioning the system to a more resilient form of production (Williams et al., Chapter 6; Woodcock et al., 2014).

Adaptive capability focuses on actions that enhance the response and recovery capacity of the agroecosystem. Actions may involve some change to system structure and function but look to introduce change incrementally while still maintaining the system's essential identity. Lengnick (2015) identified that producers utilised the full range of capital assets – financial, human, manufactured, natural, social – to enhance the adaptive capacity of their production systems. She reported a broad range of activities associated with adaptation including the use of drought-resistant species, improving soil management, reorientating the production window to avoid climate extremes, integrating livestock and crop production, direct marketing and value-added processing to diversify product range. An important element of adaptation is the development of social response capacity. This includes all activities that encourage interaction and the exchange of knowledge and ideas between producers

(Darnhofer, 2010; Lengnick, 2015). These in turn facilitate self-organisation, learning and innovation among producers, all of which are central to the functioning of agricultural production as a complex adaptive system.

Transformative capability enables a farm to undertake radical change in the face of a major shock, as illustrated by some of the producers highlighted in Box 2.1. Transformative actions help systems transition to a new identity and enhance their adaptive capacity by reorientating their structure and function. For example, the owner of Happy Cow Creamery reorganised his husbandry practices to cut out dependence on external inputs and developed a flexible regime of seasonally adapted forage crops to provide year-round, high-quality feed and improved soils. Such changes focus on altering the relationships among crops, livestock and land to promote the ecosystem services that support production (Hails et al., Chapter 4). Transformative actions need to be supported by high-levels of all five capital assets and linked to a recognition of the need for change and a vision and practical means for achieving it (Lengnick, 2015). In practice, the process needs to be undertaken in small steps, wherever possible experimenting with trial projects and products to ensure that both the production method, market and risks are well understood before undergoing transition of the whole system (e.g. Timeless Seeds in Carlisle, 2015).

Conclusions

The study of complex adaptive systems informs us that resilience is a property that emerges from the behaviour of the lower-level agents in a system. For agriculture and the food system that is built on it, these agents are the producers. They work with agroecosystems, which are themselves complex adaptive systems. Both producers and the agroecosystems they manage are key to generating the resilient patterns, behaviours and services on which the persistence of agriculture and the wider food system depend (see Hails et al., Chapter 4). In complex adaptive systems, resilience relies on the interactions of numerous heterogeneous and autonomous agents to generate the diverse forms, patterns and pathways that enable a system to persist under conditions of uncertainty (Ladyman et al., 2013). For resilience to emerge in agricultural production, these characteristics of numerosity, heterogeneity and autonomy must be present among producers and agroecosystems to enable them to self-organise and generate system properties such as resilience.

Maintaining heterogeneity within food crops, livestock and production systems is difficult for both agricultural producers and the wider food system to manage, particularly on a large scale. Heterogeneity complicates agricultural production and potentially reduces the efficiency of large-scale food processing. Producers working within today's food supply chains tend to be large and to specialise in a narrow range of outputs, with the latter managed intensively using external inputs to reduce variability within the production environment. Production is tightly controlled, making innovation and experimentation with new ideas difficult and costly to introduce. Such systems remain efficient but 'brittle', lacking resilience.

A challenge for agriculture is the need to balance resilience and efficiency. Both are needed to enable a system to, respectively, adapt to changing conditions and to grow over time (Ulanowicz et al., 2009). A tension arises between these properties because efficiency tends to drive systems towards greater uniformity among agents, while resilience requires such agents to be heterogenous (Goerner et al., 2009). This tension lies at the heart of agriculture and today's food system, because a focus on productivity has emphasised the importance of efficiency at the expense of resilience (Mazoyer & Roudart, 2006). This imbalance between efficiency and resilience is exacerbated by the structure and function of many of today's food supply chains. In these chains, the occurrence of a relatively small number of processors and retailers handling large volumes of goods has accelerated the drive towards uniformity in production practice. This reduces heterogeneity within the production environment, undermining the resilience of agroecosystems and the ecosystem services associated with them. The small number of processors and retailers also limits the choice of sale outlets for producers and the range of goods they can provide which, together with purchasing practices such as minimum supply volume that favour large producers over small ones, reduces the autonomy and heterogeneity of producers. With respect to the structure and functioning of a complex adaptive system such as agriculture, many of today's food supply chains do not seem geared towards generating the conditions that might foster resilience for agricultural producers, for agroecosystems or for the food system overall.

Are there other approaches that could be considered that might help strengthen agricultural resilience? In France, for example, the number of Food and Drink companies is an order of magnitude greater than in the UK and Germany and there is a strong culture advocating local and regional food. Support for small farmers and new entrants is strong, encouraging them to innovate and add value to their products (see Ferme de l'Or Blanc in Box 2.1). Such diversity among producers may enhance the resilience of agriculture and the associated food system to external shocks such as volatility in commodity prices. Regionalisation of the food system is being considered in some US states in order to enhance resilience by diversifying the range of agents and pathways. The aim is to develop a more modular structure for the food supply chain that might be more resilient to shocks and environmental uncertainty (Ackerman-Leist, 2013; Lengnick, 2015).

Rebuilding resilience into highly efficient agricultural systems is difficult but not impossible as examples in Box 2.1 affirm. Capital reserves of all types are required, with human, natural and social capital being as important for enabling adaptation and transformation in response to change, as financial capital. Capital reserves provide a buffer that helps producers innovate and experiment with new ideas. Adaptation and the development of new production pathways is aided by building heterogeneity into the production system and/or product range; this provides flexibility to a producer in dealing with an uncertain production environment and supports agroecosystem resilience. Maintaining autonomy among producers is also important as this enables them to act independently and take decisions based on future drivers and trends, and not just on the current stimulus alone (Holland, 2014).

Social–ecological systems such as agriculture challenge management and policy because of the high level of complexity and multiple approaches needed to understand them (Levin et al., 2013). This is certainly the case for the development of resilience in agricultural production, which arises from the characteristics and behaviour of producers and the agroecosystem they manage. Attempts to manage or control these characteristics and behaviours will constrain the system properties that emerge. Because resilience cannot be designed for, measures are needed that are indicative of resilience and that might be used to nudge or incentivise producers towards practices that enhance the characteristics of producer autonomy, heterogeneity and the development of capital reserves. For a complex adaptive agricultural system, nested within a larger complex adaptive food system and operating in an uncertain environment, these characteristics will help build resilience.

References

Ackerman-Leist, P. (2013). *Rebuilding the Foodshed: how to create local sustainable and secure food systems*. White River Junction, VT: Chelsea Green Publishing.

Becton, L. (2014). *Update on PEDV Research*. Minneapolis, MN: Minnesota Pork Congress.

Bennett, E., Carpenter, S.R., Gordon, L.J., et al. (2014). Towards a more resilient agriculture. *Solutions*, **5**, 65–75.

Biggs, R., Schlüter, M. & Schoon, M.L. (2015). *Principles for Building Resilience: sustaining ecosystem services in social–ecological systems*. Cambridge: Cambridge University Press.

Cabell, J.F. & Oelofse, M. (2012). An indicator framework for assessing agroecosystem resilience. *Ecology and Society*, **17**(1), 18. http://dx.doi.org/10.5751/ES-04666-170118

Carlisle, L. (2015) *Lentil Underground: renegade farmers and the future of food in America*. New York, NY: Gotham Books, Penguin Group (USA) LLC.

CNBC. (2017). Brace for higher orange prices as USDA expected to forecast smallest crop since 1940s (reporter Jeff Daniels, CNBC, 11 October 2017). www.cnbc.com/2017/10/11/usda-forecast-on-florida-citrus-could-show-smallest-crop-since-1940s.html (accessed January 2018).

Comptroller and Auditor General. (2002). *The 2001 Outbreak of Foot and Mouth Disease*. HC 939 Session 2001–2002. London: National Audit Office.

Crépin, A.-S. (2007). Using fast and slow processes to manage resources with thresholds. *Environmental and Resource Economics*, **36**, 191–213.

Darnhofer, I. (2010). Strategies of family farms to strengthen their resilience. *Environmental Policy and Governance*, **20**, 212–222.

Darnhofer, I. (2014). Resilience and why it matters for farm management. *European Review of Agricultural Economics*, **41**(3), 461–484.

Darnhofer, I., Fairweather, J. & Moller, H. (2010). Assessing a farm's sustainability: insights from resilience thinking. *International Journal of Agricultural Sustainability*, **8**(3), 186–198.

Defra. (2016). Agriculture in the UK 2015. https://assets.publishing.service.gov.uk/government/uploads/system/uploads/attachment_data/file/557993/AUK-2015-05oct16.pdf

Dobson, P.W., Waterson, M. & Davies, S.W. (2003). The patterns and implications of increasing concentration in European food retailing. *Journal of Agricultural Economics*, **54**(1), 111-126.

Ernst & Young. (2008). Food for thought: how global prices will hit UK inflation and employment. www.criticaleye.com/inspiring/insights-detail.cfm?id=561

Eurostat. (2013). Farm structure survey. https://ec.europa.eu/agriculture/sites/agriculture/files/statistics/facts-figures/farm-structures.pdf

Eurostat. (2015). Eurostat NewsRelease, 124/2015 – 10 July 2015. https://ec.europa.eu/eurostat/documents/2995521/6903510/3-10072015-AP-EN.pdf

Folke, C. (2006). Resilience: the emergence of a perspective for social–ecological systems analyses. *Global Environmental Change*, **16**, 253–267.

Folke, C., Carpenter, S., Walker, B., et al. (2004). Regime shifts, resilience and biodiversity in ecosystem management. *Annual Review of Ecology, Evolution and Systematics*, **35**, 557–581.

FoodDrinkEurope. (2016). Data & Trends. EU Food and Drink Industry 2016. www.fooddrinkeurope.eu/uploads/publications_documents/Data_and_trends_Interactive_PDF_NEW.pdf

Friedland, W.H., Barton, A.E. & Thomas, R.J. (1981). *Manufacturing Green Gold: capital, labor, and technology in the lettuce industry*. Cambridge: Cambridge University Press.

Gardner, S.M., Waterhouse, A. & Critchley, C.N.R. (2009). Moorland management with livestock: the effect of policy change on upland grazing, vegetation and farm economics. In: *Drivers of Change in Upland Environments*, edited by A. Bonn, K. Hubacek, J. Stewart & T. Allott. Abingdon: Routledge, pp. 186–207.

Goerner, S.J., Lietaer, B. & Ulanowicz, R.E. (2009). Quantifying economic sustainability: implications for tree-enterprise theory, policy and practice. *Ecological Economics*, **69**, 76–81.

Holland, J.H. (1992). *Adaptation in Natural and Artificial Systems: an introductory analysis with applications to biology, control and artificial intelligence*. Cambridge, MA: MIT Press.

Holland, J.H. (1998). *Emergence: from chaos to order*. Oxford: Oxford University Press.

Holland, J.H. (2014). *Complexity: a very short introduction*. Oxford: Oxford University Press.

Holling, C.S. (1973). Resilience and stability of ecological systems. *Annual Review of Ecology and Systematics*, **4**, 1–23.

Hoogervorst, N., van Veen, M. & Dassen, T. (2012). *Assessment of the Human Environment 2012*. The Hague: PBL Netherlands Environmental Assessment Agency. http://themasites.pbl.nl/balansvandeleefomgeving/wp-content/uploads/PBL_2012_Assessment-of-the-Human-Environment-2012.pdf

Johnstone, S. & Mazo, J. (2011). Global warming and the Arab Spring. *Survival*, **53**, 11–17. doi:10.1080/00396338.2011.571006

Kogan, F. & Guo, W. (2014). Early twenty-first-century droughts during the warmest climate. *Geomatics, Natural Hazards and Risk*, **7**, 127–137. doi:10.1080/19475705.2013.878399

Ladyman, J., Lambert, J. & Wiesner, K. (2013). What is a complex system. *European Journal for Philosophy of Science*, **3**(1), 33–67.

Lang, T. & Ingram, J. (2013). Food security twists and turns: why food systems need complex governance. In: *Addressing Tipping Points for a Precarious Future*, edited by T. O'Riordan & T.

Lenton. Oxford: Oxford University Press for the British Academy, pp. 81–103.

Lang, T., Barling, D. & Caraher, M. (2009). *Food Policy: integrating health, environment and society*. Oxford: Oxford University Press.

Lengnick, L. (2015). *Resilient Agriculture: cultivating food systems for a changing climate*. Gabriola Island. Toronto: New Society Publishers.

Levin, S.A. (1999). *Fragile Dominion: complexity and the commons*. Cambridge, MA: Perseus Publishing.

Levin, S.A. (2005). Self-organization and the emergence of complexity in ecological systems. *Bioscience*, **55**(12), 1075–1079.

Levin, S.A. (2009). Complex adaptive systems and the challenge of sustainability. In: *Toward a Science of Sustainability*, edited by S.A. Levin & W.C. Clark. Warrenton, VA: Report from Toward a Science of Sustainability Conference, pp. 123–128.

Levin, S.A., Xepapadeas, T., Crépin, A-S., et al. (2013). Social–ecological systems as complex adaptive systems: modelling and policy implications. *Environment and Development Economics*, **18**, 111–132.

Lewin, R. (1992). *Complexity: life at the edge of chaos*. 2nd edition. Chicago, IL: University of Chicago Press.

Macfayden, S., Tylianakis, J.M., Letourneau, D.K., et al. (2016). The role of food retailers in improving resilience in global food supply. *Global Food Security*, http://dx.doi.org/10.1016/j.gfs.2016.01.001

Mazoyer, M. & Roudart, L. (2006). *A History of World Agriculture: from the Neolithic age to the current crisis*. London: Earthscan.

Miewald, C., Ostry, A. & Hodgson, S. (2013). Food safety at the small scale: the case of meat inspection regulations in British Columbia's rural and remote communities. *Journal of Rural Studies*, **32**, 93–102. doi: 10.1016/j.jrurstud.2013.04.010

Parsonson-Ensor, C. & Saunders, C. (2011). Exploratory research into the resilience of farming systems during periods of hardship. New Zealand Agricultural & Resource Economics Society (Inc): 2011 NZARES Conference. https://ageconsearch.umn.edu/bitstream/115511/2/2492-Exploring%20the%20resilience%20of%20farming%20systems%20-%20NZARES%20Conference%20formatted.pdf

Pigou, A.C. (1920). *The Economics of Welfare*. London: Macmillan.

Portugal, S.J., Hubel, T.Y., Fritz, J., et al. (2014). Upwash exploitation and downwash avoidance by flap phasing in ibis formation flight. *Nature*, **505**, 399–402.

Rotz, S. & Fraser, E.D.G. (2015). Resilience and the industrial food system: analyzing the impacts of agricultural industrialization on food system vulnerability. *Journal of Environmental Studies and Sciences*, **5**(3), 459–473. doi: 10.1007/s13412-015-0277-1

Sengupta, A. (2006). Chaos, nonlinearity, complexity: a unified perspective. In: *Chaos, Nonlinearity, Complexity: the dynamical paradigm of nature*, edited by A. Sengupta. Berlin: Springer-Verlag, pp. 270–352.

Ulanowicz, R.E., Goerner, S.J., Lietaer, B. & Gomez, R. (2009). Quantifying sustainability: resilience, efficiency and the return of information theory. *Ecological Complexity*, **6**, 27–36.

Weimerskirch, H., Martin, J., Clerquin, Y., Alexandre, P. & Jiraskova, S. (2001). Energy saving in flight formation. *Nature*, **413**, 697–698.

Winter, M. & Lobley, M. (2016). *Is There a Future for the Small Family Farm in the UK?* Report to the Prince's Countryside Fund. London: Prince's Countryside Fund.

Woodcock, B.A., Harrower, C., Redhead, J., et al. (2014). National patterns of functional diversity and redundancy in predatory ground beetles and bees associated with key UK arable crops. *Journal of Applied Ecology*, **51**, 142–151.

World Bank. (2017). Data for 2014. https://data.worldbank.org/indicator/AG.LND.AGRI.ZS)

CHAPTER THREE

Biodiversity and agriculture

DAVID TILMAN

University of Minnesota and University of California, Santa Barbara

Introduction

Earth's biodiversity, by which I mean both the vast number of species on Earth and the genetic variability within each species, has long been of great importance to agriculture (Zimmerer, 2010). From this biodiversity were chosen the several thousand species of plants, animals and fungi that now form the global food supply. Earth's biodiversity is also the source of the pollinators that are required by many crops, of predator and parasitoid insects that control many crop pests, of soil microorganisms and fauna that create soils and their fertility, and of biologically active chemicals that we use to control crop pests and many livestock diseases (Daily, 1997; Tilman et al., 1999). Our ability to successfully produce a crop year after year depends on access to genetic diversity that is needed to overcome the impacts of emerging strains of pathogens (Cassman et al., 2003). Genetic diversity is also needed to adapt existing crop varieties to new habitats, or to changing climate, and to increase crop yields (Duvick, 1984).

For these reasons alone, biodiversity will always be essential to agricultural productivity and sustainability. However, in the past two decades there have been fundamental advances in our understanding of the numerous effects that biodiversity has on ecosystem functioning (Tilman et al., 2001a; Cardinale et al., 2012; Tilman et al., 2014). This new information is of potentially great agricultural relevance, but it has not yet been incorporated into agricultural practices. In this chapter, I first review these recent findings, their conceptual bases, and then discuss how ecosystem-level effects of biodiversity might be used as tools for advancing agriculture.

Biodiversity and ecosystem functioning

Early naturalists and agriculturalists noted that ecosystems containing more plant species often seemed to have greater productivity. For instance, Darwin (1859) stated that it was widely known that pastures planted with many different grass species produced more hay, or herbage, than those planted with one or a few grass species (Hector & Hooper, 2002). This is among the first known references to the effect of plant diversity on the amount of plant biomass produced, which ecologists call the 'primary productivity' of an ecosystem, often shortened as its 'productivity' (such as dry grassland plant mass harvested per hectare). A century later, Charles Elton (1958), based on his observations of patterns in a wide range of habitats and ecosystems, suggested that ecosystems containing more species seemed to be more stable, have lower disease incidence and be more resistant to invasion by exotic species. These ideas were widely accepted and appeared in the major textbooks of that era, such as Odum and Odum (1961). However, ecology then began a transition into being a more experiment- and theory-based discipline that sought mechanistic explanations for patterns in nature based on the dynamics and outcomes of interactions among the dominant species. Some of the initial theoretical work on diversity led to predictions that contradicted Elton's assertion that greater species diversity led to greater stability (May, 1973). The ensuing diversity–stability debate was never fully resolved (McNaughton, 1977), but the new consensus in the discipline was that there likely were no simple or general relationships between diversity and stability (Goodman, 1975). Although there remained great interest in the causes of diversity, that is the forces that allowed so many competing species to coexist with each other, few ecologists continued to study the potential effects of changes in diversity on community and ecosystem processes.

Multi-species coexistence

A large number of experiments and related mathematical theory have shown that competing species coexist with each other only if the species have appropriate interspecific trade-offs. Many such trade-offs have been identified, such as an interspecific trade-off between competitive ability and dispersal ability. If each species can only gain (or lose) competitive ability by giving up (or gaining) dispersal ability, the net result is that species fall at different points on a trade-off curve from being good competitors but poor dispersers, or, at the other end of the gradient, to being poor competitors but good dispersers. A large number of such species can stably coexist in a spatial habitat (Hastings, 1980; Tilman, 1994). Interspecific trade-offs between the abilities of species to compete for nitrogen versus water, or for phosphate versus nitrate, or silicate versus phosphate (Tilman et al., 1982), or for pulses of resources versus their mean availability can also allow stable multi-species coexistence (Armstrong & McGehee, 1980; Levins, 1979). Interspecific trade-offs between competitive ability versus resistance to predation can also allow many species to coexist on one, or a few, limiting resources when there are predators present (Levin et al., 1977).

Concerns about the potential impacts of human-caused losses of biodiversity, such as from nutrient pollution or habitat destruction, reignited interest in the possibility that diversity might influence various aspects of ecosystem functioning (Ehrlich & Ehrlich, 1981; Schulze & Mooney, 1993; Wilson, 1988). New ideas arose that suggested linkages between plant diversity and ecosystem productivity, diversity and stability, and plant diversity and the level to which limiting resources are reduced by plant consumption (McNaughton, 1993; Swift & Anderson, 1993). These ideas were followed by a trickle of field studies that grew into a torrent over the next 20 years. Simple verbal assertions were rigorously re-explored mathematically, which led to the development of mechanistic theories of why and how changes in biodiversity should impact various ecosystem processes.

Before I detail these studies, let me summarise their major findings. The body of work that resulted from the efforts of hundreds of ecologists worldwide documents that biodiversity is a surprisingly strong determinant of ecosystem functioning. Greater plant biodiversity led to greater ecosystem stability, greater productivity, more complete use of the limiting resources, lower disease incidence, less invasion by exotic competing species, and to an above-ground food web dominated by predaceous and parasitoid insects, and thus to low densities of herbivorous insects (Cardinale et al., 2012; Tilman et al., 2014).

Diversity and stability
Field studies and experiments
McNaughton (1977, 1993) presented many examples, often from his work in the Serengeti, of grassland plots being more stable when they contained more plant species. Although each example might be questioned for its low or no replication, the overall set of examples suggested that greater plant diversity led to greater stability of primary productivity. In 1994, a well-replicated field experiment containing 207 plots showed that grassland plots containing one to three plant species had their ecosystem productivity fall to about one-tenth of their usual level during a major drought, but plots containing 15 to 25 plant species had their productivity fall to only half of their usual level (Tilman & Downing, 1994; Figure 3.1a). Thus, plots containing high plant species numbers were about five times more resistant to drought than plots that were very species-poor. This greater ecosystem stability did not carry over to the stability of individual species, a result that is consistent with May (1973). Long-term observations in Mongolian grasslands (Bai et al., 2004) and a Chinese wetland (Li et al., 2013) have also suggested that ecosystem stability is greater at higher plant diversity.

None of the diversity–stability studies mentioned above were based on an actual randomised and replicated biodiversity experiment. Rather, those studies used plots that differed in their plant diversity for a variety of other reasons. A plant biodiversity experiment in Minnesota was the first study to examine the diversity-dependence of the stability of ecosystem productivity. In that experiment, there were 30 to 40 replicate plots at each diversity level

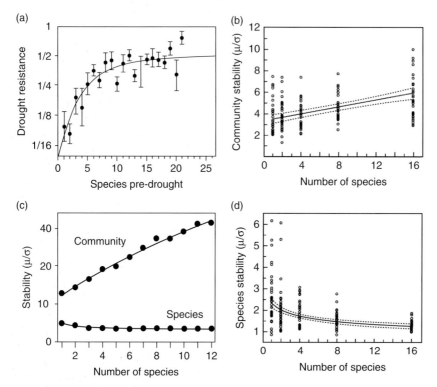

Figure 3.1 (a) The effect of the pre-drought number of plant species in each of 207 plots on drought resistance, measured as total plant biomass produced in a plot during the drought (1988) divided by plot biomass produced the year before drought began (1986). See Tilman and Downing (1994). (b) The effects of the number of plant species planted in each of 152 plots on stability of total plant biomass produced for the Minnesota biodiversity experiment. Stability is the ratio of the 10-year average biomass production of a plot to the standard deviation of these 10 numbers. See Tilman et al. (2006a). (c) Theoretical dependence of community and species stability on the number of plant species as predicted by a model of multi-species competition. See Lehman and Tilman (2000). (d) The same as for (b), except this shows species stability in the Minnesota biodiversity experiment.

(one, two, four, eight and 16 perennial prairie plant species), with the species composition of each plot determined by a separate, random draw of a predefined number of species from a pool of 18 species. Annual data on the abundances of each species in each plot for the first decade of that experiment showed that total above-ground production (the sum of all above-ground plant production by all species in a plot) was significantly more temporally stable at higher diversity (Figure 3.1b), and that the abundance of each species was slightly less temporally stable at higher diversity (Figure 3.1d; Tilman et al., 1996). Temporal stability is measured as mean abundance (in this case, annual abundances averaged across 10 years) divided by its temporal standard deviation. A review and meta-analysis showed that, across 27 different grassland biodiversity experiments, ecosystem productivity was more stable and population sizes were less stable at higher plant diversity (Gross et al., 2014). On a genetic scale, greater genetic diversity in the seagrass, *Zostera*, led to greater stability of its population size. Other studies of marine ecosystems showed that greater species diversity led to greater ecosystem stability, much like terrestrial ecosystems (Stachowicz et al., 2007).

Theory
May's (1973) classic theoretical treatment of diversity–stability relationships showed that greater diversity tends to destabilise the dynamics of individual species. This led many in ecology to conclude that greater diversity should also destabilise the whole community or ecosystem, which is essentially the sum of all of the individual species. However, in the 1974 'Afterthoughts' discussion that May inserted upon the reprinting of his book, he noted that the stability of the full community of competing species would increase with diversity even though individual species would be less stable. That insight lay unappreciated until discovered during the 1990s. Indeed, the more recent biodiversity and ecosystem functioning theory of diversity and stability (Yachi & Loreau, 1998; Tilman, 1999; Lehman & Tilman, 2000; Loreau et al., 2003) supports the two stability predictions of May (1973, 1974). For example, Lehman and Tilman (2000) found that community stability, measured as the mean community biomass divided by its temporal standard deviation, was an essentially linearly increasing function of the number of species, but that the temporal stability of individual species was a slightly declining function of diversity (Figure 3.1c).

Diversity and productivity
Experimental studies
Darwin asserted that agriculturalists of his era knew that pastures planted with many grass species produced more herbage than those planted with fewer species. Similarly, Harper (1977) showed from several experiments that two competing plant species will overyield[1] when growing together, if they are coexisting, and offered a theoretical explanation for this by using isocline diagrams for the Lotka–Volterra competition model. The first replicated experiment to show that higher plant diversity led to greater primary

---

[1] Overyielding of a species is defined as 'the relative yield of a species (when grown) in mixture compared with that in monoculture where the null expectation is the monoculture yield times the starting proportion in mixture (e.g. in a two-species mixture where species were planted or seeded at equal density the expected yield of each species is 50 per cent of its monoculture value)' (Hector, 2006).

productivity (Naeem et al., 1994) was a short-term laboratory study in which plant diversity and trophic diversity were varied together. Two field experiments were planted in 1994 in which plant diversity was the only experimentally manipulated variable. Both experiments showed that ecosystem productivity was greater at higher diversity (Figure 3.2a shows results for

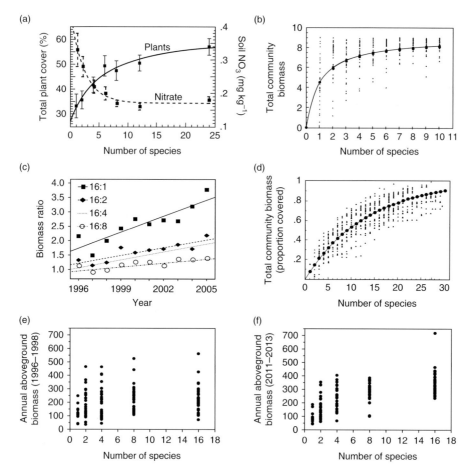

Figure 3.2 (a) Results of the 'small' Minnesota biodiversity experiment. Total community plant abundance ('Plants') increased with the number of planted species, while the concentration of unconsumed soil nitrate declined (Tilman et al., 1996). (b) Predictions of the sampling effect model of biodiversity and productivity (Tilman et al., 1997a). (c) Long-term effects of plant biodiversity on ecosystem productivity, showing the ratio of biomass in 16-species plots relative to that of monocultures (16:1), to two-species plots (16:2), etc. (Tilman et al., 2006b). (d) Predictions of a niche differentiation model of biodiversity and productivity (Tilman et al., 1997a). (e) Results of the 'big' Minnesota biodiversity experiment averaged across its first three years after plants reached adult size (1996–1998), and (f) averaged across the recent three years, 2011–2013, with results for each of 152 plots shown (unpublished results).

one experiment), and that the limiting soil nutrient was more fully used at higher diversity. Furthermore, having a greater range of functional traits[2] represented in a community, which corresponded with greater plant diversity, was associated with greater productivity (Tilman et al., 1997a, 1997b). Two other experiments used grassland species but did not include any legumes in the species pool, and observed similar results (van Ruijven & Berendse, 2003; Wilsey & Polley, 2004), thus showing that overyielding by high-diversity mixtures of species was not merely the result of nitrogen-fixing legumes. More than 100 additional diversity experiments have since been performed. Reviews of such work have reported trends like those observed in the earliest biodiversity experiments and have given deeper insights into the mechanisms causing greater productivity at higher diversity (Loreau et al., 2001; Hooper et al., 2005). A review and meta-analysis of the results of >100 experiments showed consistent effects of diversity on productivity, much like those of Figure 3.2a (Cardinale et al., 2006), with high-diversity treatments being about double the productivity of the average species grown in monoculture.

Longer-term studies have shown that the magnitude of such diversity effects can increase through time. The world's oldest biodiversity experiment had 16 species plots that were about 70 percent more productive than the average single-species plot in the third year of the experiment. The productivities of the monoculture plots did not change much through time. In contrast, the productivity of the higher-diversity plots increased through time (Figure 3.2c). After 11 years, the 16-species plots were about 250 percent more productive than the average of all monocultures (Figure 3.2c; Reich et al., 2012). A multi-experiment meta-analysis that included the age of an experiment as a variable found that across the full suite of biodiversity experiments, the positive effect of biodiversity on productivity became greater through time (Cardinale et al., 2007).

An issue immediately arose when the first biodiversity and productivity experiments were being done. Is the higher productivity of higher-diversity plots simply the result of a 'sampling effect' resulting from the greater chance that a highly productive species will be present in a high-diversity plot (Aarssen, 1997; Huston, 1997), or is it caused by interspecific 'niche complementarity' or other trade-off-based interspecific differences (Tilman et al., 1997a)?

Theory predicts that these two processes would have very different signatures (Figure 3.2b,d). Both cases assume that a pool of many species is

---

[2] 'Functional traits are morphological, biochemical, physiological, structural, phenological, or behavioural characteristics that are expressed in phenotypes of individual organisms and are considered relevant to the response of such organisms to the environment and/or their effects on ecosystem properties (Violle et al., 2007)' in Díaz et al. (2013).

randomly sampled to determine which species would be in a community. They differ in the type of interspecific interactions that are assumed. The sampling effect model assumes that species differ greatly in the productivity that they can achieve in monoculture, and that a more productive species will always competitively displace a less productive species when growing together. The sampling effect model generated Figure 3.2b. Note the great differences among the species in how productive each species is in monoculture. Note also that whenever the most productive species happens to be present, it imposes its productivity on the system because it outcompetes all other species. The net result of such interactions is that the upper bound of the variation in productivity across all diversity levels is flat (Figure 3.2b). This occurs because no species combination is ever more productive than the productivity of the most productive species that it contains. Thus, were one to wish to obtain the greatest possible productivity, all that would be required would be to find and plant the single most productive species (or variety).

In marked contrast, if there are interspecific trade-offs that make each species be a better competitor for some types of conditions and a poorer competitor for others, multi-species coexistence could occur, as already discussed. In this situation, in which niche complementarity leads to multi-species coexistence, the upper bound of the diversity–productivity relationship is predicted to be an increasing curve (Figure 3.2d). This means that there would be some combination of two species that was more productive than any of these species grown in monoculture, some combinations of 10 species that was more productive than the best combination of nine species, and so on. Niche complementarity and limiting resources thus lead to the prediction that high-diversity plots would provide much greater productivity than could any of those same species grown in monocultures.

This theoretical prediction is born out in long-term plant biodiversity experiments (Tilman et al., 2001a; Tilman et al., 2014). The 16-species plots became much more productive, on average, than the single most productive species grown in monoculture. On average, across three recent years of data collection in the Minnesota biodiversity experiment (2011–2013), every single 16-species plot was more productive than the most productive species in monoculture (Figure 3.2f) and the average 16 species plot was >200 percent more productive than the average of all species in monoculture. Compared to the most productive monoculture species (the legume *Petalostemon purpureum*), the average 16-species plot was 84 percent more productive. However, it took time for this pattern to emerge. On average, for the first three years that this experiment was sampled, 16-species plots were 82 percent more productive than the average of all the monocultures, but the most productive monoculture species at that time, *Lupinus perennis*, was as productive as the average 16-species plot (Figure 3.2e).

Mathematical models of competition among niche-differentiated plant species for limiting resources predict that ecosystem productivity is an increasing function of plant species numbers, that the levels of unconsumed limiting resources decline with diversity and that the upper bound of the diversity–productivity relationship is an increasing function of diversity (Lehman & Tilman, 2000; Loreau, 1998; Tilman et al., 1997a). These same models also predict that ecosystem stability increases with diversity, and that invasion by exotic species declines with diversity (Thébault & Loreau, 2003; Tilman, 2004).

Biodiversity and invasibility

Theory predicts, and experiments show, that limiting resources are consumed down to lower levels in ecological communities that have higher diversity of consuming species. A new consumer species that is invading into an occupied site only has available to it whatever limiting resources are left unconsumed by the established species. Invaders thus must be able to grow and survive on such leftover resources. It is for this reason that models of species invasions predict that fewer species should be able to invade, and that they would attain lower abundances once established, when invading into higher-diversity ecosystems.

These predictions are well supported by the results of biodiversity experiments. For instance, in a Minnesota biodiversity experiment, Knops et al. (1999) found that fewer exotic plant species (mainly weeds and pasture plant species of Eurasian origin) invaded into plots planted with a higher diversity of native prairie perennial plant species. A similar but more spatially complex pattern was observed when Naeem et al. (2000) mapped out the locations and sizes of planted and invading individuals in a biodiversity experiment. Invasions were less likely, and invading individual plants were smaller, when their immediate spatial neighbourhood had higher diversities and densities of the species planted in the biodiversity experiment. Such neighbourhoods also had lower light levels and lower levels of soil nitrate. In a direct test of the biodiversity–invasibility hypothesis, Fargione et al. (2003) added seed of novel plant species that had different functional traits to the plots of the Minnesota biodiversity experiment. The three major findings of this experiment were that (1) a smaller number of the added species became established in the higher-diversity treatments; (2) the added species that did become established had lower biomasses in higher-diversity treatments; and (3) the potential invaders, which differed in their functional traits, were most inhibited in plots containing high abundances of functionally similar established species (Fargione et al., 2003). Invasion experiments in a Californian grassland (Zavaleta & Hulvey, 2004), in a European grassland in Jena, Germany (Petermann et al., 2010) and in a community of marine sessile organisms found similar effects of resource levels and biodiversity on invader success (Stachowicz et al., 1999).

In total, these experiments show that high-diversity plantings are highly resistant to invasion by exotic species. On the practical side, this means that high-diversity combinations of species are self-maintaining, and thus need little if any inputs of chemicals, energy or labour to persist in their planted state. In stark contrast, agricultural monocultures, or, for that matter, monocultures of pasture plants, trees, or perhaps even fish, are much more readily invaded and thus can require considerable inputs of chemicals, energy or labour to persist as monocultures.

Biodiversity, soil fertility and sequestration of carbon

During the first 25–50 years of annual crop cultivation on newly created farmland, about 30–40 percent of the original organic carbon content of the soil is lost to the atmosphere as CO_2 released by microbial respiration. Results of the Minnesota biodiversity experiment suggest that this lost carbon can be captured and resequestered if tilling is stopped and native perennials are planted on the depleted soils (Tilman et al., 2006b). Moreover, this experiment suggests that the restoration of soil fertility and soil carbon is highly dependent on plant biodiversity. For instance, there was no detectable increase in soil carbon across all planted monocultures of prairie perennials during the first decade of the Minnesota biodiversity experiment (Figure 3.3a), but the observed annual rate of carbon accumulation in the 16-species plots was about 0.9 tonnes ha^{-1} $year^{-1}$ of C (equivalent to 3.2 tonnes ha^{-1} $year^{-1}$ of CO_2 removed from the atmosphere). The roots of the 16-species plots also contained 0.3 more tonnes ha^{-1} of organic carbon after 10 years of growth than those of the average monoculture. These feedback effects of higher plant diversity also increase soil nitrogen content (Figure 3.3b), and thus soil fertility, and are a major reason why productivity of higher-diversity plots increased through time (Fornara & Tilman, 2009).

Food web effects

In grassland biodiversity experiments, higher diversity of plant species led to greater insect diversity, greater soil microbial diversity, lower incidences of species-specific fungal diseases and greater diversity of mycorrhizal fungi diversity (Knops et al., 1999; Haddad et al., 2001; Mitchell et al., 2002; Antoninka et al., 2011). Perhaps the most intriguing food web effect of plant diversity is the shift in food web structure reported for the Minnesota biodiversity experiment. Haddad et al. (2009) found that low-diversity plots had food webs dominated by herbivorous insects, while high-diversity plots had food webs dominated by predatory and parasitoid arthropods, with low numbers of herbivorous arthropods. These food web differences likely mean that

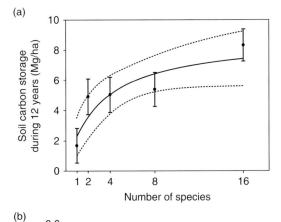

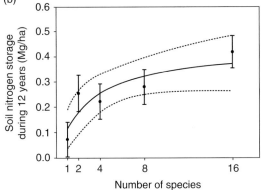

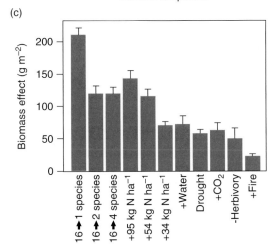

Figure 3.3 (a) Effects of plant species diversity on soil carbon storage in the Minnesota biodiversity experiment. Results show total C stored during a 12-year period (see Fornara & Tilman, 2008). (b) Soil nitrogen storage during this same period. (c) Effect of each item shown on the x-axis on annual production of aboveground plant biomass in grassland ecosystems of Minnesota. 16->1 species is the difference in the productivities of 16-species plots versus one-species plots in the Minnesota biodiversity experiment. 16->2 and 16->4 compare 16-species plots to those planted with two or four species. +95 kg N ha^{-1} compares productivity of plots receiving this amount of N fertiliser with plots receiving none, and so on for other treatments shown (Tilman et al., 2012).

arthropod herbivory is lessened at higher plant diversity, which could contribute to greater productivity, as could the lower disease incidence and greater abundances of mycorrhizal mutualists observed at higher plant diversity.

Biodiversity as a major controller of ecosystem functioning

As noted above and in many other studies (Tilman et al., 2014), greater plant biodiversity leads to greater primary productivity, greater use of limiting resources, greater ecosystem stability, greater rates of accumulation of soil carbon and nitrogen and thus higher soil fertility, and lower rates of invasion by exotic species. In one study, higher plant diversity also led to shifts in food web structure that favoured predators and parasitoids over herbivores. Some of these effects of diversity had been proposed half a century to even a century and a half ago, but a broad consensus on most of these impacts of biodiversity was not reached until recently.

Three decades ago, nutrient availability, climate, disturbance, herbivory, fire and drought were considered to be the major factors determining ecosystem functioning. Research done during the past two decades has shown not only that biodiversity must be added to this list, but also that biodiversity may be near the top of this list (Hooper et al., 2012; Tilman et al., 2012). These studies showed that human-caused loss of diversity, of the magnitude resulting from habitat simplification and habitat destruction, had as large – or larger – an effect on ecosystem primary productivity as changes in atmospheric carbon dioxide, fire, drought, herbivory and the highest rates of atmospheric nitrogen deposition observed on Earth (Figure 3.3c). It is especially for these reasons that the potential uses of biodiversity should be given careful attention as we search for ways to make agriculture both more productive and more sustainable.

Relevance of biodiversity to agriculture

Some types of biodiversity are widely known to be of central importance to the sustainability of agriculture (Zimmerer, 2010), such as genetic diversity within each crop species, which is often preserved via seed banks that contain thousands to hundreds of thousands of genetic variants (land races) of crops such as wheat, rice and maize. These genetic variants are essential for controlling crop diseases, developing new varieties for different soils or climates and for increasing crop yields (Simmonds, 1990). The need to preserve and use genetic variation applies equally well to animals grown for food.

Diversity and crop disease

The resistance of a crop to disease is a major determinant of its yield, and the ability of many pathogens to rapidly evolve ways to overcome such resistance is a continuing threat to yields. One way that such threats are overcome in modern agriculture is by using genetic diversity to breed new crop varieties through time, which is a strategy of temporal deployment of resistant crop varieties. A major epidemic of wheat stem rust greatly decreased yields in the 1950s but has been controlled since then by using the wheat's genetic diversity to continually find varieties that are resistant to the evolving traits of stem rust (Roelfs, 1988). The need for genetic diversity to overcome evolving pathogens is also well-illustrated by rice. The yield of the first major Green Revolution rice variety, IR8, declined 25 percent over the subsequent 30 years. The actual rice yields of farmers, though, did not decline because 11 major new rice strains were introduced during that time period, each of which overcame the yield impacts of then-current pathogens (Cassman et al., 2003).

An alternative way to use genetic diversity to control crop diseases is the simultaneous planting of intermingled mixtures of multiple variants of a crop that have differing genetic defences against disease. This is a strategy of the simultaneous spatial deployment of genetic resistance. The greater the diversity of defences that are deployed spatially, the lower the probability that a pathogen can spread from one infected individual to another one that is also susceptible to that particular strain of the pathogen. The use of this strategy has successfully increased wheat yields (Kiær et al., 2009). Similarly, Zhu et al. (2000) found that planting alternating rows of two different rice varieties controlled a fungal disease that previously had required application of fungicide. Because it seems plausible that spatial deployment may greatly slow the rate of pathogen evolution, the spatial deployment strategy merits much further exploration. For such evolutionary issues, a Darwinian perspective on agriculture will be essential (Denison, 2012).

Because any given crop or agricultural animal could potentially be lost if a novel disease or other pest emerged for which no genetic solution could be found, the long-term ability of the global food system to sustainably feed 9–10 billion people may well depend on continual efforts to discover and develop new crops. It is especially important that we have viable alternative grain and legume crops 'in reserve' because four such crops – wheat, rice, maize and soybeans – currently provide about 75 percent of global calories and protein. The sudden loss of any one of these four crops would be devastating if viable alternatives were not readily available.

Below are some other agricultural uses of biodiversity that merit deeper exploration and potentially broader adoption.

Biodiversity and crop yields

This juxtaposition of 'crop yields' and 'biodiversity' is not meant to imply that any or all of the findings of the effects of biodiversity on ecosystem functioning, as illustrated in Figures 3.1–3.3, can be simply or easily applied to agriculture. The potential applicability of these findings is the subject of ongoing discussion (Swift et al., 2004; Snapp et al., 2010; Zimmerer, 2010). Major sources of funding are needed that can turn these discussions into global field research. Only much additional experimental work in a variety of agricultural systems can do so. It seems likely that biodiversity may be found for some conditions to be an important new tool for assuring greater crop yields and greater stability of yields, and for reducing the environmental impacts of feeding a world of 9–10 billion people.

It is well known that growing two or more appropriately chosen crops together, which is called intercropping, can lead to increased yields (Vandermeer, 1992). Despite this, the vast majority of annual crops are grown in monocultures in developed and many developing nations. Intercropping, though, is practiced in many parts of the world, including Latin America, India, Africa, Southeast Asia and China (Vandermeer, 1992; Li et al., 2007), and is common in some nations. For instance, more than 28 million hectares of cropland are currently dedicated to intercropping in China (Li et al., 2007).

If two or more crops have differing environmental requirements (that is, are niche-differentiated), the total yield obtained from a given area of cropland via intercropping can be greater than the yields obtained from any of the crop species grown in monoculture, which is called overyielding (Rao & Willey, 1980; Vandermeer, 1992; Gliessman, 1998; Zhang & Li, 2003; Li et al., 2007; Kiær et al., 2009; Snapp et al., 2010). The yield benefits of intercropping can be large. For instance, when faba bean and maize were grown in alternating rows on phosphorus deficient soils, faba bean overyielded by 26 percent and maize by 43 percent compared to their respective monocultures. This overyielding was mainly caused by the ability of faba bean to release organic acids that mobilised otherwise unavailable soil phosphorus, and by the different growing seasons of these two crop species. Yield advantages have been reported when nitrogen-fixing legumes and cereal grains are intercropped on nitrogen deficient soils (Rao & Mathuva, 2000).

Other types of differences among crop species, which ecologists would call niche differentiation, merit fuller exploration to determine the yield benefits they might provide. For instance, some crops have physiologies that cause them to grow best in spring and early summer when conditions are cooler and wetter, while other crops grow best when it is warmer and drier. Some crops have shallow root systems and others deep. Such differences could allow the two species when grown in mixture to better exploit the full growing season and/or the full soil volume, and thus to overyield. It is also worth determining whether intercropping could inhibit weeds, and to use it to lower disease incidence (e.g. Zhu et al., 2000). The overall benefit for production would need to consider any additional costs, for example of sowing and/or harvesting a mix of crops compared to a species grown in monoculture.

In some cases, inputs of limiting resources can provide the same yield benefits as intercropping. However, such inputs have economic and environmental costs that intercropping would not have, although intercropping can have added costs associated with harvesting.

In total, the wider exploration of the potential benefits of intercropping merits serious attention. For instance, currently, 780 million hectares of cropland, which is 61 percent of total global cropland, are used to produce grains and legumes, for which a large number of studies have demonstrated significant yield increases from intercropping. An even higher percentage of the ~800 million hectares of land that may be cleared in the next 50 years will likely be dedicated to grains and legumes because these crops are also major animal feeds, and meat consumption is increasing globally. Widespread adoption of intercropping could provide a major environmental benefit by potentially negating the need to convert ~100 million hectares of intact ecosystems into cropland while also decreasing nutrient pollution and input costs. These possibilities merit serious international attention.

Biodiversity, pasture and hay

About 25 percent of the ice and permafrost-free land surface of the Earth is used to graze livestock. Given the results of all the grassland biodiversity experiments that have been performed, it seems plausible that appropriate high-diversity mixtures of pasture grasses, forbs and legumes might produce about double the forage per hectare per year of monocultures (Cardinale et al., 2012; Tilman et al., 2014). Work done on other grassland perennials suggests that, relative to monocultures, such mixtures may also be more stable (less year-to-year variation in yields), remove and sequester more atmospheric carbon, increase soil fertility, and improve water quality by removing more nutrients and thus decreasing nutrient leaching. However, there are many unanswered questions. How might different intensities and methods of grazing impact the potential benefits of biodiversity? Might intensive rotational grazing allow the long-term persistence of high plant diversity, which mowing for hay is known to do? Work over the past 150 years in the Park Grass Experiment (Rothamsted, UK) might seem to suggest that switching from grazing (which is selective removal of plant species, including legumes) to haying (which is not selective) may be the cause of a long-term threefold increase in annual biomass production in diverse (control) grassland plots that have never been fertilised but have been harvested for hay twice each year since 1856 (Tilman et al., 1994).

Biodiversity and nutrient abatement by buffers

One of the environmental costs of intensive crop production is nutrient loading into groundwater, streams, rivers, lakes and marine ecosystems associated with fertiliser application. As discussed above, intercropping might increase the amount of fertiliser taken up by crops and thus reduce this loading. Buffer strips planted orthogonal to the flow of surface and subsurface water can intercept some of the nutrients and remove them. Might buffer strips planted with high-diversity mixtures of native perennials be better at nutrient removal? Might they accrue soil C and N through time, and thus increase soil fertility? If so, locations of buffers could be slowly rotated back into cropland to increase the fertility of croplands as new areas are dedicated to serve as buffers. Such a practice might help increase stores of soil carbon and help mitigate atmospheric greenhouse gases. Further work is needed to explore such possibilities.

Conclusions

Agriculture is the foundation of modern societies, whose functioning and stability are dependent on its productivity and stability. It is also the greatest cause of extinction risks and of aquatic and groundwater pollution (Tilman et al., 2017). Based on projected increases in global population and income, global demand for crops may double within the next 50 years (Tilman et al., 2011). If agriculture continues to develop along its current trajectories, crop yields would be insufficient to meet global food demand without massive increases in land clearing, the resultant risks of extinctions of species in the developing world, especially sub-Saharan Africa, and great increases in nutrient pollution of lakes, rivers, groundwater and the oceans (Tilman et al., 2017).

As discussed above, the creative incorporation into agricultural research of the possibilities raised by recent advances in our understanding of biodiversity may help accelerate the rate at which crop yields increase, may increase fertiliser efficiency, and decrease the need for pesticides. The potential ability of intercropping to increase yields is clear, but its adoption awaits agronomic field trials, development of agricultural machinery for harvesting, discovery of the crop combinations that provide the best returns to farmers and society, and development of crop varieties that take the greatest advantage of intercropping. Agriculture also needs to incorporate evolutionary mechanisms (Denison, 2012) to find ways to better control both the evolution and outbreaks of new pathogens. One possibility is the simultaneous spatial deployment of many alternative genetic pest-resistant genotypes, which also merits exploration, as does the simultaneous planting of several different high-yielding varieties of a single crop. Pasture management already often takes advantage of the benefits of planting a diversity of forage species, but given the immense amount of land used for pasture and the value of animal protein produced, major advances in pasture management are also of crucial importance globally.

Agriculture, broadly defined, accounts for at least a third of current global greenhouse gas emissions, and these emissions are on an accelerating trajectory. There are many ways that these impacts of expanding population and consumption can be addressed (e.g. Tilman et al., 2001b; Foley et al., 2011;

Tilman et al., 2011; Matson, 2012). The work summarised here suggests that using biodiversity can contribute to creating a more environmentally sustainable world.

References

Aarssen, L.W. (1997). High productivity in grassland ecosystems: effected by species diversity or productive species? *Oikos*, **80**, 183-184.

Antoninka, A., Reich, P.B. & Johnson, N.C. (2011). Seven years of carbon dioxide enrichment, nitrogen fertilization and plant diversity influence arbuscular mycorrhizal fungi in a grassland ecosystem. *The New Phytologist*, **192**, 200-214.

Armstrong, R.A. & McGehee, R. (1980). Competitive exclusion. *American Naturalist*, **115**, 151-170.

Bai, Y., Han, X., Wu, J., Chen, Z. & Li, L. (2004). Ecosystem stability and compensatory effects in the Inner Mongolia grassland. *Nature*, **431**, 181-184.

Cardinale, B.J., Srivastava, D.S., Duffy, J.E., et al. (2006). Effects of biodiversity on the functioning of trophic groups and ecosystems. *Nature*, **443**, 989-992.

Cardinale, B.J., Wright, J.P., Cadotte, M.W., et al. (2007). Impacts of plant diversity on biomass production increase through time because of species complementarity. *Proceedings of the National Academy of Sciences of the United States of America*, **104**, 18123-18128.

Cardinale, B.J., Duffy, J.E., Gonzales, A., et al. (2012). Biodiversity loss and its impact on humanity. *Nature*, **486**, 59-67.

Cassman, K.G., Dobermann, A., Walters, D.T. & Yang, H. (2003). Meeting cereal demand while protecting natural resources and improving environmental quality. *Annual Review of Environment and Resources*, **28**, 315-358.

Daily, G. (ed.). (1997). *Nature's Services: societal dependence on natural ecosystems*. Washington, DC: Island Press.

Darwin, C. (1859). *On the Origins of Species by Means of Natural Selection*. London: Murray.

Denison, R.F. (2012). *Darwinian Agriculture: how understanding evolution can improve agriculture*. Princeton, NJ: Princeton University Press.

Díaz, S., Purvis, A., Cornelissen, J.H.C., et al. (2013). Functional traits, the phylogeny of function, and ecosystem service vulnerability. *Ecology and Evolution*, **3**, 2958-2975.

Duvick, D.N. (1984). Genetic diversity in major farm crops on the farm and in reserve. *Economic Botany*, **38**, 161-178.

Ehrlich, P.R. & Ehrlich, A.H. (1981). *Extinction: the causes and consequences of the disappearance of species*. New York, NY: Random House.

Elton, C.S. (1958). *The Ecology of Invasions by Animals and Plants*. London: Methuen.

Fargione, J., Brown, C. & Tilman, D. (2003). Community assembly and invasion: an experimental test of neutral versus niche processes. *Proceedings of the National Academy of Sciences of the United States of America*, **101**, 8916-8920.

Foley, J.A., Ramankutty, N., Brauman, K.A., et al. (2011). Solutions for a cultivated planet. *Nature*, **478**, 337-342.

Fornara, D.A. & Tilman, D. (2008). Plant functional composition influences rates of soil carbon and nitrogen accumulation. *Journal of Ecology*, **96**, 314-322.

Fornara, D.A. & Tilman, D. (2009). Ecological mechanisms associated with the positive diversity-productivity relationship in an N-limited grassland. *Ecology*, **90**, 408-418.

Gliessman, S.R. (1998). *Agroecology: ecological processes in sustainable agriculture*. Chelsea, MI: Ann Arbor Press.

Goodman, D. (1975). The theory of diversity-stability relationships in ecology. *The Quarterly Review of Biology*, **50**, 237-266.

Gross, K., Cardinale, B.J., Fox, J.W., et al. (2014). Species richness and the temporal stability of biomass production: a new analysis of recent biodiversity experiments. *The American Naturalist*, **183**, 1–12.

Haddad, N.M., Tilman, D., Haarstad, J., Ritchie, M. & Knops, J.M.H. (2001). Contrasting effects of plant richness and composition on insect communities: a field experiment. *The American Naturalist*, **158**, 17–35.

Haddad, N.M., Crutsinger, G.M., Gross, K., et al. (2009). Plant species loss decreases arthropod diversity and shifts trophic structure. *Ecology Letters*, **12**, 1029–1039.

Harper, J.L. (1977). *Population Biology of Plants*. London: Academic Press.

Hastings, A. (1980). Disturbance, coexistence, history, and competition for space. *Theoretical Population Biology*, **18**, 363–373.

Hector, A. (2006). Overyielding and stable species coexistence. *New Phytologist*, **172**, 1–3.

Hector, A. & Hooper, R. (2002). Darwin and the first ecological experiment. *Science*, **295**, 639–640.

Hooper, D.U., Chapin, III, F.S., Ewel, J.J., et al. (2005). Effects of biodiversity on ecosystem functioning: a consensus of current knowledge. *Ecological Monographs*, 75, 3–35.

Hooper, D.U., Adair, E.C., Cardinale, B.J., et al. (2012). A global synthesis reveals biodiversity loss as a major driver of ecosystem change. *Nature*, **486**, 105–108.

Huston, M.A. (1997). Hidden treatments in ecological experiments: re-evaluating the ecosystem function of biodiversity. *Oecologia*, **110**, 449–460.

Kiær, L.P., Skovgaard, I.M. & Østergard, H. (2009). Grain yield increase in cereal variety mixtures: a meta-analysis of field trials. *Field Crops Research*, **114**, 361–373.

Knops, J.M.H., Tilman, D., Haddad, N., et al. (1999). Effects of plant species richness on invasion dynamics, disease outbreaks, insect abundances and diversity. *Ecology Letters*, **2**, 286–293.

Lehman, C.L. & Tilman, D. (2000). Biodiversity, stability, and productivity in competitive communities. *American Naturalist*, **156**, 534–552.

Levin, B.R., Stewart, F.M. & Chao, L. (1977). Resource-limited growth, competition, and predation: a model and experimental studies with bacteria and bacteriophage. *American Naturalist*, **111**, 3–24.

Levins, R. (1979). Coexistence in a variable environment. *American Naturalist*, **114**, 765–783.

Li, L., Li, S-M., Sun, J-H., et al. (2007). Diversity enhances agricultural productivity via rhizosphere phosphorus facilitation on phosphorus-deficient soils. *Proceedings of the National Academy of Sciences of the United States of America*, **104**, 11192–11196.

Li, W., Li, W., Tan, R., et al. (2013). Effects of anthropogenic disturbance on richness-dependent stability in Napahai plateau wetland. *Chinese Science Bulletin*, **58**, 4120–4125.

Loreau, M. (1998). Biodiversity and ecosystem functioning: a mechanistic model. *Proceedings of the National Academy of Sciences of the United States of America*, **95**, 5632–5636.

Loreau, M., Naeem, S., Inchausti, P., et al. (2001). Biodiversity and ecosystem functioning: current knowledge and future challenges. *Science*, **294**, 804–808.

Loreau, M., Mouquet, N. & Gonzalez, A. (2003). Biodiversity as spatial insurance in heterogeneous landscapes. *Proceedings of the National Academy of Sciences of the United States of America*, **100**, 12765–12770.

Matson, P.A. (ed.). (2012). *Seeds of Sustainability: lessons from the birthplace of the green revolution in agriculture*. Washington, DC: Island Press.

May, R.M. (1973). Time-delay versus stability in population models with two and three trophic levels. *Ecology*, **54**, 315–325.

May, R.M. (1974). *Stability and Complexity in Model Ecosystems*. 2nd edition. Princeton, NJ: Princeton University Press.

McNaughton, S.J. (1977). Diversity and stability of ecological communities: a comment on the role of empiricism in ecology. *Ecology*, **111**, 515-525.

McNaughton, S.J. (1993). Biodiversity and function of grazing ecosystems. In: *Biodiversity and Ecosystem Function*, edited by E.D. Schulze & H.A. Mooney. Berlin: Springer-Verlag.

Mitchell, C.E., Tilman, D. & Groth, J.V. (2002). Effects of grassland and plant species diversity, abundance, and composition on foliar fungal disease. *Ecology*, **83**, 1713-1726.

Naeem, S., Thompson, L.J., Lawler, S.P., Lawton, J.H. & Woodfin, R.M. (1994). Declining biodiversity can alter the performance of ecosystems. *Nature*, **368**, 734-737.

Naeem, S., Knops, J.M.H., Tilman, D., et al. (2000). Plant diversity increases resistance to invasion in the absence of covarying extrinsic factors. *Oikos*, **91**, 97-108.

Odum, E. & Odum, H. (1961). *Fundamentals of Ecology*. Philadelphia, PA: W. B. Saunders Co.

Petermann, J.S., Fergus, A.J.F., Roscher, C., et al. (2010). Biology, chance, or history? The predictable reassembly of temperate grassland communities. *Ecology*, **91**, 408-421.

Rao, M.R., & Willey, R.W. (1980). Evaluation of yield stability in intercropping: studies on sorghum/pigeonpea. *Experimental Agriculture*, **16**, 105-116.

Rao, M.R. & Mathuva, M.N. (2000). Legumes for improving maize yields and income in semi-arid Kenya. *Agriculture, Ecosystems & Environment*, **78**, 123-137.

Reich, P.B., Tilman, D., Isbell, F., et al. (2012). Impacts of biodiversity loss escalate through time as redundancy fades. *Science*, **336**, 589-592.

Roelfs, A.P. (1988). Genetic control of phenotypes in wheat stem rust. *Annual Reviews of Phytopathology*, **26**, 351-367.

Schulze, E.D. & Mooney, H.A. (eds.). (1993). *Biodiversity and Ecosystem Function*. Berlin: Springer-Verlag.

Simmonds, N.W. (1990). Uniformity and yield in hybrid crop cultivars. *Tropical Agriculture (Trinidad)*, **68**, 198-199.

Snapp, S.S., Blackie, M.J., Gilbert, R.A., Bezner-Kerr, R. & Kanyama-Phiri, G.Y. (2010). Biodiversity can support a greener revolution in Africa. *Proceedings of the National Academy of Sciences of the United States of America*, **107**, 20840-20845.

Stachowicz, J.J., Whitlatch, R.B. & Osman, R.W. (1999). Species diversity and invasion resistance in a marine ecosystem. *Science*, **286**, 1577-1579.

Stachowicz, J.J., Bruno, J. & Duffy, J.E. (2007). Understanding the effects of marine biodiversity on communities and ecosystems. *Annual Review of Ecology, Evolution and Systematics*, **38**, 739-766.

Swift, M.J. & Anderson, J.M. (1993). Biodiversity and ecosystem function in agricultural systems. In: *Biodiversity and Ecosystem Function*, edited by E.D. Schulze & H.A. Mooney. Berlin: Springer-Verlag, pp. 15-42.

Swift, M.J., Izac, A.-M.N. & van Noordwijk, M. (2004). Biodiversity and ecosystem services in agricultural landscapes – are we asking the right questions? *Agriculture, Ecosystems & Environment*, **104**, 113-134.

Thébault, E. & Loreau, M. (2003). Food-web constraints on biodiversity–ecosystem functioning relationships. *Proceedings of the National Academy of Sciences of the United States of America*, **100**, 14949-14954.

Tilman, D. (1994). Competition and biodiversity in spatially structured habitats. *Ecology*, **75**, 2-16.

Tilman, D. (1999). The ecological consequences of changes in biodiversity: a search for general principles. *Ecology*, **80**, 1455-1474.

Tilman, D. (2004). Niche tradeoffs, neutrality, and community structure: a stochastic theory of resource competition, invasion, and community assembly. *Proceedings of the National Academy of Sciences of the United States of America*, **101**, 10854-10861.

Tilman, D. & Downing, J.A. (1994). Biodiversity and stability in grasslands. *Nature*, **367**, 363-365.

Tilman, D., Kilham, S.S. & Kilham, P. (1982). Phytoplankton community ecology: the role of limiting nutrients. *Annual Review of Ecology and Systematics*, **13**, 349–372.

Tilman, D., Dodd, M.E., Silvertown, J., et al. (1994). The Park Grass Experiment: insights from the most long-term ecological study. In: *Long-term Experiments in Agricultural and Ecological Sciences*, edited by R.A. Leigh and A.E. Johnston. Wallingford: CAB International, pp. 287–303.

Tilman, D., Wedin, D. & Knops, J. (1996). Productivity and sustainability influenced by biodiversity in grassland ecosystems. *Nature*, **379**, 718–720.

Tilman, D., Lehman, C.L. & Thomson, K.T. (1997a). Plant diversity and ecosystem productivity: theoretical considerations. *Proceedings of the National Academy of Sciences of the United States of America*, **94**, 1857–1861.

Tilman, D., Knops, J., Wedin, D., et al. (1997b). The influence of functional diversity and composition on ecosystem processes. *Science*, **277**, 1300–1302.

Tilman, D., Duvick, D.N., Brush, S.B., et al. (1999). *Benefits of Biodiversity*. Task Force Report No. 133. Ames, IA: Council for Agricultural Science and Technology.

Tilman, D., Reich, P.B., Knops, J., et al. (2001a). Diversity and productivity in a long-term grassland experiment. *Science*, **294**, 843–845.

Tilman, D., Fargione, J., Wolff, B., et al. (2001b). Forecasting agriculturally driven global environmental change. *Science*, **292**, 281–284.

Tilman, D., Reich, P.B. & Knops, J.M.H. (2006a). Biodiversity and ecosystem stability in a decade-long grassland experiment. *Nature*, **441**, 629–632.

Tilman, D., Hill, J. & Lehman, C. (2006b). Carbon-negative biofuels from low-input high-diversity grassland biomass. *Science*, **314**, 1598–1600.

Tilman, D., Balzer, C., Hill, J. & Befort, B.L. (2011). Global food demand and the sustainable intensification of agriculture. *Proceedings of the National Academy of Sciences of the United States of America*, **108**, 20260–20264.

Tilman, D., Reich, P.B. & Isbell, F. (2012). Biodiversity impacts ecosystem productivity as much as resources, disturbance, or herbivory. *Proceedings of the National Academy of Sciences of the United States of America*, **109**, 10394–10397.

Tilman, D., Isbell, F. & Cowles, J.M. (2014). Biodiversity and ecosystem functioning. *Annual Review of Ecology, Evolution and Systematics*, **45**, 471–493.

Tilman, D., Clark, M., Williams, D., et al. (2017). Future threats to biodiversity and pathways to their prevention. *Nature*, **546**, 73–81.

Van Ruijven, J. & Berendse, F. (2003). Positive effects of plant species diversity on productivity in the absence of legumes. *Ecology Letters*, **6**, 170–175.

Vandermeer, J.H. (1992). *The Ecology of Intercropping*. Cambridge: Cambridge University Press.

Violle, C., Navas, M.L., Vile, D., et al. (2007). Let the concept of trait be functional! *Oikos*, **116**, 882–892.

Wilsey, B.J. & Polley, W.H. (2004). Realistcally low species evenness does not alter grassland species-richness–productivity relationships. *Ecology*, **85**, 2693–2700.

Wilson, E.O. (1988). The current state of biological diversity. In: *Biodiversity*, edited by E.O. Wilson & F.M. Peter. Washington, DC: National Academy Press.

Yachi, S. & Loreau, M. (1999). Biodiversity and ecosystem productivity in a fluctuating environment: the insurance hypothesis. *Proceedings of the National Academy of Sciences of the United States of America*, **96**, 1463–1468.

Zavaleta, E. & Hulvey, K. (2004). Realistic species losses disproportionately reduce grassland resistance to biological invaders. *Science*, **306**, 1175–1177.

Zhang, F. & Li, L. (2003). Using competitive and facilitative interactions in intercropping systems enhances crop productivity and nutrient-use efficiency. *Plant and Soil*, **248**, 305–312.

Zhu, Y., Chen, H., Fan, J., et al. (2000). Genetic diversity and disease control in rice. *Nature*, **406**, 718–722.

Zimmerer, K.S. (2010). Biological diversity in agriculture and global change. *Annual Review of Environment and Resources*, **35**, 137–166.

CHAPTER FOUR

Determining the value of ecosystem services in agriculture

ROSEMARY S. HAILS[1]
Centre for Ecology & Hydrology
REBECCA CHAPLIN-KRAMER
Natural Capital Project, Stanford University
ELENA BENNETT
McGill University
BRIAN ROBINSON
McGill University
GRETCHEN DAILY
Natural Capital Project, Stanford University
KATE BRAUMAN
University of Minnesota
and
PAUL WEST
University of Minnesota

Introduction

Ecosystem services and agriculture

The Millennium Ecosystem Assessment (2003) provided the foundations for characterising the links between ecosystems and the benefits they provide to humans. This conceptual framework has since been extended to include the abiotic elements of the natural environment which combine with biotic components to produce goods and services. Ecosystems are therefore a nested component of 'natural capital', which has been defined as 'a configuration of natural resources and ecological processes, which contributes through its existence and/or in some combination to human welfare' (Natural Capital Committee, 2013). This framework has continued to advance through the International Panel for Biodiversity and Ecosystem Services (IPBES; Pascual et al., 2017; Diaz et al., 2015).

The concept of natural capital, as with other forms of capital, is that it produces goods and services of value to people. Therefore, ecosystem services are the outputs of ecosystems which flow from the underlying stocks of natural capital, to produce goods and services that people value and benefit

---

[1] Rosemary Hails now works for the National Trust, UK.

from. The Millennium Ecosystem Assessment (MA) classified ecosystem services into four groups: (i) supporting services, which include fundamental processes such as soil formation, primary production, nutrient and water cycling; (ii) regulating services, which include processes such as climate regulation, detoxification in soil, air and water, disease and pest regulation and pollination; (iii) provisioning services that produce goods such as food, fuel and fibre; and (iv) cultural services that connect people to nature, spiritually, aesthetically, educationally, recreationally or otherwise.

Some ecosystem services are directly linked to the production of goods and these are 'final ecosystem services', while others underpin those final services and are 'intermediate ecosystem services' (UK National Ecosystem Assessment, 2011). This distinction is important when valuing the benefits derived from agricultural systems, to avoid double counting (Fisher & Turner, 2008). For example, pollination is an intermediate service, necessary for the production of some foods; its value is therefore contained within the final value of the food itself. Furthermore, when valuing final goods (such as food), frequently other capital inputs such as machinery (manufactured capital) and expertise (human capital) have also contributed to the production. Final values should attempt to tease apart the value of these different contributions if the aim is to understand the specific value that is attributable to an element of natural capital or an ecosystem service.

The role of biodiversity in this framework is particularly complex, as it has roles at every level of the process (Mace et al., 2012). Tilman (Chapter 3) explores the role of biodiversity in agricultural productivity, and also in the accumulation of soil carbon and nitrogen (thus enhancing soil fertility and benefitting other ecosystem services). Biodiversity is also a regulator of other important intermediate ecosystem services, such as pollination and pest control, where the enhancement of biodiversity has been shown to enhance those services and ultimately crop yield (e.g. Tschumi et al., 2016). Biodiversity in its broadest sense (from within species to variation in landscapes) is thought to be important in enhancing the resilience of ecosystem functions and services, with agriculture being an important beneficiary (Oliver et al., 2015). In addition, biodiversity is a valued product of the agricultural landscape, with rare birds, bees and plants being sensitive to change in agricultural management (Pywell et al., 2012). However, it is also the case that biodiversity may bring dis-benefits in the form of pests and diseases, and any potential costs should be weighed against the benefits (Mulder et al., 2015).

Therefore, biodiversity has a central role to play in the ecosystem service chain, from an underpinning natural asset through to valued product. It contributes to agricultural productivity, to conferring ecological resilience to the agroecosystem, and in and of itself is important for the conservation agenda. Agriculture and biodiversity are therefore intimately linked in many landscapes, and perhaps because of this, concerns are sometimes expressed about the implications of using the natural capital framework in this context.

On the one hand, a limited number of 'environmental goods' have a market value. The natural capital framework promotes economic valuation of all environmental goods, so that those without a market value can be included (and the total economic value of those with market values can also be considered). It has been suggested, however, that in decision-making all values should be considered, from intrinsic to instrumental, including different types of value that are harder to quantify (Tallis & Lubchenco, 2014).

The dependency of agriculture on ecosystem services that flow from natural capital

Regulating and supporting services are fundamental to any form of agricultural production because they generate and maintain the resources (e.g. soil, nutrients, plant biomass, pollinators) needed for food production and ultimately human health and well-being (Figure 4.1). Agricultural practices modify and supplement these environmental resources to maximise the provisioning service capability of agroecosystems to meet human food requirements.

While food production is the acknowledged aim of agriculture, an example of a valued but indirect outcome of agricultural practices is the formation of characteristic agricultural landscapes that invoke a strong sense of place, space and beauty among local residents and visitors. This positive externality, which contributes to the cultural services derived from agriculture (Figure 4.1), is often an unacknowledged societal benefit from agricultural production which is not valued or included in the financial accounts of agriculture (although there are exceptions to this through agricultural zoning and heritage protection; Nolan, 2001). By contrast, negative externalities from agriculture that impact on supporting and regulating services (e.g. reduced flow or pollution of water, soil erosion), are well recognised and are increasingly addressed via regional, national and international policy and legislation.

As agriculture is both dependent on, and a mediator of, ecosystem services in agricultural landscapes, it is important to be clear about which goods and services we aim to value and how they are generated. In this chapter we consider ecosystem services upon which agriculture depends, and the role of agriculture in influencing both the underlying natural capital and the flow of ecosystem services for human health and well-being (Figure 4.1). Our focus is not just on valuing these goods and services, but also in how that value may be considered within the broader context of society. In what ways does society benefit from the management of ecosystem services by agriculture and how might such benefits be recognised and used to help enhance agricultural resilience?

In the first category, agriculture requires a range of supporting services (such as soil formation and nutrient cycling) that are inputs to the production of consumable goods like crops and livestock; in the second, agricultural landscapes mediate the delivery of other services that contribute to many

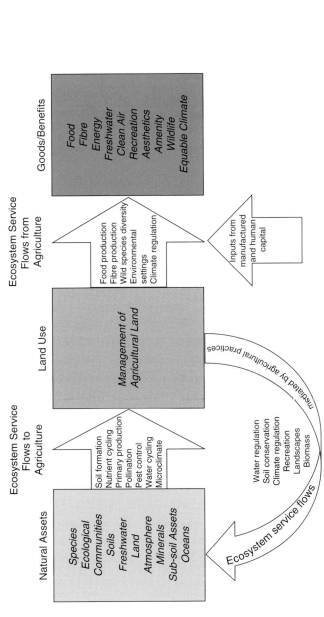

Figure 4.1 The influence of agriculture on the flow of ecosystem services. (A black and white version of this figure will appear in some formats. For the colour version, please refer to the plate section.)

dimensions of human well-being through impacting upon the underlying natural capital (e.g. water quality, climate stability, flood protection) as well as delivering a range of benefits (e.g. wine, scenic beauty) that are either traded as market goods or are available as Public Goods to society.

The second category is especially important to acknowledge because the management of agricultural landscapes can have a dramatic impact on 'downstream' regional or global services, determining whether these services are enhanced, maintained or reduced. While it is true that the management of all habitats influences the flow of services to a certain extent, the range of potential impacts is very significant for agricultural systems for a number of reasons, including the proportion of land that is under agricultural management, and the focus of subsidies on food production rather than other benefits arising from agricultural land. Environmental regulations in agricultural landscapes often focus on the negative impacts of agriculture's mediating role, such as reducing water quality from increasing fertiliser application or erosion causing tillage practices; however, downstream impacts of agricultural management can easily be positive: for example, management of land to intercept pollutants arising from an upstream area.

The role of natural capital and delivery of ecosystem services

The impact of agricultural management on the assets supplying the supporting, provisioning, regulating and cultural services can result in either positive or negative feedbacks on ecosystem service flows. Here we consider how agriculture influences the flow of non-provisioning ecosystem services.

Supporting services

Soil fertility and structure maintained by invertebrates and microorganisms living in the soil are critical to agriculture and yet are also directly influenced by agricultural practices. Growing food benefits from fertile, uncompacted soil, yet the process of growing food can improve or undermine future soil productivity, depending on the management practices used and the nature of the soil. Practices such as ploughing and fertiliser use are intended to raise soil productivity in the short term, but can have impacts on future soil productivity by affecting the ability of the soil to retain nutrients and/or its susceptibility to erosion. Fertiliser applied on fields can increase nutrient concentration downstream beyond the level acceptable for drinking water, especially if water flows from the field directly into streams without any interception. If, however, elements in the agricultural landscape, such as the addition of riparian buffer strips, reduce nutrient loading, such elements would contribute to the service of maintaining water quality for drinking water or other uses downstream (Brauman et al., 2014). Thus, the choice of soil and vegetation management practice has important consequences not only for food production, but also for other ecosystem services, including water quality and flood management.

Irrigated agriculture is the dominant consumer of water on the planet and agriculture is thus an important beneficiary of hydrologic ecosystem services. The volume of water available for irrigation, the water quality, the time of year it is available, and whether it is surface water or groundwater are all influenced by upstream land cover and land management. Because both irrigation and rainwater pass through agricultural fields, agriculture is also a mediator of water-related services downstream as illustrated above, with too much abstraction for irrigation being a cause of anthropogenic salinity in some parts of the world. Reliance on voluntary action to prevent anthropogenic salinity for public benefit has proved insufficient in some cases as the costs to private individuals are greater than private benefits (see Harris, Marshall & Pannell, Chapter 14).

Regulating services
Climate regulation includes the ability of ecosystems to sequester greenhouse gases (GHGs) and to influence microclimate, e.g. by modifying evapotranspiration rates, buffering winds and providing shade.

When land conversion is included, agriculture accounts for approximately 20–35 percent of global GHG emissions (Vermeulen et al., 2012), and reducing this contribution is critical. To put this range in perspective, GHG emissions from agriculture are double the emissions from all sources of transportation. Deforestation in the tropics is the primary source of these emissions, accounting for 12–18 percent of global GHGs (DeFries & Rosenzweig, 2010; Vermeulen et al., 2012; van der Werf et al., 2009) and yet the resulting agricultural yields are typically very low (West et al., 2010). Ewbank (Chapter 7) considers small-scale farming as a means of reducing deforestation.

The next major source of GHGs is methane from rice and livestock production (52–66 percent of global methane emissions; Smith et al., 2008) and nitrous oxide emissions from synthetic and organic nitrogen fertiliser (61–84 percent of global nitrous oxide emissions; Montzka et al., 2011). GHG emissions from agriculture can be reduced through changing management practices to reduce methane (Smith et al., 2008) and nitrous oxide emissions (Hillier et al., 2012; Venterea et al., 2012). No-till cropping and perennial crops can also be effective by increasing carbon storage on farmlands relative to cropping systems with higher soil disturbance or unmanaged systems with low productivity.

Pollination and pest control are regulating services provided to agriculture by diverse habitat elements in agricultural landscapes. Pollinators are essential to the productivity of certain agricultural crops, e.g. fruit and nuts, and the role of wild pollinators is becoming increasingly important as colony collapse disorder and other threats posed to honeybee colonies compromise their availability for crop pollination on many farms (Potts et al., 2010).

Indeed, in several crop systems, wild pollinators have been found to be more effective at pollination (per visit) than managed pollinators (Garibaldi et al., 2013). Farming practice (e.g. use of pesticides, companion cropping, crop diversity) and farm topology (e.g. field size, connectedness and the proportion of cropped to uncropped land) are important factors influencing the effectiveness of both pollination and pest control services. The proportion of non-crop habitat in the landscape is critical for maintaining effective pollination services (Kennedy et al., 2013; Ricketts et al., 2008) and the diversity and abundance of natural enemies (Chaplin-Kramer et al., 2011a). Measures to enhance pollinator habitat have been found to deliver other benefits such as improved soil and water quality, enhanced biodiversity and natural pest control and improved aesthetic value of the agricultural land (reviewed in Wratten et al., 2012), as well as reducing crop injury in many systems (Thies & Tscharntke, 1999; Werling & Gratton, 2010; O'Rourke et al., 2011; Karp et al., 2013) and improving or at least maintaining yields overall at the farm scale (Pywell et al., 2016). However, it is important to recognise that non-crop habitat is a resource for pests and pathogens as well as pollinators and natural enemies, although in many cases, the types or qualities of habitat are very different. It may be possible to establish non-crop habitats that promote the services (pollinators and natural enemies) and constrain the disservices (pests and pathogens) (Chaplin-Kramer et al., 2011b).

Cultural services and biodiversity
Agricultural landscapes can include a wide variety of land use types – including not only cropland and pastures, but also woodlots, larger natural areas (wetlands, forest parks), streams and housing. While some agricultural settings are nearly entirely intensive cropland, far more of the world's agriculture takes place in mixed landscapes (Ellis & Ramankutty, 2008), which often evoke an intensely loyal sense of place for local residents. Aesthetic quality and recreational opportunities such as walking, cycling and hiking are an important part of many agricultural landscapes and people derive enjoyment, spirituality or pride from the beauty of the countryside. These services are not always considered as being provided by agricultural landscapes, but they are important, not only for the wealthier countries with leisure time for recreation, but for many agricultural economies to maintain their cultural identity and sense of place. These cultural services are strongly mediated by the specific type of agriculture practised and its intensity, as well as by the mix of land-use types and land management that happens in an agricultural setting.

Farmland also supports a number of species of cultural value ('wildlife'). As the intensity of agricultural management has increased, so has the abundance and diversity of some valued groups declined (Baulcombe et al., 2009). However, recent evidence supports the idea that targeted measures within farmland (forms of 'wildlife-friendly farming') can benefit rare species of birds, bees and plants (Pywell et al., 2012). For example, decreasing the intensity of hedgerow management improves the diversity of Lepidoptera (Staley et al., 2016).

A fundamental question when considering the links between the agricultural production and biodiversity conservation agendas is the extent to which biodiversity enhances the ecological resilience of agricultural production through the services delivered. The closer this connection, the more benefit we can expect from integrating strategies for biodiversity and agricultural production – both in terms of service delivery today and options for the future. Initial work, in Costa Rica (Karp et al., 2012) and also in Central California (Winfree & Kremen, 2009) shows relatively high stability and resilience in low- and intermediate-intensity production systems, and the reverse in high-intensity production systems. This relationship between biodiversity in agricultural landscapes and the resilience of services they generate has been conceptualised as having different forms for different types of services, with the benefits being most evident in lightly used or extensive systems (Braat & ten Brink, 2008).

A second question which then follows is how can we manage landscapes to enhance biodiversity and which species are most likely to respond. Habitat elements within farms can provide important refuges for a diversity of species. Maintaining such habitat, or farming in a way that is more hospitable to the species passing through it, may reduce yields, which has led to calls by some scientists to consolidate and intensify agriculture to use as little space for growing food as possible (Green et al., 2005; Phalan et al., 2011). If, however, the biodiversity-enhancing attributes of landscapes do not involve great trade-offs for farmers and, indeed, if those species yield valuable services to agriculture, enhancing yields, then there is considerable scope for integrating conservation and production strategies (Abson, Sherren & Fischer, Chapter 11). Continuing with the Costa Rican example, the ribbons and dots of forest that span the landscape are good predictors for a wide range of vertebrate and invertebrate taxa, across local to landscape and regional scales (Karp et al., 2012; Mendenhall et al., 2011). This also correlates with the enhanced pest control and pollination reported above for coffee farms.

Why value and what is value? The challenge of valuing public goods
Why value?
Why do we need to quantify the extent to which we value something, rather than merely knowing that we value it and acting accordingly to preserve it?

Decisions rarely come without trade-offs, as resources are limited. Valuation helps to make the costs and benefits of such trade-offs more explicit, and *economic* valuation converts values into a common unit of measurement – literally a common currency. However, it is clear that decisions are often made without this explicit accounting. There is also the issue that some environmental goods defy economic valuation methods, that other quantitative and qualitative measures frequently resonate better with people, and that social and political debate should integrate economic with societal and personal values. The most important thing in valuation is to translate biophysical changes in ecosystem function resulting from a decision to an endpoint that the people making the decision care about. Crucially, this requires good understanding of the biophysical changes as well as valuation methods – and it is here, because of our uncertainty about what effect decisions will have in agroecological systems (see Martin, Chapter 13), that the concept of resilience becomes important.

Agriculture, as with any other business (arguably more than many businesses), has dependencies upon and impacts upon natural capital and the ecosystem services that flow from that capital. Only some of the environmental goods produced by those processes have market values (food, fibre and energy); others, such as biodiversity, water quality and enhanced quality of landscapes, do not have market values but certainly have economic value. Understanding those values will help farmers make management decisions that enhance their economic sustainability, and at regional and national levels help set policy and investment priorities to enhance the resilience of agricultural systems. In 2016 a 'Natural Capital Protocol' was launched by a coalition of organisations (the Natural Capital Coalition, NCC), to provide a framework to generate trusted and consistent information around impacts on natural capital by business, as a guide to inform business decisions.[2] In some instances such information may also reveal opportunities for business, or coalitions of public and private investors, to provide economic incentives to farmers to alter management practices in ways that reduce impacts on ecosystem services. For example, water funds are emerging as a viable strategy for enhancing water quality downstream, while restoring and conserving forest within agricultural landscapes because, if investments are targeted to the most beneficial places, they can provide a return on investment to downstream water users (see Box 4.1). If we wish to conserve the ecosystem services that support and flow from agricultural landscapes, we must align incentives such that doing so is more attractive than short-term gains made at the expense of long-term degradation (Heal, 2000) and consequent reduction in resilience.

[2] http://naturalcapitalcoalition.org/protocol/

Box 4.1. Case study for managing ecosystem services in agricultural landscapes: Latin America

The Northern Andean region (Colombia, Peru, Ecuador), like much of South America, faces mounting pressure for agricultural conversion of native forest as a growing population competes for increasingly marginal lands and productivity declines in previously deforested lands. Farmers and ranchers are practising agriculture on steeper slopes than many would deem possible, with consequent erosion problems causing loss of soil fertility and water quality problems downstream. Meanwhile, demand for a clean and stable water supply is growing with the urban population, and many cities look to these same watersheds undergoing so much transformation for their water security. Natural areas that are deemed important for water production are often protected, but in many cases in Latin America this presents a social justice problem, denying rural inhabitants their only opportunities for improving their livelihoods.

In response to this growing resource conflict, Quito, Ecuador, established the first 'water fund', a financial mechanism for watershed management in predominantly agricultural systems, promoting habitat conservation, restoration and improved agricultural practices in the upper reaches of a watershed (Arias et al., 2010). Downstream water users such as city drinking water municipalities, hydropower companies, bottling corporations and other large commercial entities like agribusiness invest in these funds, often administered by local watershed management associations, with diverse and multiple goals including securing ample and clean water, recharging groundwater supplies, protecting against floods, landslides and other natural disasters, and enhancing biodiversity (Goldman-Benner et al., 2012). The concept has quickly taken off throughout the Andean region and spread to the rest of the continent. Latin America has formed a platform for the creation of water funds, to establish a community of practice for sharing lessons learned on science and governance and developing standardised methods and metrics to design water funds and measure their performance.

Once a fund has been established, however, an open question is how to invest the money – in which activities, and where in the watershed? Best management practices could obviously be employed everywhere that agriculture is practised, but if there are trade-offs to yields or areas under production, as there often are, spatial targeting can find the most cost-effective places to focus efforts. Likewise, conservation and restoration of native habitat should be prioritised for the places that will provide the most benefits at the least opportunity cost to production. Strategically locating a

> **Box 4.1. (cont.)**
>
> cover crop, lower tillage, a buffer strip, or a restored tract of forest in places that will make the greatest difference to explicit water fund goals (e.g. improving overall water quality or flood protection) can provide other benefits as well (e.g. nesting habitat for native birds, floral resources for pollinators or natural enemies of agricultural pests) and gains can be made in many ecosystem services with minimal losses to agricultural production. One pressing problem with water funds is that both the biophysical and economic value of various land management activities are often unclear, especially in the high-elevation tropics like the Andes, where the impact of land-use change remains poorly understood (Ponette-González et al., 2014). As a result, many projects are premised on upstream residents conforming with required actions, not on changes to delivered water services (Farley et al., 2011).
>
> Improved targeting of these watershed investments can provide enhanced and more resilient ecosystem services in agricultural landscapes. Approaches to prioritise investments, whether focused on a single service or multiple services, can improve the production of those services up to fivefold over a random investment (Figure 4.2). Considering both current and future environmental conditions, including climate extremes, can help identify priority areas for protecting ecosystem services to enhance resilience in the system. For example, in the Putomayo region of Colombia, areas with the highest levels of water yield today overlap with areas most susceptible to soil erosion in future climates (Suarez et al., 2013). In Nicaragua, modelling of agricultural productivity and hydrological ecosystem services is helping to guide climate adaptation planning to identify climate 'hot spots', where adaptation measures are not likely to be effective because the expected change is too great; 'adaptation spots', where different agricultural management can enhance agricultural production and other ecosystem services amidst climate change; and 'pressure spots', where growing conditions will improve in the future and trade-offs in ecosystem services should be considered as part of development (Girvetz et al., 2014). Based on the evidence of these projects, the water fund approach is now being replicated across the world, with several under consideration in as diverse contexts as the Midwestern US, Kenya and China.

Considering ecosystem services in agricultural systems illustrates that there are costs and benefits to every decision which should be weighed in different contexts. Where agriculture is located in a landscape – and how and where specific practices are employed – can have an enormous effect on the total set of services produced and how equitably they are delivered (Vejre & Brandt, 2003; Fischer et al., 2008; Mitchell et al., 2013). For example, stream buffers and low tillage regimes may secure water quality only along certain slopes or on certain soils, and may outweigh the opportunity cost of their implementation only in some areas and not in others (Zheng et al., 2016). Defining multiple social goals for agricultural landscapes thus helps guide which trade-offs should be assessed and valuation can assist in understanding their relative importance (Goldstein et al., 2012; Polasky et al., 2011). Box 4.2 describes a case study in Hawai'i where a consideration of the multiple benefits arising from agricultural landscapes allowed the full environmental and financial implications of different management options to be explored. When trade-offs are recognised as including not only delivery of food and water but also a larger suite of services such as climate regulation, flood mitigation, recreation opportunities, cultural heritage and provision of goods such as timber and wild game cogenerated in the landscape, the implications of alternative choices are clearer, and this enables design of agricultural systems that can better meet society's full needs.

Box 4.2. Case study for managing ES in agricultural landscapes: Hawai'i

The State of Hawai'i is a microcosm of pressures on land and water globally, with a growing human population and ongoing conversion of forest to agricultural uses and of agricultural land to residential and other development. Food and energy security are major concerns: at any moment, the state has an estimated 10-day food supply and, in addition to food, it imports over 90 percent of its energy in the form of oil and coal. There is also deep concern over sustaining critical mass in the farming and ranching communities, to maintain at least some food production in the state. Given these pressures, and a feeling of crisis, there is demand for exploring the implications of alternative land-use scenarios, especially on private land. We illustrate such scenarios and their societal implications in two cases.

First, on the Kona coast of leeward Hawai'i Island, potential conversion between native forest and cattle pasture has raised concerns about impacts on water resources and other ecosystem services (Brauman et al., 2015). Evaluation of a variety of potential transitions showed that only in a scenario of dense forest restoration did agricultural production trade off against a wide variety of ecosystem services, including increased groundwater recharge, carbon sequestration and native bird habitat. In other transitions, some services were increased while others decreased. When some areas of cattle pasture were converted to native timber plantation, for

> **Box 4.2. (cont.)**
>
> example, carbon sequestration and bird habitat increased while groundwater recharge decreased (Goldstein et al., 2008; Brauman et al., 2012). This case highlights the importance of identifying and evaluating specific agricultural transitions of interest and assessing changes in the delivery of each service.
>
> Second, ecosystem-service values helped the largest private landowner in Hawai'i, Kamehameha Schools, design a land-use development plan that balances multiple private and public values on its North Shore land holdings (Island of O'ahu) of ca. 10,600 ha (Goldstein et al., 2012). One of the earliest applications of the InVEST (Integrated Valuation of Ecosystem Services and Trade-offs) toolkit, the decision process evaluated the environmental and financial implications of seven planning scenarios encompassing contrasting land-use combinations that included biofuel feedstocks, food crops, forestry, livestock and residential development. All scenarios were characterised in biophysical terms, as well as in terms of their overall financial return. While each scenario had positive financial return, relative to the status quo of negative return, trade-offs existed between carbon storage and water quality as well as between environmental improvement and financial return (Goldstein et al., 2012).
>
> Based on several iterations of analysis and community input, Kamehameha Schools is now implementing a plan to support diversified agriculture and forestry. This plan generates a positive financial return ($10.9 million; figure 3 in Goldstein et al., 2012) and improved carbon storage (0.5 percent increase relative to status quo) with negative relative effects on water quality (15.4 percent increase in potential nitrogen export relative to status quo). The effects on water quality will be mitigated partially (reduced to an estimated ca. 5 percent increase in potential nitrogen export) by establishing vegetation buffers on agricultural fields. This plan contributes to policy goals for climate change mitigation, food security and diversifying rural economic opportunities. More broadly, this approach illustrates how information can help guide local land-use decisions that involve trade-offs between private and public interests.

What is value?

There are many ways to estimate value, and economic valuation is one, but this is not always the most appropriate or effective. For example, the contribution of high-quality water to human health can be valued as the reduced risk of mortality (Keeler et al., 2012), which can be monetised, but can also be valued simply in terms of the percentage of population with access to clean drinking

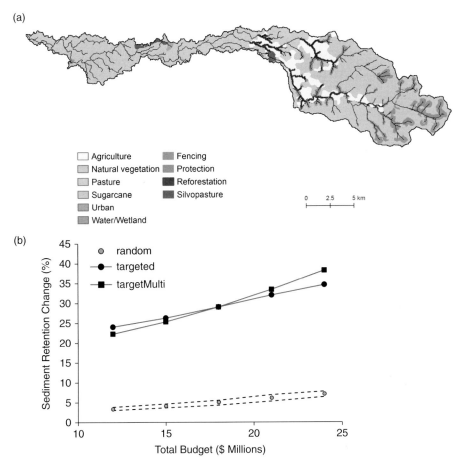

Figure 4.2 (a) Investment portfolio for where to prioritise different conservation activities in the Cauca Valley of Colombia. (b) Estimated improvement in sediment retention resulting from conservation activities that were either targeted to sediment specifically (targeted), targeted to multiple ecosystem services (targetMulti), or randomly allocated across the watershed. Figure credit: S. Wolny and A. Vogl. (A black and white version of this figure will appear in some formats. For the colour version, please refer to the plate section.)

water. Using such health metrics implicitly places greater emphasis on expanding human rights for access to safe water, as stated in national or international goals (e.g. the Sustainable Development Goals). Likewise, changes in crop production, fisheries landings and non-timber forest products are typically valued in economic terms, but could also be valued in terms of their contribution to alleviating malnutrition (Tallis et al., 2012).

There are instances, however, where economic valuation is desirable, one reason being it allows direct comparison between different scenarios. Food, timber and fuel are environmental goods produced from agricultural land which have a market price, which may fluctuate due to the quality and quantity of the goods, and due to other external drivers. In general, agricultural prices only reflect current supply and demand conditions (i.e. marginal values), and do not reflect change in value resulting from large changes in the provision of agricultural products, such as potential non-linear changes in ecosystems (Heal, 2000). Market prices as a basis for valuation may be adequate for informing decisions that are made for small, marginal changes. However, this price should not be confused with the full economic value of these goods, which should include the costs or benefits of wider impacts on the ecosystem as a result of producing those goods.

Furthermore, biodiversity poses particular challenges in valuation, yet clearly is of value in underpinning ecosystem services, in delivering wildlife benefits through the management of agricultural land, and in conferring resilience to ecosystem functions and to agriculture, providing buffering to environmental change.

Approaches for valuing ecosystem services and for integrating ecosystem service values into agricultural systems

Provisioning services

Environmental goods such as food, timber and fibre provided to people from agricultural landscapes are thought to be the easiest to value because a market is generally already established. However, there are many issues with using market prices as proxies for the marginal value that ecosystems provide to people. First, agricultural products like crops and livestock have many inputs. These can include seeds, fertilisers, machinery and human labour in addition to biophysical inputs like water, solar energy and soil fertility. The market price reflects contributions from all these inputs, and thus other capital inputs are conflated with natural capital inputs (Fezzi et al., 2013). Only a portion of this value is attributable to natural capital and the ecosystem services that flow from them. Second, economic value will include the value of positive and negative externalities, such as the removal of pesticide residues from water, and these costs are not reflected in market values but are borne externally. In this vein, the influence of taxes and subsidies should also be taken into account. Third, when valuing environmental goods it is necessary to not only tease apart the value of other human inputs, but also to ensure that intermediate services (such as pollination and pest control) are not double-counted in the final value (of food, for example), although services such as pollination can also be considered as a final ecosystem service depending upon context.

Production function approaches provide a method by which to isolate the contribution of the natural environment from other human inputs.

Ecosystem inputs are viewed as 'factors of production', and the loss or reduction in the environmental good quantified as the level of ecosystem input is varied: in other words, the 'value added' by the ecosystem service over and above the other factors involved in the production of the good. Good overviews of this approach are available (see, for example, McConnell & Bockstael, 2005; Barbier, 2007).

Supporting services
Soil
For naturally occurring soil fertility and structural suitability, as can result from agroecological practices that build or maintain soil biota and soil organic matter, there are several ways to value the contribution to agricultural production. First, the production function approach can be applied to test how field-based measurements of soil properties (e.g. soil moisture, nitrogen or phosphorus content, etc.) correlate with variation in agricultural yield (McConnell & Bockstael, 2005). Alternatively, production functions may be fitted to long-term field experiments using different agricultural treatments (Brady et al., 2015), where soil organic carbon correlates with changes in soil biodiversity and therefore the associated ecosystem services. Second, the cost of importing and spreading organic matter, fertilisers or soil conditioners to replace native soil conditions can serve as a surrogate for the cost of maintaining soil fertility. This is the replacement value method (Heal, 2000). Finally, hedonic valuation may be used to test whether variation in soil properties has an effect on agricultural land prices (Heal, 2000), assuming potential purchasers have a knowledge of the state of the soil and a competitive market for land exists.

Water
Whether the upstream watershed or the farm itself is transformed to forest, monoculture cropping, or even pavement, rain will still fall and rivers will still flow (Brauman et al., 2012; Keeler et al., 2012). The value of a particular management option lies in the value of a change in an aspect of water resulting from a change in land use or land management: a change in water availability, in water quality, in the time of year water is available, or in whether groundwater is recharged. For irrigated agriculture, the value of upstream land management to ensure dry-season flows could be considered not as the additional yield from irrigated crops compared to rain-fed crops but as the additional yield *and reduced risk* given a managed versus unmanaged upstream watershed (Ponette-González et al., 2014). Downstream, the value of reduced sedimentation because cattle are moved away from riverbanks could be calculated as a reduction in water treatment costs (Dearmont et al., 1998).

Regulating services
Pollination and pest control
Valuation of pollination and pest control could take several forms. Pollination services have been valued in terms of the replacement costs (of renting honeybees or hand-pollinating) to perform the equivalent service provided by wild pollinators (de Groot et al., 2002). The replacement cost for pest control services could be considered as the amount spent on pesticides to replace the ecosystem service of pest control, although it is somewhat controversial whether the two methods provide equivalent levels of control (Pimentel et al., 1992). A more commonly used approach for valuing pollination services estimates the value of crop production attributable to pollination (i.e. a production function method). Crop yield reduction in the absence of pollinators has been approximated for all major crop types (Klein et al., 2007); this yield reduction can then be multiplied by the market value of production (Morse & Calderone, 2000; Ricketts et al., 2004). Recently, a new approach – the 'attributable net income method' – to valuing these services in agriculture has been developed in the context of pollination; this deducts the cost of inputs to crop production from the value of the ecosystem service and does not attribute value to the service in excess of plant requirements (Winfree et al., 2011). The analogous approach in pest control is to quantify the amount that yield is reduced by pest damage in the absence of natural enemies, and value the avoided damage (Losey & Vaughan, 2006). Arguably, a more complete approach would be to consider not only changes in yield, but also the cost of changes in water treatment required as a consequence of pesticide use, and also any impacts on wider biodiversity.

Climate mitigation: carbon
The management of soil organic carbon (SOC) will influence not only the productive capacity of land, but also has the potential to deliver other ecosystem services, including climate change mitigation via soil-based carbon sequestration, improvement of water quality downstream and enhancement of biodiversity above- and belowground. As Public Goods, the value to the individual farmer needs to be demonstrated (e.g. through policy instruments such as subsidy). Farming context will also dictate which management measures will improve farm margins, and to what extent (Glenk et al., 2017).

Cultural services
As well as aesthetic meaning and recreational opportunities, agricultural landscapes also provide cultural value to populations. Most often, these services relate to communities' cultural heritage or spiritual significance (Daniel et al., 2012). For example, backyard gardens are not often economically viable, but homeowners derive a certain amount of enjoyment from them. In some cases,

cultures choose to grow economically inferior crops to maintain cultural traditions and foods. Therefore, the benefits provided by agricultural landscapes can come in the form of the diversity of crops grown or the longstanding production practices and associated community activities supported. The use and management of agricultural landscapes may also relate to communities' spiritual values. For example, forested 'sacred sites' that are often alongside agricultural landscapes in Kampa Tibet, providing wild medicines, fuel wood and aiding water storage (Anderson et al., 2005; Salick et al., 2007).

As these examples demonstrate, communities can derive cultural value from agricultural landscapes in various and complex ways that may not be consistent across populations, making valid quantitative measures of cultural ecosystem service value difficult to calculate and often controversial. Two methods frequently practiced are Willingness to Pay (WTP) and Willingness to Accept (WTA). In the former case, the extent to which individuals are willing to pay for a particular benefit is estimated, and in the latter, the extent to which they are willing to accept compensation to forgo the benefit. Three types of data may be used to estimate these quantities: market prices, observed consumer behaviour and statements made by individuals about their WTP. For example, agri-tourism may in some instances have a market value through ticket prices, but as with provisioning services this price may not properly reflect reliance on services supporting agricultural production and landscapes and the damage costs of potential impairment of services flowing through agricultural landscapes. For a second example, the effort and expense individuals will expend to enjoy a landscape can be used to reflect the values individuals place on those landscapes. Methods for quantifying the value of these kinds of cultural or recreational opportunities include travel cost methods (which measure the costs incurred to travel to a place where the ecosystem service is enjoyed, be it the cost of fuel to a trailhead or the price of a holiday package to climb Kilimanjaro). Another type of 'observed consumer behaviour' is revealed in hedonic pricing, which uses variation in observable values of property that affect the value of land, some of which may be due to the proximity to natural landscapes. The issue here is to tease apart the contribution that the natural landscape plays in influencing price alongside many other factors.

All the valuation methods described previously require data, and the quality of the data will determine the quality of the valuation. In the absence of other data, carefully designed surveys may be used to elicit people's responses in terms of how much they would be willing to pay. These methods are referred to as contingent valuation, and the questionnaires use hypothetical situations to elicit directly stated values from people about what they may be willing to pay for a service in the form of choices. Clearly, the design of such surveys is crucial in obtaining the best possible data, and can be used to investigate values of any environmental good or service; however, it is the *only* quantitative method that can be used to estimate non-use values like existence value, the value someone derives from simply knowing something exists (Heal, 2000). Our ability to measure non-use value through methods like contingent valuation is improving, but is still rooted in public perception and understanding of ecosystem connections and interdependences (Daily et al., 2000).

The valuation of cultural services is often the context in which the uncertainties in the methods used, and the juxtaposition of very different environmental benefits (beauty of landscapes compared with the production of food, for example), causes most unease. One potential resolution to these issues is to highlight the cultural benefits from landscapes in an open dialogue between scientist, policymakers and stakeholders (Daniel et al., 2012). This is also the context in which non-monetary forms of valuation can be especially useful, and alternative metrics can be estimated alongside economic valuations (Thompson & Segerson, 2009). For example, biodiversity is often reported in species or habitat loss, which better reflects the non-use value of that service (e.g. Naidoo et al., 2008) or as a diversity index (Bateman et al., 2013). New and novel ways of assessing the psychological value of a landscape are emerging as we begin to better understand the benefits of how human experience in nature relates to cognitive function, including concentration, impulse inhibition, short-term and working memory and mood (Bratman et al., 2012). Some of these measures are not monetary, so expanding beyond conventional economic notions of 'value' will help more thoroughly capture the full benefits of ecosystem services to people.

An alternative approach is to constrain the set of possible decisions to include only those which do not reduce the biophysical metric in question (such as biodiversity indices). This was one approach adopted by Bateman et al. (2013), who explored a set of six future policy scenarios for Great Britain, estimated the consequences for the economic values of a range of ecosystem services and compared the outcomes with those produced with the additional constraint of no biodiversity loss. The difference between these two sets of outcomes represents the opportunity cost of conserving biodiversity at the expense of other ecosystem services. This cost was discernible, although relatively minor, and caused a shift in some areas from enhancing greenbelt areas for recreation to preserving and extending existing areas of conservation value (Bateman et al., 2013). The principle illustrated here could be applied in other contexts to incorporate other impacts that cannot be readily monetised, or when there are other policy or political reasons for imposing constraints.

Managing trade-offs

Given the range of ecosystem services that flow from the management of agricultural land, there will be choices to be made and trade-offs to be considered. The most fundamental trade-offs include the use of land to produce food rather than unmanaged ecosystems that provide biodiversity and the intensity with which the land is managed.

Trade-offs are often characterised by the delivery of ecosystem services on different spatial or temporal scales, as a consequence of the need to reduce variability in the supply of a particular service, and by the provision of private versus public goods. For example, farmers will have a direct interest in managing soil fertility and soil retention; however, the methods used may impact on water quality downstream (for example, the over-application of fertilisers). This is an example where the impacts of management to enhance one ecosystem service locally are felt in a more distant location. Similarly, farmers would benefit from local agents of pest control; however, the consistency and magnitude of this service may not be sufficient, causing substitution with insecticides that may ultimately reduce the provision of that service supplied naturally. Furthermore, measures to enhance habitat for pollinators and predators will benefit not only the local farmer, but also the neighbours. A farmer acting in isolation may lack sufficient incentive.

Greater understanding of the ecosystem service flows may lead to more frequent identification of 'win–win' scenarios, although reviews have identified that trade-offs are three times more frequent than synergies (Howe et al., 2014); the involvement of provisioning services and stakeholders having a private interest were two features that were associated with trade-offs. Progress can be made in the context of 'sustainable intensification' (Baulcombe et al., 2009) by reducing trade-offs through management, as has been empirically demonstrated in various case studies in developing and developed countries (Pretty et al., 2006; Badgley et al., 2007; Pywell et al., 2016). In this way, yields may be maintained or even enhanced while delivering a broad range of other ecosystem services.

The overarching goal for valuing different ecosystem services flowing from agricultural land is to inform policies, regulations and incentives to enable the better management of natural resources when trade-offs are inevitable. Valuation methods can bring greater clarity to the consequences of different options.

Valuation of ecosystem services as a tool for enhancing the resilience of agriculture
Future benefits and values

Resilience, *the capacity to buffer change, learn and develop*, has both ecological and socioeconomic aspects, to which the valuation of ecosystem services could contribute (Folke, 2006). Understanding the value of the multiple benefits that may be delivered from agricultural land, and appropriate acknowledgement of that value, could result in new markets, behaviours, subsidies or regulation. The Millenium Ecosystem Assessment (2003) and other national ecosystem assessments that have followed have shown that taking account of the values of the benefits for which there is no market would profoundly alter policies and behaviours around land use, and the management of agricultural land is a key example (National Ecosystem Assessment, 2011). Appropriate management to deliver multiple benefits could also enhance the ecological (and ultimately economic) resilience of agriculture. It is necessary, however, to understand how agricultural production and the delivery of other environmental goods are expected to change under future climate scenarios, and how the values of those goods may change. While progress has been made in developing methods to estimate those future values, considerable uncertainties remain.

Changes in climate can have effects that are spatially heterogeneous in magnitude, scale and direction of impact on agricultural productivity. Thus, quantifying these changes to agriculture is complex, and further complicated because farmers may or may not adapt to (sufficiently slow) change by changing crop mixes or land uses altogether. Many studies predict changes in agricultural productivity in light of expected climate changes based on biophysical processes (e.g. Piao et al., 2010). To account for the complex land-use and crop choice dynamics, some have employed integrated modelling, coupling climate change projections with econometric land-use modelling to incorporate potential farmer responses. Such modelling approaches enable the potential contribution of particular factors (biophysical or socioeconomic) to be explored by comparison with the *ceteris paribus* (or 'other things being equal') situation. This was the strategy of the UK National Ecosystem Assessment (Bateman et al., 2011), which valued changes in precipitation and temperature on agricultural production across the UK. The same approach may explore how human activities may influence the delivery of a range of ecosystem services in the future. Fezzi and colleagues (2013) review several other techniques, but ultimately apply similar methods to the UK NEA; the conclusions reached are that changes in climate may benefit agriculture in the UK's north but are likely to be costly in the UK's south.

There is also considerable uncertainty in the nature of the relationships between the underpinning natural capital stocks and the delivery of benefits. It is widely recognised that natural capital stocks are in decline (through depletion of water resources, soil erosion, biodiversity decline, for example) and this has been linked to declines in benefits. In some instances, non-linear relationships may exist, for which stock depletion can lead to a sudden and abrupt change in the delivery of a benefit (a 'tipping point'). Ecologists have long demonstrated that many ecosystems have non-linear dynamics that can exhibit multiple stable states. Economists have begun to try to place a value on the ability of a system to remain in its current state versus being subject to some disturbance, exemplifying resilience in terms of resistance to flipping into an alternative and undesired stable regime. Such transitions can be costly to livelihoods and welfare, so avoiding these 'flips' has economic value (Mäler, 2008; Mäler & Li, 2010).

The valuation of ecosystem services can incorporate future benefits through using standard discounting methods; 'net present value' should consider how and why the delivery of the benefit may change over time. For example, if the underlying ecological dynamics describing the probability of flips to alternate states are known (as in, for example, Carpenter et al., 1999), we can incorporate the risk of a flip in our evaluation (Mäler, 2008). A good understanding of the ecological science is necessary to predict these changes, and this will often not be the case (see above); even where understanding is good, an ecological system may exhibit unpredictable (chaotic) characteristics. However, it will still be useful to use sensitivity analysis to test different probabilities to establish a range over which a system remains resilient. A 'small' range would suggest a lack of resilience, or at least vulnerability to unexpected change.

Responding to future uncertainty

Gunderson and Holling (2002) identify three, potentially interlinked, strategies for coping with variability: (i) living passively, but with some form of adaptation; (ii) reducing variability and its influences; and (iii) creating or manipulating variability, within the system itself, to beneficial effect.

The first strategy has been employed in agricultural systems since their inception, through selective breeding that has allowed domestication to track shifting environmental conditions, through the development of new agricultural practices (e.g. the terracing of hillsides or nomadic lifestyles: see Homewood et al., Chapter 9). Technological advances may in some instances enhance the potential pace of adaptation; for example, the development of new plant breeding techniques which allow more precise manipulation of the genome, in theory allowing pest and disease resistance to be integrated into locally adapted varieties (see Meyer, Chapter 8). Adaptations may also include less-intensive farming systems that actively manage biodiversity on-farm, to enhance pollination and natural pest control. Nevertheless, these strategies have their limitations, and it may be increasingly challenging to match the pace of change in the environment as anthropogenic stressors intensify.

The second strategy has been a central pillar of modern twentieth-century agriculture through the use of inorganic fertilisers, crop protection products and irrigation (although the latter is well recorded in ancient societies) to reduce variability in soil fertility, pest and disease outbreaks, weed populations and water availability. In essence, ecosystem services such as soil fertility and pest control have been enhanced/substituted for by man-made products. Understanding to what extent technology and other forms of human, built and financial capital can substitute for natural capital is an integral part of managing for resilience in agricultural systems. The difficult aspect of understanding this substitutability is that we may not always choose to replace a service that is lost or impaired if the replacement cost is too high (Heal, 2000). A system that builds resilient food production is not well represented by market valuation, because, as we have argued, the potential impacts of large shocks that resilience buffers against cannot be understood through the marginal changes best represented by market values. If all we value in agricultural systems is production, we are only measuring the final or outcome variable in the system, which is the one that tends to respond rapidly to a wide variety of underlying changes. Without also measuring the services that are inputs to agricultural production, we are ignoring the more slow-changing variables (e.g. phosphorus regulation) that are likely to drive the agricultural system and connected systems (e.g. downstream lakes and rivers) over thresholds (Carpenter et al., 1999). So while this focus on food production has yielded major production successes, this has been at the cost of unanticipated and undesirable loss of ecosystem services (Foley et al., 2011), potentially undermining long-term resilience.

The final strategy offers an opportunity for building resilient agricultural systems for the longer term: creating variability through the establishment of multi-functional agricultural landscapes with a variety of different crops and breeds, involving a wide range of human and natural inputs, and producing a diverse set of tangible and intangible benefits, each with a slightly different response to change. Designing agricultural systems that co-produce multiple ecosystem services rather than ones that maximise a single service (like food provision from a single commodity) at the expense of others could enhance resilience by increasing the diversity of ecological processes that maintain a regime in the face of disturbance (Gordon et al., 2008) and by diversifying income streams for farmers (see below). This not only requires the quality and quantity of provisioning and cultural services to be enhanced, but also demands management of other services supporting and influenced by agriculture. In particular, a decline in regulating services can reduce resilience even when they do not initially reduce the level of other services because regulating services can buffer the impacts of shocks on agricultural systems (Bennett et al., 2009). Furthermore, ecosystem service outputs from agriculture can feed back to affect future ecosystem service inputs to agriculture (Figure 4.1), especially if agriculture competes with other uses for the services that support its production. Sustainable agriculture is concerned with the capacity for an agricultural system to produce the desired level of all benefits through time and, while resilience is necessary to achieve sustainability and ecosystem services are a large part of defining resilience in a system, these approaches are often not well integrated.

Financial instruments to enhance the resilience of agriculture

At present, farming enterprises that deliver positive ecosystem service benefits are often not rewarded for these goods. Where payments are offered, these are often at a 'flat-rate', taking little account of the quality, quantity or appropriateness of the benefit produced and not incentivising farmers to do more than the absolute minimum to achieve the payment. Such payments can encourage agricultural management practices that are inappropriate for the local ecological context and poorly targeted at the desired ecosystem service outcome. Getting the incentives right for sustaining biodiversity and the many ecosystem service values to which it contributes is challenging. Even in Western Europe, with substantial study and policy support for farmland biodiversity, there remains considerable uncertainty and several examples of outright and costly failure (Kleijn & Sutherland, 2003). In the Netherlands, for instance, over 20 years of biodiversity management schemes for plants and birds yielded little perceptible benefit. This failure was not caused by a lack of compliance by farmers, but instead, by a lack of understanding of the constraints on the conservation and restoration of biodiversity at both landscape and local scales (Kleijn et al., 2001, 2004). There are indications that the implementation of agri-environment schemes is improving, as recent assessments have been more positive (Batáry et al., 2011), and the co-benefits such as improved opportunities for recreation and reduced pollution should not be overlooked. Nevertheless, as the Common Agricultural Policy is currently constructed, subsidies that indirectly incentivise production and payments to help prevent damage to the environment are two instruments that potentially work against each other. Reform and realignment could provide (at least) the same income opportunities for farmers, while providing other environmental benefits that over a period of time could enhance the environmental and economic resilience of farming systems (Natural Capital Committee, 2015).

In some senses, agri-environment stewardship schemes run by government agencies can be thought of as an example of 'payment for ecosystem services', although crucially, these agri-environment payments are made for on-going management practices rather than the delivery of ecosystem services. Spatial targeting of such payments, as well as payment by delivery of ecosystem service, could fundamentally improve value for money for the tax payer (see Drechsler & Wätzold, Chapter 12). Such schemes seek to address market failure, providing financial incentives to farmers where markets do not. Commercial companies are now also recognising the opportunities in engaging with landowners and farmers in this manner. One notable example is

ScaMP (Sustainable Catchment Management Programme) developed by a UK water company. In this pilot study in north-west England, a water company has developed a partnership approach to improving water quality within its catchments by incentivising tenant farmers to implement certain land management practices. The financial incentives provided by the company are more than repaid by improvement in water quality, which reduces the cost of downstream treatment.[3]

Conclusion

In this chapter we have provided an overview of a common 'toolkit' with which practitioners and researchers can begin to think about valuing ecosystem services in agriculture as a gateway toward better inclusion of processes, such as resilience, that operate across different scales. Challenges for the future include the development of the environmental and ecological science alongside the development of effective policy and financial instruments. Work that addresses or incorporates concepts of resilience and potential links with policy and management have a long history (Levin et al., 1998; Adger, 2000; Walker et al., 2002; Martin-Breen & Anderies, 2011), but valuing resilience as an entity itself is relatively new in the economics literature. Some work has been done on the theoretical side (Baumgärtner & Strunz, 2014) but even less has been explored empirically from an economic point of view. This is perhaps understandable given the challenges of linking data, developing an appropriately designed study (which would require a robust counterfactual), and the long time horizons often needed for measuring resilience concepts. We anticipate increased interest and growth in this area of interdisciplinary research in the near future.

References

Adger, W.N. (2000). Social and ecological resilience: are they related? *Progress in Human Geography*, **24**, 347–364.

Anderson, D.M., Salick, J., Moseley, R.K. & Xiaokun, O. (2005). Conserving the sacred medicine mountains: a vegetation analysis of Tibetan sacred sites in Northwest Yunnan. *Biodiversity and Conservation*, **14**, 3065–3091.

Arias, V., Benitez, S. & Goldman, R. (2010). TEEBcase. Water fund for catchment management, Ecuador. Available at TEEBweb.org

Badgley, C., Moghtader, J., Quintero, E., et al. (2007). Organic agriculture and the global food supply. *Renewable Agriculture and Food Systems*, **22**(2), 86–108.

Barbier, E.B. (2007). Valuing ecosystem services as productive inputs. *Economic Policy*, **22**, 177–229.

Batáry, P., Báldi, A., Kleijn, D. & Tscharntke, T. (2011). Landscape-moderated biodiversity effects of agri-environmental management: a meta-analysis. *Proceedings of the Royal Society B*, **278**, 1894–1902.

---

[3] http://www.gov.uk/government/uploads/system/uploads/attachment_data/file/200901/pb13932a-pes-bestpractice-annexa-20130522.pdf/

Bateman, I.J., Abson, D., Beaumont, N., et al. (2011). Economic values from ecosystems. In: *UK National Ecosystem Assessment*. The UK National Ecosystem Assessment Technical Report. Cambridge: UNEP-WCMC, pp. 1067–1152.

Bateman, I., Harwood, A., Mace, G., *et al.* (2013). Bringing ecosystem services into economic decision-making: land use in the United Kingdom. *Science*, **341**, 45–50.

Baulcombe, D., Crute, I., Davies, B., et al. (2009). *Reaping the Benefits: science and the sustainable intensification of global agriculture*. London: The Royal Society.

Baumgärtner, S. & Strunz, S. (2014). The economic insurance value of ecosystem resilience. *Ecological Economics*, **101**, 21–32.

Bennett, E.M., Peterson, G.D. & Gordon, L.J. (2009). Understanding relationships among multiple ecosystem services. *Ecology Letters*, **12**, 1394–1404.

Braat, L.C., Brink, P. ten, (eds.). (2008). *The Cost of Policy Inaction: the case of not meeting the 2010 biodiversity target*. Report to the European Commission Under Contract: ENV.G.1./ETU/2007/0044. Wageningen, Brussels: Alterra Report 1718/.

Brady, M.V., Hedlund, K., Cong, R.-G., et al. (2015). Valuing supporting soil ecosystem services in agriculture: a natural capital approach. *Agronomy Journal*, **107**, 1809–1821. doi:10.2134/agronj14.0597

Bratman, G.N., Hamilton, J.P. & Daily, G.C. (2012). The impacts of nature experience on human cognitive function and mental health. *Annals of the New York Academy of Sciences*, **1249**, 118–136.

Brauman, K.A., Freyberg, D.L. & Daily, G.C. (2012). Land cover effects on groundwater recharge in the tropics: ecohydrologic mechanisms. *Ecohydrology*, **5**, 435–444.

Brauman, K., Meulen, S. & Brils, J. (2014). Ecosystem services and river basin management. In: *Risk-Informed Management of European River Basins*, edited by J. Brils, W. Brack & D. Müller-Grabherr. Berlin: Springer.

Brauman, K., Freyberg, D. & Daily, G. (2015). Impacts of land-use change on groundwater supply: ecosystem services assessment in Kona, Hawaii. *Journal of Water Resources Planning and Management*, **141**(12), A4014001.

Carpenter, S.R., Ludwig, D. & Brock, W.A. (1999). Management of eutrophication for lakes subject to potentially irreversible change. *Ecological Applications*, **9**, 751–771.

Champ, P.A., Boyle, K.J. & Brown, T.C. (2017). *A Primer on Non-market Valuation (The Economics of Non-Market Goods and Resources)*. 2nd edition. Dordrecht: Springer.

Chaplin-Kramer, R., O'Rourke, M.E., Blitzer, E.J., *et al.* (2011a). A meta-analysis of crop pest and natural enemy response to landscape complexity. *Ecology Letters*, **14**, 922–932.

Chaplin-Kramer, R., Kliebenstein, D.J., Chiem, A., et al. (2011b). Chemically-mediated tritrophic interactions: opposing effects of glucosinolates on a specialist herbivore and its predators. *Journal of Applied Ecology*, **48**, 880–887.

Daily, G.C., Söderqvist, T., Aniyar, S., et al. (2000). The value of nature and the nature of value. *Science*, **289**, 395–396.

Daniel, T.C., Muhar, A., Arnberger, A., et al. (2012). Contributions of cultural services to the ecosystem services agenda. *Proceedings of the National Academy of Sciences*, **109**, 8812–8819.

Dearmont, D., McCarl, B. & Tolman, D. (1998). Costs of water treatment due to diminished water quality: a case study in Texas. *Water Resources Research*, **34**, 849–853.

DeFries, R. & Rosenzweig, C. (2010). Toward a whole-landscape approach for sustainable land use in the tropics. *Proceedings of the National Academy of Sciences*, **107**, 19627–19632.

Diaz, S., Demissew, S., Carabias, J., et al. (2015). The IPBES conceptual framework – connecting nature and people. *Current Opinion in Environmental Sustainability*, **14**, 1–16.

Ellis, E.C. & Ramankutty, N. (2008). Putting people in the map: anthropogenic biomes of the world. *Frontiers in Ecology and the Environment*, **6**, 439–447.

Farley, K.A., Anderson, W.G., Bremer, L.L. & Harden, C.P. (2011). Compensation for ecosystem services: an evaluation of efforts to achieve conservation and development in Ecuadorian páramo grasslands. *Environmental Conservation*, **38**, 393–405.

Fezzi, C., Bateman, I., Askew, T., et al. (2013). Valuing provisioning ecosystem services in agriculture: the impact of climate change on food production in the United Kingdom. *Environmental and Resource Economics*, **57**, 197–214.

Fischer, J., Brosi, B., Daily, G., et al. (2008). Should agricultural policies encourage land-sparing or wildlife-friendly farming? *Frontiers in Ecology and the Environment*, **6**, 380–385.

Fisher, B. & Turner, R.K. (2008). Ecosystem services: classification for valuation. *Biological Conservation*, **141**(5), 1167–1169.

Foley, J.A., Ramankutty, N., Brauman, K.A., et al. (2011). Solutions for a cultivated planet. *Nature*, **478**(7369), 337–342.

Folke, C. (2006). Resilience: the emergence of a perspective for social–ecological systems analyses. *Global Environmental Change*, **16**, 253–267.

Garibaldi, L.A., Steffan-Dewenter, I., Winfree, R., et al. (2013). Wild pollinators enhance fruit set of crops regardless of honey bee abundance. *Science*, **339**, 1608–1611.

Girvetz, E., Valle, A.M., Laderach, P. & Vogl, A. (2014). *Technical Support to Apply the RIOS Analytical Modeling Tool to Guide Climate Adaptation Planning*. Report in 158 collaboration with The Nature Conservancy, International Center for Tropical Agriculture and Natural Capital Project.

Glenk, K., Shrestha, S., Cairstiona, F.E., et al. (2017). A farm level approach to explore farm gross margin effects of soil organic carbon management. *Agricultural Systems*, **151**, 33–46.

Goldman-Benner, R.L., Benitez, S., Boucher, T., et al. (2012). Water funds and payments for ecosystem services: practice learns from theory and theory can learn from practice. *Oryx*, **46**, 55–63.

Goldstein, J., Pejchar, L. & Daily, G. (2008). Using return-on-investment to guide restoration: a case study from Hawaii. *Conservation Letters*, **1**, 236–243.

Goldstein, J.J.H., Caldarone, G., Duarte, T.K.T., et al. (2012). Integrating ecosystem service tradeoffs into land-use decisions. *Proceedings of the National Academy of Sciences*, **109**, 7565–7570.

Gordon, L.J., Peterson, G.D. & Bennett, E.M. (2008). Agricultural modifications of hydrological flows create ecological surprises. *Trends in Ecology and Evolution*, **23**, 211–219.

Green, R.E., Cornell, S.J., Scharlemann, J.P.W. & Balmford, A. (2005). Farming and the fate of wild nature. *Science*, **307**, 550–555.

De Groot, R.S., Wilson, M.A. & Boumans, R.M. (2002). A typology for the classification, description and valuation of ecosystem functions, goods and services. *Ecological Economics*, **41**, 393–408.

Gunderson, L.H. & Holling, C.S. (2002). *Panarchy: understanding transformations in systems of humans and nature*. Washington, DC: Island Press.

Heal, G. (2000). Valuing ecosystem services. *Ecosystems*, **3**, 24–30.

Hillier, J., Brentrup, F., Wattenbach, M., et al. (2012). Which cropland greenhouse gas mitigation options give the greatest benefits in different world regions? Climate and soil specific predictions from integrated empirical models. *Global Change Biology*, **18**(6), 1880–1894.

Howe, C., Suich, H., Vira, B. & Mace, G. (2014). Creating win–wins from trade-offs? Ecosystem services for human well-being: a meta-analysis of ecosystem service trade-offs and synergies in the real world. *Global Environmental Change*, **28**, 263–275.

Karp, D.S., Rominger, A.J., Zook, J., et al. (2012). Intensive agriculture erodes β-diversity at large scales. *Ecology Letters*, **15**, 963–970.

Karp, D.S., Mendenhall, C.D., Sandí, R.F., et al. (2013). Forest bolsters bird abundance, pest control and coffee yield. *Ecology Letters*, **16**, 1339–1347.

Keeler, B.L., Polasky, S., Brauman, K.A., et al. (2012). Linking water quality and well-being for improved assessment and valuation of ecosystem services. *Proceedings of the National Academy of Sciences*, **109**, 18619–18624.

Kennedy, C.M., Lonsdorf, E., Neel, M.C., et al. (2013). A global quantitative synthesis of local and landscape effects on wild bee pollinators in agroecosystems. *Ecology Letters*, **16**, 584–599.

Kleijn, D. & Sutherland, W.J. (2003). How effective are European agri-environment schemes in conserving and promoting biodiversity? *Journal of Applied Ecology*, **40**, 947–969.

Kleijn, D., Berendse, F., Smit, R. & Gilissen, N. (2001). Agri-environment schemes do not effectively protect biodiversity in Dutch agricultural landscapes. *Nature*, **413**, 723–725.

Kleijn, D., Berendse, F., Smit, R., et al. (2004). Ecological effectiveness of agri-environment schemes in different agricultural landscapes in the Netherlands. *Conservation Biology*, **18**, 775–786.

Klein, A.-M., Vaissière, B.E., Cane, J.H., et al. (2007). Importance of pollinators in changing landscapes for world crops. *Proceedings of the Royal Society B: Biological Sciences*, **274**, 303–313.

Levin, S.A., Barrett, S., Aniyar, S., et al. (1998). Resilience in natural and socioeconomic systems. *Environment and Development Economics*, **3**, 221–262.

Losey, J.E. & Vaughan, M. (2006). The economic value of ecological services provided by insects. *Bioscience*, **56**, 311–323.

Mace, G.M., Norris, K. & Fitter, A.H. (2012). Biodiversity and ecosystem services: a multilayered relationship. *Trends in Ecology and Evolution*, **27**(1), 19–26.

Mäler, K.-G. (2008). Sustainable development and resilience in ecosystems. *Environment and Resource Economics*, **39**, 17–24.

Mäler, K.-G. & Li, C.-Z. (2010). Measuring sustainability under regime shift uncertainty: a resilience pricing approach. *Environment and Development Economics*, **15**, 707–719.

Martin-Breen, P. & Anderies, J.M. (2011). *Resilience: a literature review*. Brighton: Bellagio Initiative, IDS.

McConnell, K.E. & Bockstael, N.E. (2005). Valuing the environment as a factor of production. *Handbook of Environmental Economics*, **2**, 622–666.

Mendenhall, C.D., Sekercioglu, C.H., Brenes, F.O., et al. (2011). Predictive model for sustaining biodiversity in tropical countryside. *Proceedings of the National Academy of Sciences*, **108**, 16313–16316.

Millennium Ecosystem Assessment. (2003). *Ecosystems and Human Well-Being: a framework for assessment*. Washington, DC: Island Press.

Mitchell, M., Bennett, E.M. & Gonzalez, A. (2013). Linking landscape connectivity and ecosystem service provision: current knowledge and research gaps. *Ecosystems*, **16**, 894–908.

Montzka, S.A., Dlugokencky, E.J. & Butler, J.H. (2011). Non-CO_2 greenhouse gases and climate change. *Nature*, **476**, 43–50.

Morse, R.A. & Calderone, N.W. (2000). The value of honey bees as pollinators of U.S. crops in 2000. *Bee Culture Magazine*, **128**, 2–15.

Mulder, C., Bennett, E.M., Bohan, D.A., et al. (2015). 10 years later: revisiting priorities for science and society a decade after the millenium ecosystem assessment. *Advances in Ecological Research*, **53**, 1–54.

Naidoo, R., Balmford, A., Costanza, R., et al. (2008). Global mapping of ecosystem services and conservation priorities. *Proceedings of the National Academy of Sciences*, **105**, 9495–9500.

Natural Capital Committee. (2013). The State of Natural Capital: towards a framework for measurement and valuation. www.naturalcapitalcommittee.org

Natural Capital Committee (2015). The State of Natural Capital: protecting and improving natural capital for prosperity and wellbeing. www.gov.uk/government/publications/natural-capital-committees-third-state-of-natural-capital-report

Nolan, J. (2001). *Well Grounded: using local land use authority to achieve smart growth*. Washington, DC: Environmental Law Institute.

O'Rourke, M.E., Rienzo-Stack, K. & Power, A.G. (2011). A multi-scale, landscape approach to predicting insect populations in agroecosystems. *Ecological Applications*, **21**, 1782–1791.

Oliver, T.H., Heard, M.S., Isaac, N.J.B., et al. (2015). Biodiversity and resilience of ecosystem functions. *Trends in Ecology and Evolution*, **30**(11), 673–684.

Pascual, U., Balvanera, P., Diaz, S., et al. (2017). Valuing nature's contributions to people: the IPBES approach. *Current Opinion in Environmental Sustainability*, **26**, 7–16.

Phalan, B., Onial, M., Balmford, A., et al. (2011). Reconciling food production and biodiversity conservation: land sharing and land sparing compared. *Science*, **333**, 1289–1291.

Piao, S., Ciais, P., Huang, Y., et al. (2010). The impacts of climate change on water resources and agriculture in China. *Nature*, **467**, 43–51.

Pimentel, D., Acquay, H., Biltonen, M., et al. (1992). Environmental and economic costs of pesticide use. *Bioscience*, **42**, 750–760.

Polasky, S., Nelson, E., Pennington, D. & Johnson, K.A. (2011). The impact of land-use change on ecosystem services, biodiversity and returns to landowners: a case study in the State of Minnesota. *Environmental and Resource Economics*, **48**, 219–242.

Ponette-González, A.G., Marín-Spiotta, E., Brauman, K.A., et al. (2014). Hydrologic connectivity in the high-elevation tropics: heterogeneous responses to land change. *Bioscience*, **64**, 92–104.

Potts, S.G., Biesmeijer, J.C., Kremen, C., et al. (2010). Global pollinator declines: trends, impacts and drivers. *Trends in Ecology and Evolution*, **25**, 345–353.

Pretty, J.N., Noble, A.D., Bossio, D., et al. (2006). Resource-conserving agriculture increases yields in developing countries. *Environmental Science & Technology*, **40**(4), 1114–1119.

Pywell, R.F., Heard, M.S., Bradbury, R.B., et al. (2012). Wildlife-friendly farming benefits rare birds, bees and plants. *Biology Letters*, **8**, 772–775.

Pywell, R.F., Heard, M.S., Woodcock, B.A., et al. (2016). Wildlife-friendly farming increases crop yield: evidence for ecological intensification. *Proceedings of the Royal Society B*, **282**, 20151740.

Ricketts, T.H., Daily, G.C., Ehrlich, P.R. & Michener, C.D. (2004). Economic value of tropical forest to coffee production. *Proceedings of the National Academy of Sciences*, **101**, 12579–12582.

Ricketts, T.H., Regetz, J., Steffan-Dewenter, I., et al. (2008). Landscape effects on crop pollination services: are there general patterns? *Ecology Letters*, **11**, 499–515.

Salick, J., Amend, A., Anderson, D., et al. (2007). Tibetan sacred sites conserve old growth trees and cover in the eastern Himalayas. *Biodiversity and Conservation*, **16**, 693–706.

Smith, P., Martino, D., Cai, Z., et al. (2008). Greenhouse gas mitigation in agriculture. *Philosophical Transactions of the Royal Society of London B: Biological Sciences*, **363**, 789–813.

Staley, J.T., Botham, M.S., Chapman, R.E., et al. (2016). Little and late: how reduced hedgerow cutting can benefit Lepidoptera.

Agriculture, Ecosystems and Environment, **224**, 22–28.

Suarez, C.F., West, N., Naranjo, L.G., et al. (2013). *Climate Adaptation in Colombia: designing an adaptive compensation and rewards program for ecosystem services.* World Wildlife Fund Report.

Tallis, H. & Lubchenco, J. (2014). Working together: a call for inclusive conservation. *Nature*, **515**, 27–28. doi:10.1038/515027a

Tallis, H., Polasky, S., Lozano, J.S.S. & Wolny, S. (2012). Inclusive wealth accounting for regulating ecosystem services. In: *Inclusive Wealth Report 2012. Measuring progress toward sustainability*, edited by UNEP and UNU-IHDP. Cambridge: Cambridge University Press.

Thies, C. & Tscharntke, T. (1999). Landscape structure and biological control in agroecosystems. *Science*, **285**, 893–895.

Thompson, B.H. & Segerson, K. (2009). Valuing the Protection of Ecological Systems and Service. Report of the EPA Science Advisory Board.

Tschumi, M., Albrecht, M., Bartschi, C., et al. (2016). Perennial, species-rich wildflower strips enhance pest control and crop yield. *Agriculture Ecosystems and Environment*, **220**, 97–103.

UK National Ecosystem Assessment. (2011). The UK National Ecosystem Assessment: Synthesis of the Key Findings. Cambridge: UNEP-WCMC.

Van der Werf, G.R., Morton, D.C., DeFries, R.S., et al. (2009). CO_2 emissions from forest loss. *Nature Geoscience*, **2**, 737–738.

Vejre, H. & Brandt, J. (2003). *Multifunctional Landscapes: theory, value and history*. Southampton: WIT Press.

Venterea, R.T., Halvorson, A.D., Kitchen, N., et al. (2012). Challenges and opportunities for mitigating nitrous oxide emissions from fertilized cropping systems. *Frontiers in Ecology and the Environment*, **10**, 562–570.

Vermeulen, S.J., Campbell, B.M. & Ingram, J.S.I. (2012). Climate change and food systems. *Annual Review of Environmental Resources*, **37**, 195–222.

Walker, B., Carpenter, S.R., Anderies, J.M., et al. (2002). Resilience management in social-ecological systems: a working hypothesis for a participatory approach. *Conservation Ecology*, **6**, 14.

Werling, B.P. & Gratton, C. (2010). Local and broadscale landscape structure differentially impact predation of two potato pests. *Ecological Applications*, **20**, 1114–1125.

West, P.C., Gibbs, H.K., Monfreda, C., et al. (2010). Trading carbon for food: global comparison of carbon stocks vs. crop yields on agricultural land. *Proceedings of the National Academy of Sciences*, **107**, 19645–19648.

Winfree, R. & Kremen, C. (2009). Are ecosystem services stabilized by differences among species? A test using crop pollination. *Proceedings of the Royal Society B: Biological Sciences*, **276**, 229–237.

Winfree, R., Gross, B.J. & Kremen, C. (2011). Valuing pollination services to agriculture. *Ecological Economics*, **71**, 80–88.

Wratten, S.D., Gillespie, M., Decourtye, A., et al. (2012). Pollinator habitat enhancement: benefits to other ecosystem services. *Agriculture, Ecosystems and Environment*, **159**, 112–122.

Zheng, H., Li, Y., Robinson, B.E., et al. (2016). Using ecosystem service trade-offs to inform water conservation policies and management practices. *Frontiers in Ecology and Environment*, **14**, 527–532.

CHAPTER FIVE

Resilience in agricultural systems

STEPHEN J. RAMSDEN
University of Nottingham
and
JAMES GIBBONS
Bangor University

Introduction

In this chapter, we assess the extent to which resilience concepts have informed studies that use 'bio-economic' optimisation models of various forms, where the bio-economy described by the model is generally a farm or collection of farms. We use the term 'bio-economic' here to specifically refer to models that have optimisation and constraint components; terminology in this area is somewhat confusing, but this generally means that models are based on some form of mathematical programming such as Linear Programming, Integer Programming, Dynamic Programming or Multiple Objective Programming. In this volume, Drechsler and Wätzold use Linear Programming to model the effect of financial incentives on farm biodiversity. From an economic perspective, the great advantage of these models is the facility to capture features and processes of natural biological and physical systems – for example, the relationship between soils, rainfall and loss of nitrogen, in various forms, to air and water – and integrate these into the micro-economy of the farm: availability of physical and financial resources, market prices, farmer decision-making, availability of different management practices or interventions. It is this latter facility that makes the approach particularly useful when combined with the constrained optimisation framework, as models can be run to establish which interventions are optimal for different types of objective. Objectives can be economic (e.g. 'utility') or environmental (e.g. 'reduce loss of nitrogen to the environment'). However, as outlined later in the chapter, despite considerable potential, these models have not been used to represent some of the more complex dynamic and spatial aspects of environmental systems.

Bio-economic models have been widely used in disciplines ranging from Agricultural and Ecological Economics, through the Agricultural and Environmental Sciences to Operations and Management Research. There are a number of established journals that have editorial policies broadly sympathetic to the systems approach, most obviously *Agricultural Systems*. Originating in England in 1976, the first editorial for *Agricultural Systems* states boldly that 'it is now generally recognised that whole agricultural systems deserve to be studied in their own right' (Spedding, 1976). Spedding argued that research into what he termed 'component and component processes' was also needed, but that *whole* system studies were essential if we are to understand the effects of *change*: change in one part of a system, its effect on other parts of the system and the consequent (probably uncertain to some degree) effect that change has on system outputs and inputs. There are three elements of the system approach that are worth emphasising. First, we are often interested in assessing the effects of changes that are a consequence of human intervention of some form: if reducing loss of nitrate to the environment is the objective, what is the effect of intervention A (e.g. grow a 'cover crop' with the objective of storing nitrogen in plant biomass so that less is lost to the environment) in comparison to intervention B (e.g. impose farm, field and timing limits on the amount of mineral and organic nitrogen that can be applied in a year – the requirement in some European countries for receiving payments under the Common Agricultural Policy)? Second and related to the first point, bio-economic models embrace, overtly – or implicitly in the assumptions – decision-making as guided by incentives. Often, these will be profit-based: if intervention A reduces farm profit, per milligram of nitrate-N per litre reduced, by less than intervention B, it is a better intervention, other things being equal. While this raises questions (do farmers maximise profit?) it does make the approach objective. We ask 'What is the cost of change?', allowing for the existing – probably highly productive in developed countries – state of the system. This way of thinking also helps us to understand why farmers sometimes do not respond to 'expert advice': when we allow for their incentive structures, we see that the change may be detrimental to the system on which the farm manager – or the farming family – relies for income. Interestingly, as shown by Pannell (2006) and demonstrated in some of our own work (Gibbons et al., 2005), model results suggest that the negative effect may not always be that great, at least over a certain range of profitability.

The third advantage of bio-economic models is that they can be constructed for different scales. A typical formulation is to model a catchment (e.g. Gibbons & Ramsden, 2008) or an area with a particular habitat (e.g. upland farms in England). Thus, it should be possible to capture some spatial aspects of resilience, for example the relationship between decisions made by groups of farmers, land use on farms and species numbers and diversity for a particular habitat, although from an ecological perspective, 'spatial' implies a rather more complex set of relationships, including the spatial arrangement of habitats and species and the effect that this has on abundance and persistence of flora and fauna in a particular area. However, despite this potential and the other advantages discussed above, there are relatively few applications of bio-economic modelling that directly address whether agricultural systems are resilient. The emphasis has been more towards generating solutions that are optimal and efficient in the short run; where uncertainty is considered, model results tend to focus on optimal solutions to known *variability*, typically variability of commodity prices or yields, rather than to less-predictable shocks or the existence of threshold effects or tipping points.

The chapter is arranged as follows. First, we consider aspects of resilience that are of particular relevance to agricultural systems. We use the term 'agricultural systems' to include the economic, biological and physical aspects of farms and the processes that are influenced by decision-making, principally by farmer owners of land, tenants or salaried farm managers. We then conduct a brief review of relevant literature to assess the extent to which these aspects have been included in bio-economic optimisation models of agricultural systems. From these two strands, we then conclude by considering how these models should be developed in the future, focusing in particular on short-run and long-run effects, substitution possibilities between different forms of capital and the potential for building resilience thinking into existing economic techniques for modelling agricultural risks.

Resilience in agricultural systems: what should we model?

Much of the work that has been done on resilience focuses on whether disturbance effects to a system can be absorbed 'while maintaining function', where function relates to the system's capacity to return to the same equilibrium state or to move to a new stable equilibrium while behaving in a similar way to the pre-disturbed state. Given that humans rely on ecosystems for a range of products and services, most obviously food, similarity in behaviour of farm systems over time is a highly desirable attribute. This is particularly the case if substitution opportunities – alternative, non-land-based ways of producing food – are limited, as is certainly the situation for many agricultural products, at least to any reasonable degree in the face of current population and income projections and consequent demand for food. Resilience thinking has also developed to encompass the adaptive capacity of systems and human intervention. An example here is moorland habitats in the UK, which are quite diverse but function as a habitat in a similar manner, despite differences in dominant species. 'Human interventions' include learning and innovation adaptations by humans. Folke (2006) refers to these as 'social–ecological resilience'.

As in our moorland example, a system may therefore maintain function while behaving differently, where this changed behaviour is a product of human intervention; the extent to which this type of adaptation is possible in agriculture is thus also an important factor to consider. However, as we will argue, adaptation may also lead to a reduction of resilience elsewhere in the system, through trade-off effects: it is this kind of unforeseen system effect that bio-economic models are designed to capture, although the emphasis has been on economic rather than ecological trade-offs.

Some aspects of the natural and hence agricultural environment are more vulnerable than others. Despite evidence that economic growth beyond a certain level can lead to some environmental improvements, Arrow et al. (1995) conclude that the relationship is unlikely to apply to natural resources such as soil, water, natural vegetation and biodiversity because of the risk of sudden loss of biological productivity and the uncertainty caused by 'flipping' from one state to another. A lack of resilience reduces our – and future generations' – options. If resources are degraded, we have less choice about how we produce food and we have less choice of environmental benefits – whether these benefits are directly productive or, like penguins in Antarctica, 'nice to have around'. It is not that technology cannot substitute for some aspects of what the environment provides: genetic modification and traditional plant breeding techniques will lead to 'land sparing' (Phalan et al., 2011), other things being equal, not just through yield but also through improved soil and root interactions and their potential benefit for soil quality (York et al., 2016). However, with a resilient farm system we can choose from both land management interventions and technological solutions to increasing and adapting food supplies to future demands from humanity. Furthermore, while it is possible to envisage technology substituting for land to some degree (in a land-sparing sense), it is more difficult to identify how technology can sufficiently substitute for the core elements of the carbon or nitrogen cycles, or for biological biodiversity.

Walker et al. (2012) make the general point that change results from internal relationships within systems and the effects of external drivers: for example, external drivers can cause changes in controlling or 'slow' system variables and may push them towards threshold levels. In the case of soil organic matter (a 'slow' variable), as a threshold level of organic matter is approached, a given external shock, e.g. flood or drought, results in greater instability in ecosystem services such as crop production and increases the chance that the system is pushed into an alternative equilibrium with the sort of substantial reduction in biological productivity referred to by Arrow et al. (1995). Human intervention can be used in an attempt to influence the slow system variables ('incorporate plant matter such as straw') and in the UK many arable farmers use machinery to incorporate cereal straw into the soil. This is partly due to specialisation (there are fewer 'mixed' arable and straw-using livestock farms, particularly in England). However, the main reason is legislation: burning of straw in the UK has been banned since the early 1990s. There are interesting short-run efficiency/long-run resilience themes here: there have been calls for the ban to be lifted – burning is a method of controlling blackgrass, one of the most problematic grass weeds in cereal crops in the UK.

However, Powlson et al. (2011) conclude that even small reductions in organic matter, as measured by soil organic carbon, would lead to deterioration in beneficial soil properties, a deterioration that would be difficult to offset through improvements in plant breeding alone. A further complication is the uncertainty surrounding the effect of straw incorporation: long-term trials suggest that benefits for soil organic matter may be small (Powlson et al., 2011).

It is therefore clear that we do not always have a good understanding of ecological or agricultural systems. This is particularly the case once we introduce people. Peterson et al. (2003) show that uncertainty in a model of lake management and eutrophication can lead to cycles of collapse of the lake's primary ecosystem services; these threshold events are, in part, due to the rational behaviour – from a management perspective – of those managing the system. This is a known result in Economics, or at least in the fields of economics that encompass game theory. Thomas Schelling's *Micromotives and Macrobehaviour* (1978) explores the scope for individual decisions to lead to undesirable outcomes, in particular the difficulty of achieving cooperative decisions with high joint value when strong incentives exist to pursue private outcomes that have lower total social value. Uncertainty in one form *is* a feature of many bio-economic models; typically, a model is run with profit-making and variability-reducing objectives where, for a risk-averse decision maker, profit will be traded-off in return for an acceptable (utility maximising) level of variability. This decision maker essentially has three options: (i) to manage the external variability; (ii) to manage the internal farm system to reduce exposure to external variability; and (iii) to build up the capital position of the business. Examples respectively would be to use financial instruments such as futures markets ('hedging'), enterprise diversification (production of different types of crops or livestock, but also non-land based enterprises) and building net worth, a process that is very much dependent on a flow of profit over time. From the perspective of Folke's social–ecological approach, all three are ways that can increase resilience. However, leaving aside the potential for thresholds to be exceeded and seemingly rational behaviour to lead to severe system malfunction (in the Peterson sense), there are also potential trade-off effects – for example, a farmer using the futures market to lock into a known price is reducing exposure of the farm system to variability – and thus other internal forms of resilience become *less* important. This doesn't matter (farmer utility is maximised by choosing the best combination of management interventions for a given level of risk aversion) if there are no adverse effects on the biological/environmental part of the farm system, but this is unlikely to be the case. It is probable that these internal elements will have less relevance – to the rational farmer – as ways of building resilience, at least to some degree. Thus, we return to the question of how well human-created capital and methods of management can substitute for natural capital and methods of management that involve natural capital (see also discussion in Hodge, Chapter 10). This is an ideal area for research using bio-economic models, with built-in risk and profit objectives and sufficiently rich biological and environmental components. As a minimum this would need to include nutrient loss (nitrogen and potassium in various forms, including nitrate, nitrous oxide and ammonia), greenhouse gas emissions (as well as nitrous oxide, agriculture is associated with methane and carbon dioxide emissions, the latter largely 'embedded' in inputs such as nitrogen fertiliser), water (both quantity and quality), biodiversity (e.g. pollinators, natural pest predators), soils and the functional relationships that the natural world has with agricultural production – and vice versa. The framework could be extended to capture uncertainty (unpredictable, rare, large-scale events – the 'black swans' of Nassim Taleb, 2007) rather than known variability, and attempts have been made to do this by some authors; however, there are relatively few examples where this has been linked to a comprehensive range of environmental outcomes.

From this brief overview, we conclude that there are at least five aspects of resilience thinking that ought to feature within bio-economic optimisation models of farm systems.

(1) Resilience thinking puts emphasis on systems and bio-economic models are systems' models; however, existing models probably do not represent the ecological and environmental components of farms to a sufficient extent.
(2) Thresholds are important areas of the bio-economic system and environment that may not be well represented in current models, particularly for natural resources such as soil, water and biodiversity.
(3) Rational decision making may be optimal in the short run, but not resilient in the long run; current optimisation models may not reflect this sufficiently.
(4) Farms are run by rational agents, with multiple objectives and ways of doing things; there will be trade-offs between these ways of doing things (and ways of doing things developed in the future) that meet the farmers' needs (for example, risk management) but reduce resilience elsewhere in the farm system.
(5) Existing approaches to modelling risk in bio-economic models could be extended to include resilience, in particular, resilience to uncertainty rather than variability.

With respect to (1), the requirement is completeness but not necessarily complexity. Thus, models of livestock farms should include loss of nitrogen to air (nitrous oxide, ammonia) and water (nitrate), through known mechanisms, together with a complete range of other environmental impacts. Parsimonious approaches, perhaps involving the use of indicators, would be an example of how this could be done, and we give some examples in the following section. We have considered thresholds in relation to soil organic carbon and this type of relationship would be relatively straightforward to model. The term 'tipping point' is also used; if we take this to mean a major shift in the system (perhaps as a result of crossing a threshold) we are on home territory from a bio-economic modelling perspective, as often the optimal solution (the optimal farm plan) will change dramatically in response to external and internal drivers. What is needed are models that fully capture the consequences of these shifts, both environmental and economic. The third and fourth points are both optimisation versus sustainability issues; the latter allows for the idea that adaptability may just be a short-run solution that does not necessarily improve environmental resilience over time. The final point relates to the opportunity to integrate resilience thinking with thinking on risk management as exemplified in the Agricultural Economics literature: we often have information about short-run variability in prices or yields, but this tells us little about the resilience of farms to events with unknown probability and potentially large impact.

The following section considers a selection of studies that use constrained bio-economic models of farm systems that include both an agricultural and an environmental or ecological component and the extent to which the above ideas have been reflected in both model construction and more generally in the objectives set for each study. The intention is not to be comprehensive, but to cover examples of models that have been in use in the relatively recent past.

Resilience in bio-economic optimisation models

None of the 42 approaches covered in Janssen and Van Ittersum's (2007) review of bio-economic farm models directly address resilience, although as we would expect from the previous discussion, there are elements within these models that are relevant to the resilience concept, including farmer decision-making (single or multiple goals including profit and risk, operational, tactical and strategic decision-making), resource constraints (land, labour, capital; but also institutional and policy-related limitations), available activities (crop and livestock enterprises, transfers between system components, buying and selling activities, environmental activities) and dynamics (decision-making over different periods). As argued, one of the great strengths of these models is their scope for capturing different management interventions and treating them in the same way as existing elements of the farm system. An intervention designed to reduce greenhouse gas emissions ('feed animals a new diet') becomes another activity choice available to the decision-making component of the model; feeding the new diet has associated costs and benefits which, when considered within the model system as a whole, under a particular policy target ('reduce emissions by 20 percent') may or may not be chosen over existing diets. Normative and positive approaches are combined here: our normative, expert judgement that the new diet is better is qualified by the positive understanding that we gain about the farm system ('it's not better when evaluated with the existing diet and other methods for achieving the reduction target'). This understanding will be enriched further if we have information on variability of different farm systems and decision makers – to extend the example, existing diets and their effects on production will vary from farm to farm, as will farmer attitudes. The 'right' response will therefore vary from farm to farm.

Interestingly, modelling approaches most relevant to resilience thinking – tactical and strategic decision-making, incorporation of environmental activities and dynamics – are also areas that Janssen and Van Ittersum suggest are in need of additional research. Existing models are usually static (75 percent of the models reviewed were in this category) and constraints and activities are often limited to specific environmental problems with relatively abundant data and continuous response and damage functions, such as loss of nitrate to watercourses. Less attention has been devoted to diseases, pests, soil characteristics and biodiversity: characteristics that are not static and potentially subject to threshold events and discontinuities.

Turning to individual studies, Table 5.1 gives some example modelling approaches, all based around programming and optimisation techniques in some way and some example applications. Three core elements are common to all the approaches considered. First, the farm system is constrained by the availability of natural and economic inputs. Second, given the prevailing technology, there are a range of production possibilities available to the farmer (grow different crops, produce different types of livestock); production uses the constraining inputs at a rate given by technical coefficients of production; these coefficients form the third component of the bio-economic modelling approach. Optimisation gives a combination of inputs and outputs that maximise the objective function, usually but not always profit. Some aspects of agricultural system complexity can be introduced by increasing the number of constraints, production possibilities, associated technical coefficients and management interventions. Attitude to variability can be introduced into the objective function, dynamics can be introduced with formal linkages capturing the effects (for example, the benefits of fertility-building crops in a rotation) of decision-making over time and sensitivity analysis can be used to explore the response of a model to changes in key parameters. Models with multiple objectives have been developed, as have models that 'self-calibrate' (for example, Louhichi et al., 2010) to match farmers' actual production and input choices.

In part because of the relatively high risk (prices and yields are variable over time), potato production systems in Europe and North West America offer

Table 5.1. *Examples of bio-economic modelling from the literature.*

Approach	Topic and region	Authors
Multi-attribute linear programming models	Policy impact analysis, water, Italy	Bartolini et al., 2007
Multiple objectives: risk	Optimisation of soil organic carbon, Sweden	Cong et al., 2014
Integer programming	Minimum conservation targets, Australia	Crossman & Bryan, 2006
Multiple objectives: goal programming	Conservation of natural resources, Portugal	Fleskens & de Graaff, 2010
Dynamic programming	Climate change adaptation and capital investment, UK	Gibbons & Ramsden, 2008
Linear programming	Reducing soil erosion, Canada	Jatoe et al., 2008
Stochastic programming	Climate change adaptation and perennial crops, Australia	John et al., 2005
Linear programming	Crop and livestock mix, Ethiopia	Kassie et al., 1999
Positive Mathematical Programming (PMP)	Policy impact analysis in the EU	Louhichi et al., 2010
Linear programming	Reducing soil erosion, China	Lu et al., 2004
Linear programming	Bee conservation, Scotland	Osgathorpe et al., 2011
Linear programming	Policy analysis, impact on natural vegetation, Scotland	Topp & Mitchell, 2003
Linear programming	Modelling worker physical health and societal sustainability on dairy farms, Netherlands	Van Calker et al., 2007

high net returns. Profitability of the crop, combined with specialisation of production, creates a short-run incentive to increase the area and frequency that potatoes are included within a rotation within a particular area. On Prince Edward Island, Canada, production in the early 2000s accounted for nearly 25 percent of crop land (Jatoe *et al.*, 2008) with increasing concern about the impact that farm systems were having on soil erosion, pesticide toxicity and potential negative effects on tourism. Although not couched in resilience terms, this is an ideal situation for evaluating human intervention effects – in this case, regulations that limited the frequency that potato and other crops could be used in a rotation. Potatoes, together with sugar beet, are also modelled, in this case under 2020s and 2050s climate change scenarios, by Gibbons and Ramsden (2008) using a dynamic, mixed-integer programming optimisation model. Farm attributes such as machinery, crop storage and irrigation capacity were set up as discrete integer variables, whereas crop areas and other variables can respond continuously. Dynamics were captured in three stages: a planning stage, where the model farmer makes profit-maximising input and output decisions for the current year; a resolution stage within this year, where discrepancies between the plan and actual outcomes are recorded; and a second planning stage, where profit-maximising decisions for the following year are made using the actual outcomes from the resolution stage. For example, current-year plans may include irrigation water holding capacity sufficient to irrigate root crops in a normal year; variability of climate may reduce precipitation to less than normal, resolution stage yields are thus reduced and the following year's plans are based on reduced net returns for the two crops. A series of dry years will increase the incentive to invest in 'lumpy' inputs such as irrigation capacity, other things being equal. The model uses a naive expectations approach – i.e. farmers are assumed to make decisions using information from the recent past. There is also a spatial element: in the 2008 paper, 29 farms in the Nar catchment of East Anglia are used, capturing 79 percent of the farm area in the catchment. In terms of climate change impacts on cropping, potato area was robust to change whereas sugar beet area showed increased variability. As an intervention, large-scale investment in winter abstraction capacity was not financially worthwhile. In another UK-based analysis, Rigby (1997) assesses the effect of guidelines to reduce nitrate pollution across multiple farms. In line with Spedding's original ideas (that a systems approach helps to identify areas of the system that are particularly important), the stand out result is the limited availability of land for spreading animal waste in the form of slurry. Now that a version of the guidelines – through Nitrate Vulnerable Zone regulations – is in place, the availability of land for slurry disposal remains one of the main concerns of dairy farmers in England.

Crossman and Bryan (2006) also use integer programming, in this instance to ensure minimum areas for the restoration of different environment types on multiple sites in a catchment in South Australia. This is one of a number of examples where the most efficient solution produced by modelling has beneficial environmental outcomes – in this case, a better spatial prioritisation for restoration of remnant natural lands – that are suboptimal to only a small degree. The implication is that environmental restoration benefits can be achieved at relatively low cost if prioritised as indicated by model results. It is worth emphasising that lack of 'efficient' prioritisation – i.e. allocating scarce resources to where the additional environmental benefits are relatively low – will increase cost; perhaps substantially. Pannell (2013), citing data from Fuller et al. (2010) covering nearly 7000 protected areas in Australia, calculates that the average benefit–cost ratio for the top 10 percent of schemes – those giving best value for money – was 200 times greater than the average scheme. This is not an argument for saving money (although clearly it could be presented in this way); rather, it is an argument for getting more environmental benefit for existing public expenditures on natural capital in Australia.

In a study of soil-loss in Anstai County, China, Lu and Van Ittersum (2004) show that interventions such as terracing and rotational cropping increase farm performance as measured by production efficiency. These results support the findings of the Crossman and Bryan study mentioned earlier and arguments put forward by Pannell (2006) that multiple input and output combinations exist that are suboptimal for one objective, but only marginally. Where this objective is profit, it is often possible to derive substantial improvements elsewhere in the modelled system (for example, reduction in risk, increase in crop diversification) without substantially reducing profit. To an extent this again depends on the degree to which substitution options – the land-use and management interventions – are available to the model and hence the underlying farm system.

Positive financial and ecological benefits from interventions (intercropping with legumes and the use of 'cross-bred' cattle) to improve soil management are reported by Kassie et al. (1999) for highland Ethiopia. In north-west Scotland, Osgathorpe et al. (2011) also conclude that conservation objectives – in this case increasing populations of rare bee species – can be complementary with farm business objectives, although this depends on farm type and location. From a policy perspective, their recommendation is that the appropriate response is to encourage communication – 'this is a win–win scenario and change would benefit both you and the environment' – as well as measures to gain a greater understanding of the barriers that may exist to bringing about this favourable change. However, we should acknowledge that the 'barrier' (for example, unacceptable trade-offs) may be known to the farmer but not to the modeller.

A more sophisticated optimisation approach and one that is more closely aligned with resilience ideas than those discussed so far is provided by John et al. (2005). The authors use a profit-maximising stochastic programming model that allows for tactical (within-season) decision-making. An important aspect to the farm planning problem is that decisions on what to produce are made well in advance of the states of nature and markets that will affect the outcomes of a decision. A common misunderstanding is that farmers make area-based decisions at the time of crop planting; preceding this is the choice of crop, which to allow for timely delivery has to be done well in advance of planting (in the UK, for one crop, wheat, there were about 40 recommended varieties to choose between for the 2014–15 crop year). Once these decisions are made, there is some scope to tactically adjust the farm system (for example, use more or less inputs) to changing weather events, particularly within the Australian context. John and colleagues model external shocks – under differing climate change scenarios – and allow for a wide range of adaptation options including growing perennial crops. The results show that under more extreme climate assumptions, profitability drops considerably, even allowing for adaptation. The scope for tactical decision-making falls as optimal farm plans shift to a more stable pattern, with a slightly increased area of perennial crops: effectively the farm system is forced to become more resilient as a consequence of climate change. This is somewhat akin to arguments put forward by Nassim Taleb (2012): exposure to uncertainty develops resilience (or 'antifragility' in Taleb's terms). It is also worth noting that the tactical management interventions modelled here are not directly substituting for resilience, but that increased resilience arising from these interventions was insufficient to alleviate, from a profit perspective, the negative effects of greater uncertainty under a changed climate.

In a paper that explicitly tackles resilience and risk while demonstrating the interdisciplinary benefits of building a bio-economic model (the authors are from Centres and Departments covering Economics, Agricultural Economics, Biology, Environmental and Climate Science), Cong et al. (2014) conclude that greater soil organic capital reduces exposure to risk as measured by yield variability. This increased resilience of the 'slow' variable is argued to be due to factors such as improved water-holding capacity and reduced requirements for purchased mineral fertiliser. The authors also emphasise the need for knowledge exchange, as it is argued that farmers are unlikely to be aware of the potential benefits of increased soil organic carbon. Data used are from long-term experiments and thus the authors assume that the benefits will occur within the – very different – commercial farm system.

Fleskens and de Graaff (2010) address uncertainty in decision-making where there are multiple objectives using goal programming. Agricultural (for example, production, labour use) and environmental (for example, biodiversity index value, water use) goals are considered by running an optimisation procedure that reduces the sum of absolute deviations from all the goals considered. Validation (one of the neglected areas in model development identified by Janssen and van Ittersum (2007), and Ramsden and Gibbons, 2009) was used to assess the extent to which the model replicated the initial farm plans. The authors conclude that policy instruments were beneficial for the 'goals' of soil conservation, water use, wildfire control, biodiversity conservation and landscape value.

In a paper that addresses some of the concerns of Janssen and van Ittersum (2007), Louhichi et al. (2010) introduce a model that is 'generic and re-usable for different bio-physical and socioeconomic contexts, facilitating the linking of micro and macro analysis'. FSSIM is a variant of the positive mathematical programming approach, a major benefit of which is 'self-calibration' – input and output combinations of baseline runs of a model will exactly match base year conditions (farm inputs and outputs). Coupling FSSIM with APES (the Agricultural Production Externalities Simulator) allows different interventions to be linked to environmental impacts, such as nutrient loss, in a way that will now be familiar.

A feature of the studies considered so far is that they use a form of approximation to link management practice to environmental outcome. In the stochastic programming work of John et al., environmental degradation from increased salinity is accounted for by assuming that the area of saline soils increases to encompass a greater area of the lowland (valley floor) region considered. The authors acknowledge that 'complex interactions' involving nutrient cycling in soils, pests and diseases are not considered and an ecological critique of the approach would probably centre on the lack of process-based modelling in the study. For example, the probability that heather will recover on degraded UK moorlands is strongly determined by its initial spatial distribution on the ground. Change in heather abundance requires reliable modelling of its competitive ability, spatial distribution and animal grazing preferences. There are approaches that do link process-based models to the farm system model; examples include Gibbons et al., 2005 (nitrate loss) and Neufeldt et al., 2006 (carbon and nitrogen cycles of soils). More typically, an indicator type approach is used, for example Gómez-Limón and Sanchez-Fernandez (2010) develop 16 sustainability indicators covering economic, social and environmental components of agriculture in Spain. Van Calker et al. (2007) combine indicators of this type with a bio-economic model for Dutch agriculture; indicators are used to capture food safety, animal welfare, animal health and landscape quality. Bartolini et al. (2007) use similar approaches, with indicator methodologies based on OECD (2001). Topp and Mitchell (2003) use a bio-economic model to predict the effect of changes to the Common Agricultural Policy and link this with multivariate and probabilistic techniques for estimating the effect that this change has on natural vegetation. Results show that only 2.6 percent of the area considered changes its vegetation type, but this is unsurprising given the relatively modest nature of the policy change involved: the shock to the environmental component of the system was relatively small, in part because the policy involved – 'Agenda 2000' – incorporated compensation payments for farmers, in return for reduced market price support. It would be interesting to set the model up to respond to an environmental, rather than a policy shock: for example, a series of years with relatively high precipitation at important points in the farming year.

This brief review suggests that while some elements of resilience are captured in these models, much is not. In particular, there is little attention given to modelling thresholds and environmental tipping points. The dynamic stochastic portfolio model of Cong et al. (2014) is perhaps closest in spirit to Folke's ideas: the potential benefits of increased resilience, in this case through improvement of a slow variable, soil organic carbon, to achieve both improvement in yield and reduced variability of yield to system shocks, are clearly demonstrated by the model employed. However, despite the sophistication in modelling of price and yield variability, the complexity of the farm system is captured in a relatively simple way: through rotational constraints. In contrast, the stochastic model of John et al. (2005) has 1400 rows and 1700 activities representing multiple land, labour and capital constraints and crop, livestock and management activities.

How would we build and use a better model?
Constrained, bio-economic optimisation models, where the optimal solution is unique and meets single or multiple objectives, are an effective framework for understanding the complex and unpredictable changes that occur when we perturb an agricultural system. They can and have been used to capture the way that these systems behave when exposed to variability, particularly market variability and gradual change such as climate change. By their constrained nature, they can also capture the inevitable trade-offs that occur in reaching the solutions that meet decision-maker objectives. Improved models would give greater emphasis to feedback loops between management interventions and the biological part of the farm system, thresholds (for example, crop failure under climate-shifted temperature distributions), environmental tipping points (for example, where system processes either fail or degrade at a much faster rate) and poorly understood uncertainty rather than reasonably well-known variability. With respect to thresholds and tipping points, dynamic programming models would appear to be more appropriate, as exceeding a threshold will tip the system into a different state: this weakened state – and its consequences – would then be 'available' to the farmer in the next time period.

With this kind of model specification, we can test the resilience of different farm systems. For example, in the UK, livestock production ranges from low input, sometimes organic, systems, to those characterised by high inputs and outputs. We can construct 'perturbance scenarios' that may occur, but about which we have little or no understanding of the probability of their occurring. Factors that such scenarios ought to consider include market and policy effects, export and import embargoes, changing perceptions relating to human health and animal welfare relating to milk and meat consumption, withdrawal of farm support payments, disease outbreak and spread, climate effects, particularly the effect of extreme rainfall events on the productivity of upland and lowland pastures and the effect of extreme temperature events on livestock health and welfare. It is interesting to speculate whether very specialised high-input systems, featuring one type of high-yielding breed as is often the case in milk production in the UK, requiring a very specific management and feed system, are resilient to disturbance: a priori, the substitution and adaptation possibilities – without major system change and cost – appear limited. Resilience modelling suits techniques such as Monte Carlo Analysis, where multiple runs of a model are exposed to multiple shocks and combinations of shocks. As we have argued, rather than capturing the risk of short-run variability, the emphasis should be more towards the consequences of combined events with large negative outcomes and uncertain probability of occurring.

In many ways, resilience is an argument about the validity of short-run optimal solutions. In a UK context, studies that attempt to measure efficiency (for example, Wilson et al., 2001) show that although there is considerable variation, many farmers are relatively efficient at using their land, labour and capital resources at a point in time. Farm Management teaching in UK universities also emphasises efficient resource use, built as it is on principles developed by Farm Management economists in the 1960s and 1970s (a good example of an undergraduate text of this time is Barnard & Nix, 1980). Rational decision-making may be optimal in the short run, but not resilient in the long run, and current optimisation models may not reflect this sufficiently. The question bio-economic models should be used to address is whether this approach to management is resilient to the type of uncertainty described above and able to deliver the mixed basket of goods that higher-income country populations increasingly desire from agriculture: biomass for various uses, environmental benefits such as biodiversity and landscape amenity. An interesting lower-income country example would be to compare the internal system resilience of pastoralist agriculture in sub-Saharan Africa, as reported by Homewood et al. (Chapter 9), with the alternatives that are replacing it.

We have seen that bio-economic models are designed to capture trade-offs and we have argued that some adaptations (for example, the use of futures markets by farmers) may increase the social aspect of 'social–ecological resilience' but reduce the ecological component. This is partly an argument about semantics, which is always an unproductive pastime, as people from different disciplines frame and define things differently. However, there is an important point about how we develop resilience and what we want to be resilient. Uncertainty will tend to encourage resilience; human or economic capital when it arises and is adopted will in some circumstances reduce the original resilience of the system. The same may be true of new innovations and again, in a sense, this is why resilience is seen as an important concept – technological *progress* may make us less resilient to future shock. As a natural capital stock that is difficult to replace, substitute for and which is also a store of carbon, soil resources are a prime candidate for resilience modelling. A hypothesis that could be tested with a sufficiently comprehensive model is whether short-run optimal solutions that farmers are implementing now are resilient for soils over a 10- to 20-year time horizon, under the sort of shock scenarios discussed earlier. If resilience is improved through increasing soil organic carbon, as argued by authors such as Powlson et al. (2011), management interventions – such as making use of more extended and different rotations – could be investigated and the cost of moving towards more resilient soils established. Alternatively, multi-cropping approaches, where mixtures of crops or cultivars are grown at the same time (in contrast to monocropping) of the kind advocated by Ehrmann and Ritz (2014) could be tested. The bio-economic modelling literature suggests that these costs may not be that great for some interventions (but may well be very high for others). With a sufficiently comprehensive model, other environmental effects can be monitored, policies aimed at encouraging soil resilience could be tested and the effects of changes in other aspects of the farm system (such as different approaches to marketing) investigated. A bolder move would be to shift the emphasis from measuring trade-offs that arise from the imposition of some change that has environmental benefits to restating the problem so that the objective is to achieve a pre-defined environmental outcome. The advantage here is that this could be determined by ecology experts and more closely tailored to the desired outcome. Following our theme of potentially limited economic impact, it would be interesting to know whether these more targeted approaches to developing resilience fall within the 'negligible impact' range or whether the trade-off costs are too great across our range of desired farm outputs. Resilience thinking also explicitly embraces *change* as a means of bringing about greater resilience: system collapse and regeneration can be part of an adaptive cycle of resilience (Gunderson & Holling, 2002). A similar concept exists in economics: Joseph Schumpeter (2013) discussed the idea of 'creative destruction' in the 1940s, arguing that capitalism brings about improvements in human well-being through the actions of entrepreneurs and innovators. It may be that improved resilience in agricultural systems requires some economic 'structural' changes: in many countries, farming is carried out by older generations with the young excluded (unless they can inherit) by high start-up costs. However, it has been argued – perhaps somewhat speculatively – that the destructive costs of modern capitalism may be greater than in the past, as new technologies are 'often creating products which are close substitutes for the ones they replace, whose value depreciates substantially in the process of destruction' (Komlos, 2014).

Our final point relates to existing risk modelling work and the scope for integrating this type of research with resilience thinking. Farmer attitude to risk is often captured through some form of utility function that reflects risk-reducing and profit-increasing objectives relative to the farmer decision-maker's net worth (financial assets less liabilities at a point in time). Approaches to managing risk include elements of both resilience (build net worth) and adaptation (make use of risk management interventions). The Agricultural Economics literature also distinguishes between embedded and non-embedded risk (Dorward, 1999); in the former, tactical, short-term adaptations are possible in the light of new information. In the latter, adaptation is not possible (the time-dependent biological nature of agriculture means that it is too late to adapt) and hence resilience becomes more important. Many farmers in the UK have adapted to reductions in price support by diversifying into alternative enterprises, such as small-scale service activities linked to tourism or the local economy. The emphasis in this work has been on management interventions that can reduce variability of income: an optimal bio-economic farm plan will have lower income variability on average than a suboptimal plan, although again, the suboptimal plan may not be much worse than the optimal plan across a certain range. It has been argued (Pannell et al., 2000; Hardaker et al., 2015) that greater emphasis should be given to what in resilience terms would be framed as the effect of a shock on tipping points within the system; a recent example that takes this approach is Ogurtsov et al. (2015), where 'catastrophic farm risks' are considered; it would be interesting to extend this type of approach further. However, these tipping points – for example, major system change or the effect of large but inappropriate or misplaced investments on farmer objectives – have tended to be economic rather than environmental or ecological. As we have argued, with models that better reflect the dynamic and spatial aspects of ecological resilience, it would be possible to assess the impact of these types of system change on environmental and ecological resilience and capture, through the utility function, the benefits of resilience planning for farmer decision-makers. As explored in some of the literature presented here, with a comprehensive set of management interventions, realistically modelled, results could be used through knowledge exchange mechanisms to encourage farms to move from short-run optimal solutions to longer-term solutions that may well be achievable without appreciable economic cost.

References

Arrow, K., Bolin, B., Costanza, R., et al. (1995). Economic growth, carrying capacity, and the environment. *Ecological Economics*, **15**(2), 91–95.

Barnard, C. & Nix, J. (1980). *Farm Planning and Control*. Cambridge: Cambridge University Press.

Bartolini, F., Bazzani, G.M., Gallerani, V., Raggi, M. & Viaggi, D. (2007). The impact of water and agriculture policy scenarios on irrigated farming systems in Italy: an analysis based on farm level multi-attribute linear programming models. *Agricultural Systems*, **93**(1), 90–114.

Cong, R.G., Hedlund, K., Andersson, H. & Brady, M. (2014). Managing soil natural capital: an effective strategy for mitigating future agricultural risks? *Agricultural Systems*, **129**, 30–39.

Crossman, N.D. & Bryan, B.A. (2006). Systematic landscape restoration using integer programming. *Biological Conservation*, **128**(3), 369–383.

Dorward, A. (1999). Modelling embedded risk in peasant agriculture: methodological insights from northern Malawi. *Agricultural Economics*, **21**(2), 191–203.

Ehrmann, J. & Ritz, K. (2014). Plant:soil interactions in temperate multi-cropping production systems. *Plant and Soil*, **376**(1–2), 1–29.

Fleskens, L. & de Graaff, J. (2010). Conserving natural resources in olive orchards on sloping land: alternative goal programming approaches towards effective design of cross-compliance and agri-environmental measures. *Agricultural Systems*, **103**(8), 521–534.

Folke, C. (2006). Resilience: the emergence of a perspective for social–ecological systems analyses. *Global Environmental Change*, **16**(3), 253–267.

Fuller, R.A., McDonald Madden, E., Wilson, K.A., et al. (2010). Replacing underperforming protected areas achieves better conservation outcomes. *Nature*, **466**, 365–367.

Gibbons, J.M., Ramsden, S.J., Sparkes, D.L. & Wilson, P. (2005). Modelling optimal strategies for decreasing nitrate loss with weather variation – a farm level approach. *Agricultural Systems*, **83**, 113–134.

Gibbons, J.M. & Ramsden, S.J. (2008). Integrated modelling of farm adaptation to climate change in East Anglia, UK: scaling and farmer decision making. *Agriculture, Ecosystems and Environment*, **127**, 126–134.

Gómez-Limón, J.A. & Sanchez-Fernandez, G. (2010). Empirical evaluation of agricultural sustainability using composite indicators. *Ecological Economics*, **69**(5), 1062–1075.

Gunderson, L.H. & Holling, C.S. (2002). *Panarchy: understanding transformations in systems of humans and nature*. Washington, DC: Island Press.

Hardaker, J.B., Lien, G., Anderson, J.R. & Huirne, R.B. (2015). Coping with Risk in Agriculture: applied decision analysis. Wallingford: CABI.

Janssen, S. & Van Ittersum, M.K. (2007). Assessing farm innovations and responses to policies: a review of bio-economic farm models. *Agricultural Systems*, **94**(3), 622–636.

Jatoe, J.B.D., Yiridoe, E.K., Weersink, A. & Clark, J.S. (2008). Economic and environmental impacts of introducing land use policies and rotations on Prince Edward Island potato farms. *Land Use Policy*, **25**(3), 309–319.

John, M., Pannell, D. & Kingwell, R. (2005). Climate change and the economics of farm management in the face of land degradation: dryland salinity in Western Australia. *Canadian Journal of Agricultural Economics/Revue canadienne d'agroeconomie*, **53**(4), 443–459.

Kassie, M., Jabbar, M.A., Kassa, B. & Saleem, M.M. (1999). Benefits of integration of cereals and forage legumes with and without crossbred cows in mixed farms: an *ex ante* analysis for highland Ethiopia. *Journal of Sustainable Agriculture*, **14**(1), 31–48.

Komlos, J. (2014). Has creative destruction become more destructive? National Bureau of Economic Research, NBER Working Paper No. 20379.

Louhichi, K., Kanellopoulos, A., Janssen, S., et al. (2010). FSSIM, a bio-economic farm model for simulating the response of EU farming systems to agricultural and environmental policies. *Agricultural Systems*, **103**(8), 585–597.

Lu, C.H., van Ittersum, M.K. & Rabbinge, R. (2004). A scenario exploration of strategic land use options for the Loess Plateau in northern China. *Agricultural Systems*, **79**(2), 145–170.

Neufeldt, H., Schäfer, M., Angenendt, E., et al. (2006). Disaggregated greenhouse gas emission inventories from agriculture via a coupled economic-ecosystem model. *Agriculture, Ecosystems & Environment*, **112**(2), 233–240.

OECD. (2001). *Environmental Indicators for Agriculture*. Paris: OECD Publications.

Ogurtsov, V.A., van Asseldonk, M.A.P.M. & Huirne, R.B.M. (2015). Modelling of catastrophic farm risks using sparse data. In: *Handbook of Operations Research in Agriculture and the Agri-food Industry*, edited by L.M. Plà-Aragonés International Series in Operations Research & Management Science. New York, NY: Springer, pp. 259–275.

Osgathorpe, L.M., Park, K., Goulson, D., Acs, S. & Hanley, N. (2011). The trade-off between agriculture and biodiversity in marginal areas: can crofting and bumblebee conservation be reconciled? *Ecological Economics*, **70**(6), 1162–1169.

Pannell, D.J. (2006). Flat earth economics: the far-reaching consequences of flat payoff functions in economic decision making. *Review of Agricultural Economics*, **28**, 553–566.

Pannell, D. (2013). Value for money in environmental policy and environmental economics (No. 146501). University of Western Australia, School of Agricultural and Resource Economics.

Pannell, D.J., Malcolm, B. & Kingwell, R.S. (2000). Are we risking too much? Perspectives on risk in farm modelling. *Agricultural Economics*, **23**(1), 69–78.

Peterson, G.D., Carpenter, S.R. & Brock, W.A. (2003). Uncertainty and the management of multistate ecosystems: an apparently rational route to collapse. *Ecology*, **84**(6), 1403–1411.

Phalan, B., Onial, M., Balmford, A. & Green, R.E. (2011). Reconciling food production and biodiversity conservation: land sharing and land sparing compared. *Science*, **333**(6047), 1289–1291.

Powlson, D.S., Glendining, M.J., Coleman, K. & Whitmore, A.P. (2011). Implications for soil properties of removing cereal straw: results from long-term studies. *Agronomy Journal*, **103**(1), 279–287.

Ramsden, S.J. & Gibbons, J.M. (2009). Modelling agri-environment interactions and trade-offs using Farm-adapt – integration of models, investigation of scale, dynamics and uncertainty. *Aspects of Applied Biology*, **93**, 123–130.

Rigby, D. (1997). European community guidelines to reduce water pollution by nitrates from agriculture: an analysis of their impact on UK dairy farms. *Journal of Agricultural Economics*, **48**(1–3), 71–82.

Schelling, T.C. (1978/2006). *Micromotives and Macrobehavior*. New York, NY: WW Norton & Company.

Schumpeter, J.A. (2013). Capitalism, socialism and Democracy. 2nd edition. London: Routledge.

Spedding, C.R.W. (1976). Editorial, *Agricultural Systems*, **1**, 1–3.

Taleb, N.N. (2007). *The Black Swan: the impact of the highly improbable*. New York, NY: Random House.

Taleb, N.N. (2012). **Antifragile: how to live in a world we don't understand** ... London: Allen Lane.

Topp, C.F.E. & Mitchell, M. (2003). Forecasting the environmental and socio-economic consequences of changes in the Common Agricultural Policy. *Agricultural Systems*, **76**(1), 227–252.

Van Calker, K.J., Berentsen, P.B.M., De Boer, I.J.M., Giesen, G.W.J. & Huirne, R.B.M. (2007). Modelling worker physical health and societal sustainability at farm level: an application to conventional and organic dairy farming. *Agricultural Systems*, **94**(2), 205–219.

Walker, B., Carpenter, S., Rockstrom, J., Crépin, A.S. & Peterson, G. (2012). Drivers, 'slow' variables, 'fast' variables, shocks, and resilience. *Ecology and Society*, **17**(3).

Wilson, P., Hadley, D. & Asby, C. (2001). The influence of management characteristics on the technical efficiency of wheat farmers in eastern England. *Agricultural Economics*, **24**(3), 329–338.

York, L.M., Carminati, A., Mooney, S.J., Ritz, K. & Bennett, M.J. (2016). The holistic rhizosphere: integrating zones, processes, and semantics in the soil influenced by roots. *Journal of Experimental Botany*, **67**(12), 3629–3643.

CHAPTER SIX

Building resilience into agricultural pollination using wild pollinators

NEAL M. WILLIAMS
University of California, Davis
RUFUS ISAACS
Michigan State University
ERIC LONSDORF
Franklin and Marshall College, Lancaster
RACHAEL WINFREE
Rutgers University, New Brunswick
and
TAYLOR H. RICKETTS
University of Vermont

Introduction

In many regions of the world, agricultural production depends strongly on animal-mediated pollination, most of it provided by the European honey bee (*Apis mellifera*) and a diverse set of unmanaged bee species (Aizen et al., 2009; Garibaldi et al., 2013). Although speculation that humanity would cease to exist in the absence of bees is no doubt overstated, approximately 35 percent of the crop-based global food production (measured as biomass) benefits from animal pollinators (Klein et al., 2007). Furthermore, a synthesis of crop statistics from UN-FAO shows a global trend toward increasing production of and reliance on pollination-dependent crops, particularly in the developing world. These trends coincide with recent reductions in the supply of managed honey bee colonies, leading to projected pollination deficits throughout the world if these global trends continue (Aizen et al., 2009).

Managed populations of *A. mellifera* continue to face challenges from disease, lack of reliable forage resources and environmental toxins (vanEngelsdorp et al., 2010; Vanbergen et al., 2013). As a result, increasing attention has been paid to the role of wild bee species, which can make substantial contributions to crop pollination in certain circumstances (Kremen et al., 2002; Winfree et al., 2007; Carvalheiro et al., 2012; Garibaldi et al., 2013). Diverse communities of wild bees offer an important way to add resilience to pollination in agricultural lands. Their ability to do so depends on maintaining biodiversity and management strategies designed to promote and sustain their populations. In this chapter we focus on the biodiversity of wild bees as part of a larger integrated pollination strategy to increase pollination resilience. Importantly, actions to promote robust wild bee communities within agricultural landscapes often positively affect honey bees as well (Morandin & Kremen, 2013; Williams et al., 2015). For example, reducing inputs of bee-toxic pesticides or planting additional plants to provide pollen and nectar can benefit wild and managed pollinators alike.

In the sections that follow, we discuss the inherent vulnerability of pollination in the modern world, review the mechanisms by which pollinator biodiversity can promote pollination services and its stability over space and time and suggest where a lack of data limits our ability to build resilience of agricultural pollination using wild bees and other pollinators. We then review current practices used to promote pollinator communities and resilience and the degree to which those practices have achieved their goals. Finally, we propose and demonstrate a Pollination Decision Framework that illustrates how a more formalised decision-making approach can help farmers choose among alternative management actions to promote wild pollinators and pollination on their farms.

Pollinator diversity and the resilience of pollination services
Inherent vulnerability of crop pollination

Pollinators are important for the production of many crops, and can be as essential as irrigation, fertiliser or labour. Bee-pollinated crops are therefore vulnerable to changes in the availability of their mutualist pollinators. This vulnerability stems from various intersecting issues; some of these are a function of our modern agricultural systems, some are caused by reliance on a limited set of pollinator taxa and pollination strategies, and still others are inherent to the pollination process itself.

First, modern agriculture has produced increasingly simplified and intensively managed landscapes through the removal of semi-natural habitats and consolidation of fields (Matson et al., 1997; Robinson & Sutherland, 2002; Tscharntke et al., 2005; Meehan et al., 2011). This landscape simplification is combined with regional specialisation of crops, creating large expanses of agricultural monocultures bordered by small patches of weedy or forested vegetation (Tscharntke et al., 2005; Meehan et al., 2011). Regional specialisation in agricultural production also creates a temporal homogenisation with truncated periods of bloom followed by periods of relative dearth. Ironically, such changes can create landscapes that present the greatest challenges for pollinators precisely where pollination demands are greatest. The Central Valley of California is a prime example where almond (*Prunus dulcis*) orchards cover approximately 370,000 ha, with individual orchards often spanning 50–100 ha. This intensity leaves little space for habitat to provide nesting or floral resources for most bee species. Moreover, billions of almond flowers bloom almost simultaneously, requiring intense pollination over a short period in the spring, so that there is a high flower to bee ratio that can limit pollination success. The short blooming also creates landscapes with limited resources for bees when crops are not in bloom. For crops grown in this way, the demand for pollination is great and the potential for a shortfall due to low wild bee abundance is most likely, with potentially catastrophic implications for crop yield.

Second, growers of most pollinator-dependent crops rely almost exclusively on *A. mellifera* to provide pollination, and for many crops and contexts, such as the almonds mentioned above, this reliance is expected to continue. However, honey bees face a variety of challenges, including exposure to parasites and pathogens, poor nutrition during critical times of the year and exposure to chemical toxins. Together, these have lowered supply, raised management effort and costs, and increased the vulnerability of their populations in many regions (vanEngelsdorp & Meixner, 2010). Mean annual colony loss of managed hives in the USA now regularly reaches 30 percent (vanEngelsdorp & Meixner, 2010; Steinhauer et al., 2014) and in Europe can exceed 25 percent (van der Zee et al., 2014). In order to sustain their capacity to provide pollination services, beekeepers must therefore replace these lost hives annually, adding costs to their production. This leads to higher colony rental fees and greater risk that the honey bee supply will not be able to meet agricultural demand for honey bee colonies. Farmers are increasingly searching for options to address this vulnerability and help stabilise their crop pollination. It is unlikely that wild pollinator communities will replace managed honey bees for agricultural pollination globally, particularly in intensive farm systems, but diversification of pollinators offers a chance to mitigate the inherent vulnerabilities of current pollination practices. Indeed, these insects can provide significant contributions to agriculture; wild bees accounted for 49 percent of visits to crop flowers across 41 crop systems examined worldwide (Garibaldi et al., 2013).

Third, animal-mediated pollination itself is a highly uncertain process. Variation in pollen removal, transport and deposition on flowers is compounded by the variation in frequency and patterns of flower visitation, making pollen deposition variable within and among plants (Herrera, 2002; Richards et al., 2009). Reported patterns of pollination limitation suggest this is widely true of wild plant populations (Knight et al., 2005) and many crop systems (Chacoff et al., 2008; Garibaldi et al., 2011; Benjamin & Winfree, 2014). Successful animal pollination also requires temporal and spatial synchrony between pollinators and receptive flowers, and like other interactions, pollination may be particularly sensitive to global change because multiple species with distinct physiology, life-history and resource needs are involved (Tylianakis et al., 2008). Temporal synchrony between bees and crop bloom appears to have been maintained in some regions (Bartomeus et al., 2013; Iler et al., 2013), but for individual crop species the future impacts are far from certain.

Finally, the relationship between pollen deposition and fruit or seed yield is saturating (Morris et al., 2010), including for pollinator-dependent crops (Garibaldi et al., 2011). This means that times or locations with above-average pollination result in relatively smaller increases in yield, whereas times or locations with below-average pollination result in relatively greater decreases in yield. Thus, variability in biotic pollination over time or space can cause additional reduction in mean crop production (Garibaldi et al., 2011).

Threats to wild bees

Long-term trends in the abundance of various wild bee species suggest that multiple species are declining (Biesmeijer et al., 2006; Goulson et al., 2008; Cameron et al., 2011; Burkle et al., 2013), although few robust data sets exist to quantify population or distributional changes (Bartomeus et al., 2011; Colla et al., 2012). Drivers of decline include loss of land that provides nesting habitat and foraging resources, both of which must be within the flight range of a bee to support their populations. In agricultural landscapes, even when natural habitats remain, large-scale conversion of natural areas or marginal lands into intensive agriculture can cause a loss or a decoupling of these resources in space, thereby limiting the carrying capacity of landscapes for wild bees. The recent surge in corn prices and the associated transition of marginal lands to crop production in the Midwest region of the USA exemplifies how quickly these changes can occur, even when programmes are in place to pay farmers to maintain their land in conservation programs (e.g. Wright & Wimberly, 2013).

Although wild bee populations available to pollinate crops are directly influenced by the composition and management intensity of land at local and landscape scales, the response varies among species and also depends on the intensity of agricultural land use (Winfree et al., 2011). We are only beginning to develop tools to predict how the different species within a community of bees are likely to respond to land conversion or conservation programmes (Lonsdorf et al., 2009; Kennedy et al., 2013). Developing this information is essential to project the spatial provision of pollinators and their function within agricultural landscapes (Ricketts & Lonsdorf, 2013). In the last section of the chapter we provide an example of how a similar approach can be used to predict how different types of enhancements will improve pollination within individual farms.

Pollinator diversity, buffering and resilience

The resilience concept includes the ability of the system both to absorb disturbance without loss of function and to reorganise in ways that achieve new stable states in the face of environmental change (Folke et al., 2004). Of these two elements, most attention regarding pollination has been on the relationship between diversity and temporal/spatial stability. A recent synthesis of data on wild bee communities and crop pollination shows a negative correlation between bee diversity and variation in pollination over space and time (Garibaldi et al., 2011). This finding supports the idea that bee diversity stabilises pollination services and therefore crop yields. Stability, or buffering, is a key element of long-term sustainability of agricultural pollination because it ensures reliable yields over time, which are critical to grower's economic viability. The potential for such buffering was originally pointed out by Lawton and Brown (1993), who termed it the insurance hypothesis, and proposed that species contributing to the same function, and thus superficially appearing to be redundant, could in fact stabilise function over time.

The buffering capacity provided by diversity under the insurance hypothesis is promoted by multiple mechanisms. First, buffering will be enhanced where there is variation among pollinators in their response to environmental change or disturbance (Blüthgen & Klein, 2011). The resulting *response diversity* is a primary mechanism for stabilising ecosystem function in the face of disturbance (Elmqvist et al., 2003). Specifically, response diversity occurs when species that provide the same ecosystem service have differential responses to environmental change, resulting in compensatory fluctuations in their populations and the ecosystem services they provide. For example, response diversity of wild bee species to increasing agricultural land use stabilises crop pollination services across land-use gradients, although not in all cases (Cariveau et al., 2013). The Cariveau et al. study (2013) directly measured pollination provided by different bee species and found that response diversity among species will not buffer pollination if those species that decline with disturbance are higher functioning and those that increase are lower functioning. In this case, even with perfect numerical compensation, every trade among species is for lower function. Whether such trades to lower function are general in pollination and agricultural landscapes has not been explored, but this would be critical for managing these systems to promote pollination resilience.

Response diversity also stabilises pollination under environmental variation. For example, temperature, wind and precipitation all fluctuate during the bloom season of almond. Brittain et al. (2013) documented differences in tolerance to wind speed among bee species pollinating this crop. These differential responses among species led to farms with greater bee diversity having both higher mean levels of pollination and more stable pollination among days.

This same variation in tolerance to weather conditions could provide resilience in the face of longer-term climate changes. Spring blooming orchard crops illustrate this effect. The timing of bloom in these crops has advanced due to climate change (e.g. Chmielewski, 2013) and in the systems where this has been measured such as apple (Bartomeus et al., 2013), bee activity at the community scale seems to have retained synchrony with bloom phenology. This community-level stability occurs despite individual bee species shifting their phenology faster or slower than apple. This is an example of buffering due to response diversity, in this case measured as rate of phenological advance. While the diversity of bees present in the orchards studied by Bartomeus et al. (2013) provides resilience in the face of climate-related phenological shifts, these shifts also bring the crop into greater risk of damage to sensitive flower tissues by spring frost. In this respect, resilience will be provided by achieving rapid pollination in poor weather so that the flowers are set into fruit before frost occurs.

The second mechanism by which diversity provides resilience for pollination is the *selection effect*. It occurs because species will naturally vary in the quality of their functional traits, some of which are crucial to providing the function in novel conditions associated with global change or other disturbance. Greater diversity of species in the community increases the chance that such a species will be present. For example, bee species that are able to forage at lower temperatures than honey bees may have important economic benefits for the farmer, by pollinating flowers during poor spring weather (Bosch et al., 2006). In this case, the presence of species with a particular trait, cold tolerance, matters and the probability of including a cold-tolerant species increases with the number of species in the community. Further research is needed on the economics of these interactions, for example to determine the magnitude and frequency of yield increase required to provide a long-term return on investment for different pollinator enhancement programs, such as those described below.

A third mechanism, *density compensation*, refers to compensatory or negatively correlated changes in the abundances of different pollinator species over space or time. As such, it is a phenomenological term and could encompass response diversity; that is, if species show differential responses to environmental variables (response diversity) they might show negatively correlated fluctuations in abundance (density compensation). Density compensation could occur for other reasons, though; for example, due to competitive interactions, if subdominant species increased after the decrease of a dominant competitor. Density compensation appears uncommon in real-world systems, like farms, that undergo frequent human disturbance (Houlahan et al., 2007; Winfree & Kremen, 2009). This might be because in real-world systems the disturbances affect resources for all species similarly. For example, agricultural intensification reduces floral forage plants such that locations that are poor for one bee species are relatively poor for others (Winfree, 2013).

A final mechanism that could promote resilience in pollination is *cross-scale resilience*, in which abundances of different species are affected similarly by an environmental change, but at different spatial and/or temporal scales (Holling, 1992). Evidence for the importance of cross-scale resilience in stabilising ecosystem function is lacking, but different bee species do vary in the scale at which populations respond to landscape composition (Steffan-Dewenter et al., 2002; Winfree & Kremen, 2009; Benjamin et al., 2014). This difference in response has been demonstrated by showing that within the same landscapes bee species vary in the spatial scale at which the amount or composition of habitats best predict estimated population abundances. No study has yet explored whether the cross-scale response pattern actually leads to greater resilience of pollination function in agricultural or other systems.

Whether through mechanisms of cross-scale resilience or response diversity, the differential responses of pollinators to their environment are determined by differences in suites of traits that govern individual species' responses. Because such traits also partly define species niches, we might expect diversification of these functional response traits through natural selection and assembly of regional bee species pools. Such functional trait diversity will also likely allow pollinator communities to restructure and reorganise in the face of altered landscape management and global climate change. Novel bee communities assemble in agricultural lands both on farms and in restored habitats within the larger landscape (Williams, 2011). Agriculture also filters bee communities based on functional groups, however. Several studies report changes in the relative representation of different nesting guilds in response to agricultural intensification (Grundel et al., 2010; Williams et al., 2010). For example, the number of species that nest aboveground and their absolute abundance as a group decrease on farms compared with natural areas, whereas the abundance of ground-nesting species increases in some farming contexts (Forrest et al., 2015). Such changes could be interpreted as revealing the vulnerability of certain groups; however, the robustness of ground-nesting bees in the face of land-use intensification also shows how diversity of nesting traits may be critical to maintain some elements of pollinator communities. The effect of agricultural intensification, however, appears to filter functional diversity more than species diversity (Forrest et al., 2015), so the critical diversity of functional traits that enhance resilience of pollination is eroded. This does not bode well for future resilience of pollination in agricultural systems (see Laliberte et al., 2010; Mayfield et al., 2010 for similar arguments related to plant communities).

Despite the increased attention to how pollinator diversity may stabilise pollination, most research is limited to single locations over limited time. Thus, we do not fully understand how pollinator diversity provides resilience for pollination to agriculture at the large spatial and temporal scales typical of commercial agriculture (e.g. Kremen et al., 2002, which considers interannual variation in the identity of the bee species providing the pollination). Yet the stabilising effect of biodiversity is most likely to act at larger spatial and temporal scales, as the turnover in pollinator species composition over space and time (beta diversity) results in larger sets of bee species being important for delivery of the function in some sites or years (Winfree, 2013). Thus, we expect that biodiversity might have even stronger stabilising effects on pollination in agricultural systems than is predicted by current ecological knowledge, which is built predominantly on small-scale experiments (Cardinale et al., 2012). How the biodiversity–pollination relationship scales over space and time will depend on the degree of turnover in functionally dominant pollinator species: widespread dominants – less diversity needed; turnover of dominants – more diversity needed. In any case, pollinator diversity at these larger spatial and temporal scales is likely to be of even greater importance to resilience in the face of global change, because it is the regional species pool that can allow for the 'reorganisation' component of resilience.

In addition to the greater number of pollinator species required across larger spatial and temporal scales, more bee species will likely be needed to pollinate multiple types of crops. To date, research on the benefits of pollinator diversity for pollination has focused on single crop species; however, we expect greater pollinator biodiversity to be more important where a range of different crop species are grown within a farm that each have different pollination needs, different flowering times and different response rates to visitation by different species within the bee community.

The degree to which the required pollinator diversity increases across crops may depend on whether different crop plants share the same dominant pollinator species. A recent synthesis suggests that across crops within regions there are common dominant pollinator species (Kleijn et al., 2015). In this case the full range of available native pollinator biodiversity is not required for pollination. This outcome is not surprising given that crops represent a relatively limited diversity of plant species, many of which are introduced to their cultivated region and tend to have generalised flower morphologies (Corbet, 2006). In addition, a small subset of pollinator species might persist well in the agricultural landscapes where most crops are grown. The exceptions are in cases where the crop is cultivated in its native range. Here the dominant pollinator species frequently include at least one trophic specialist (e.g. *Helianthus annuus* and *Curcurbita* species in western North America; Tepedino, 1981; Greenleaf & Kremen, 2006; Artz & Nault, 2011). Although the number of dominant visitor species is limited, it is typically more than one.

Despite this evidence for shared dominant pollinators, it is too early to conclude that there is a limited role for biodiversity, explicitly because of the importance of maintaining resilience. The dominant pollinator taxa today will not necessarily be dominant in different landscapes, under different management scenarios (Kremen et al., 2002) or under altered climate conditions (Rader et al., 2013). In uncertain futures, resilience may depend on diversity beyond the few dominant bee species currently providing the bulk of pollination services.

Practices to support pollination resilience

Strategies to support wild pollinators and enhance the resilience of pollination for agriculture seek to bolster populations, maintain diverse communities and increase the long-term sustainability of crop production. Multiple strategies are being developed to mitigate the impacts of agricultural intensification and to support the diversity and abundance of pollinators (Garibaldi et al., 2014).

Loss of forage and nesting resources are among the primary drivers of pollinator loss within intensified agricultural landscapes (Roulston & Goodell, 2011; Winfree et al., 2011). Although pathogens and pesticides have also been implicated as potential drivers of pollinator decline, mitigating their potential effects represents another set of partially overlapping actions, which we do not treat here. Species richness, community-wide abundance and pollination provided by wild pollinators all decrease with isolation from semi-natural habitat that provides bee resources (Ricketts et al., 2008). Isolation from semi-natural habitat is also associated with increasing variation (i.e. loss of stability) of pollination over time and space (Garibaldi et al., 2011). Across many study systems, the quality of the landscape contributes to bee abundance and diversity of wild bees found on farm sites (Kennedy et al., 2013).

Although landscape context is important, coordinating management to promote resilience of pollinators across whole landscapes is often beyond the control of individual growers. As such, it represents an important goal for agricultural communities to work toward. The strong benefit of local-scale habitat diversity and quality on bee communities found by Kennedy et al. (2013) and Benjamin et al. (2014) are encouraging because this indicates that enhancement of habitat as well as management choices within farms have potential to enhance the diversity and resilience of pollinator populations where mitigation practices are implemented. The simplest approaches to improve habitat quality on farms are designed to broadly support biodiversity and not pollinators specifically. These involve managing vegetation, such as through reduced mowing or autumn cultivation of field margins and fallow areas in an attempt to increase floral diversity, abundance and duration. The benefits of these basic options for pollinator conservation are mixed (Haaland et al., 2011; Pywell et al., 2011) and may reflect the variety of ways in which this approach has been implemented, as well as suitability for enhancing bee populations. Plantings of wildflowers targeted for enhancing pollinators along field margins, road verges and uncropped areas on the farm have been more successful, but are also more costly. These added costs arise from the price of seed mixes and specific management requirements. It is interesting to consider whether with clear guidance and flexibility among specific actions the added costs could be minimised. These enhancements seek to increase floral diversity and abundance so as to benefit pollinators, other beneficial organisms and the services they provide, rather than to restore a specific formerly present community of plants. Many studies of these enhancements have targeted specific pollinator guilds (e.g. *Bombus* spp.) and these actions almost always greatly augment the abundance and species richness of target pollinators compared to unmanaged areas or unenhanced areas managed for conservation (Carvell et al., 2006; Heard et al., 2007; Haaland et al., 2011; Pywell et al., 2011; Kleijn et al., 2015). Some research, mostly from North America, has considered pollinator communities more generally (Hannon & Sisk, 2009; Morandin & Kremen, 2013). For example, wildflower plantings along field margins supported over six times the species richness and 13 times the abundance of bees found on unenhanced field margins. Even when compared to the floristically diverse field margins, wildflower plantings designed to support bees supported 26 additional species, five of which are important crop pollinators (N. M. Williams & K.L. Ward, unpublished data).

Landscape and local site characteristics often interact such that the importance of local habitat quality and farm management for supporting wild pollinators depends on the quality of the surrounding landscape (Williams & Kremen, 2007; Rundlöf et al., 2008; Batáry et al., 2011). Such interactions help to dictate where mitigation actions to support pollinators may be most effective. Theoretical predictions and recent synthesis of empirical research suggest that on-farm restoration is likely to have the greatest benefit in simplified landscapes, but where at least a modest amount of natural vegetation persists (Tscharntke et al., 2005; Batáry et al., 2011; Kleijn et al., 2011). Although wild bee populations use resources from some crop fields (Westphal et al., 2003, 2009), their persistence requires some natural habitat (Potts et al., 2010; Jauker et al., 2012), because crop resources can be relatively ephemeral and bees need to move between non-crop and agricultural areas to exploit these temporally and spatially variable resources (Mandelik et al., 2012). Remnants also may be sources for pollinators that repeatedly recolonise farms on which they persist only intermittently. Such source–sink dynamics between natural and farm components of the landscape are untested.

Choices of plant species and a robust methodology for establishing plantings are critical to successful functioning of wildflower plantings. Phenological coverage appears to be key to successful plant mixes (Redpath-Downing et al., 2013; Russo et al., 2013). Mixes of species that bloom throughout the growing season support a greater diversity of pollinators and can benefit crops that bloom at different times of year. Season-long flowering also promotes pollinator species whose flight periods extend before and after the bloom of a single crop or season. For example, extended bloom is critical for the support of bumble bee species in temperate fruit and seed crops, where queens and workers pollinate during May and June, but where colonies grow through to September and require access to pollen and nectar from flowers for continued growth and reproduction (Isaacs et al., 2009; Westphal et al., 2009).

Crop pollination represents foraging from the pollinator perspective, thus crops can potentially serve as forage resources to support the pollinators on which they rely for pollination service. Several lines of investigation suggest the potential for larger-scale coordination of farming systems that would involve multiple mass-flowering crops that bloom sequentially. The presence of early mass-flowering crops did not promote greater reproductive success of bumble bees (Westphal et al., 2009). Later-season crops increased queen bumble bee abundance in some contexts, but not others (Williams et al., 2012; Rundlöf et al., 2014), suggesting that combinations of early and late crops together could promote populations. Indeed, temporal diversification of flower resources, including crops, is one likely mechanism by which smaller organic farms and locally diversified farms as well as farm gardens benefit pollinators (Holzschuh et al., 2008; Osborne et al., 2008; Rundlöf et al., 2008; Kennedy et al., 2013).

Several key aspects of practices designed to mitigate pollinator loss in intensive agricultural lands remain largely untested, yet they are critical to understanding the potential of such practices to increase the resilience of pollinator communities and pollination services on which sustainable agriculture relies. First, most assessments of habitat enhancements have measured flower visitors only on the enhancements themselves and concluded they offer important benefits to pollinators without directly quantifying effects on the pollinator population. Short-term surveys showing greater bee abundance and diversity in response to enhancements may simply represent aggregative behavioural, rather than population, responses to floral rich areas (Heard et al., 2007), at the expense of non-enhanced areas. Heard et al. (2007) noted important interactions of wildflower strips with landscape context, where more bees were found in wildflower strips in landscapes with fewer resources for pollinators. This pattern is consistent with an aggregative effect by which high-quality floral resources draw foragers from the surrounding landscape. Recent studies have monitored the abundance of bees in fields surrounding hedgerows as a better proxy for true population affects (Morandin & Kremen, 2013; Rundlöf et al., 2014). These are a step closer, although they do not directly measure population responses or associated population demographic variables *in situ* (see Williams & Kremen, 2007). More careful study would involve long-term monitoring and measurement of occupancy in enhanced and non-enhanced areas, an approach which has been explored for birds (Mackenzie et al., 2003). Alternatively, we could assess demographic parameters that underlie population dynamics as has been done for bees in other contexts (e.g. Williams & Kremen, 2007; Steffan-Dewenter & Schiele, 2008).

Second, to date nearly all investigation has focused on flower enhancement, which provides bee forage. Several pieces of information suggest that nesting resources may be just as important for bee distribution and diversity. Nesting and floral resources are non-substitutable and are required by all bee species. In addition, recent results from surveys of response of bee communities to agriculture show that nesting requirements are among the most predictive of community change (Williams et al., 2010; Forrest et al., 2015). Nest availability also appears to limit wild bee populations in some agricultural contexts (Steffan-Dewenter & Schiele, 2008). Understanding the relative importance of nesting limitation as well as the types of nesting resources used by different species and how they can be provided will be a key part of future actions to promote diversity and resilience using wild pollinators.

Third, and perhaps most important to our understanding of whether pollinator enhancements build resilience into agricultural systems, is the lack of data on their impact on pollination and yield of crops. Currently, most research on pollinator restorations is limited to measuring effects on pollinator abundance and diversity, thus leaving their presumed positive effect on pollination services untested. Evidence of such economic benefit of pollinator habitat enhancement has recently been found for mango in South Africa (Carvalheiro et al., 2012) and in blueberry in the USA (Blaauw & Isaacs, 2014). The work on blueberry showed that crop fields adjacent to these plantings had higher wild bee abundance and this was associated with higher fruit set and larger berries. Due to the time for perennial plants to establish and flower, this effect was not apparent until three years after the plantings were established, however, and the benefit was again seen in the fourth year. In relation to resilience, it is also instructive to contrast the benefits seen in the third and fourth years because the latter was characterised by early spring frost damage that reduced yield across the experimental sites. Although the absolute increase in yield observed in fields adjacent to the plantings was similar in the two years, there was a greater percentage increase in yield compared with the unenhanced fields in the season with the cold spring (31 percent) than in the year with the more typical moderate spring conditions (14 percent). Of course, from the perspective of farm economics, the *net* effect of enhancements (i.e. the yield benefits minus the costs of implementing and managing them) are the relevant factor. Empirical field studies have tended to focus instead on gross benefits, and we illustrate in the next section an approach to including costs for an assessment more relevant to farmer decisions.

Economic-based decision framework for building resilience, engaging stakeholders and informing policy

As our understanding of how to create resilient pollinator communities improves, the question remains of how to practically apply this knowledge to land management actions on farms. Previous sections focused on determining the ecological mechanisms that explain the role of pollinator diversity in promoting functional resilience of pollination and on current approaches to bolster native communities within agricultural landscapes. However, critical decisions remain about whether, when, where and how best to implement these approaches in order to ensure stable returns on investment that resilience is predicted to provide. The ability to economically evaluate benefits of increasing stability against the cost of creating it is an integral part of turning this ecological knowledge into practical on-farm decisions that will build resilience into crop pollination.

We illustrate an application of economic decision analysis to pollinator-dependent crop management by using four watermelon farms in California that differ in field size and the amount of surrounding natural habitat, a primary source of pollinators (Kremen et al., 2002; Ricketts et al., 2008). At each farm, the problems to solve are whether to create pollinator habitat and where to locate it. We first describe the characteristics and context of each farm, then present potential alternatives in detail, and finally evaluate the costs and benefits of each alternative. Although we are analysing real farm landscapes, our main intention is to demonstrate the decision process for evaluating a grower's potential options.

Farm characteristics

The four farms are located in northern California in an agricultural landscape consisting of mixed-row crop, fruit and nut orchards. All farms are organically managed, but differ in average field size and proportion of natural habitat within the surrounding landscapes (Table 6.1). Farms A and B have large fields of watermelon and are within relatively simplified landscapes. Farms C and D grow smaller areas of watermelon within more diverse fields, one in a relatively simple landscape, the other in a more complex landscape.

The farmers' management alternatives

The management alternatives we consider were developed to focus solely on habitat enhancement for pollination. To apply an economic approach, we developed five alternatives for each farm. The first alternative is the status quo or baseline, in which nothing additional is done. The others build on general strategies for habitat enhancement that are realistic options at many farms in our study region. In one, we considered replacing a strip of

Table 6.1. *Site and general pollinator enhancement information for four study sites in California.*

Site	Landscape class*	Percent natural habitat within 1.5 km	Field size (ha)	Enhancement size (ha)
A	Simple	7	31.1	0.70
B	Cleared	< 1	15.9	0.55
C	Simple	14	4.1	0.36
D	Complex	73	1.9	0.57

* Terms used match those of Tscharntke et al. (2005); cleared < 1 percent non-crop habitat, simple 1–20 per cent non-crop habitat, complex > 20 percent non-crop habitat in the surrounding landscape.

watermelon within the focal crop field with the habitat enhancement. In the next, we identified a location for a pollinator meadow at an uncropped area within the farmer's ownership boundary. Examples of such areas include tailwater ponds, equipment storage yards, or augmentation of existing grassland areas. In the final two, we considered adding habitat enhancement along two separate borders of the focal watermelon field. For this initial exploration, all alternatives provide the same flower and nesting resources per unit area. This allows us to focus on spatial placement, a primary question for growers.

Consequences of the farmers' decisions

To help evaluate the consequences of each alternative for the farmer, we applied a previously developed model that provides a landscape-wide spatially explicit index of pollinator abundance based on habitat quality (Lonsdorf et al., 2009) and yield response of a pollinator-dependent crop (Lonsdorf et al., 2009; Ricketts & Lonsdorf, 2013). The model has worked well to predict bee abundance in watermelon fields of California (Lonsdorf et al., 2009). The model has been published before, so we provide only a summary of it here.

Estimating pollinator abundance

First, the model translates a GIS land cover map into an index of suitability for bees, to create a pollinator source map. Higher scores indicate sources of greater relative bee abundance. To calculate the index, the model assumes that bees require two types of limiting resources to persist in a landscape: nesting substrates and floral resources. Given an input of land cover that describes the landscape, we assign various suitability values of each land cover type based on their ability to provide these resources. Thus, the potential supply of pollinators arising from a nest site depends on the quality of the nest and the floral resources surrounding it, and the model assumes that nearby resources contribute more than those farther away. Second, we use the index of bee abundance at nest sites to estimate the abundance of pollinators visiting the watermelon field. The model again assumes that the supply from nearby parcels contributes more than those farther away. The model produces a relative index of pollinator abundance that describes the relative quality of habitat in the landscape for pollinators.

Estimating yield and gross value
The relationship between visitation and yield is critical for this analysis, and this would ideally be tied directly to yield experiments that connect visitation with yield. Here, we employ the logic of past studies that show that as the index of abundance increases, bee abundance and pollen deposition increase (Lonsdorf et al., 2009) and that as pollen deposition increases, the size of watermelon increases, but is saturating (Stanghellini et al., 1998; Winfree et al., 2007). We used a simple saturating yield function to estimate the expected size of watermelons resulting from pollination as a function of the pollinator abundance index. This provides a proxy for how changing the land cover changes crop yield. We model the yield provided per flower within pixel i of a farmer's field, Y_i, as: $Y_i = 4.5 \frac{P_i}{P_i + K}$ where P_i is the pollinator abundance index at pixel i within a farmer's field, and K is the half-saturation constant of watermelon yield with increasing abundance. The value, 4.5, is the maximum size in kilograms for a fruit resulting from a flower saturated with pollen. We do not currently have an empirical yield curve so we set K to 0.05 for illustrative purposes. This value provides a curve similar to the relationship between index of abundance and estimates of pollen deposition for watermelon at these farms (Lonsdorf et al., 2009). The value of watermelon in pixel i, V_i, is equal to the product of the yield per flower, the number of flowers per plant, the number of plants planted per 30-metre pixel (0.09 ha) and the value of watermelon per kilogram (Baumer et al., 2009). The gross value is thus the sum of yield values across all planted pixels that represent the watermelon field. Gross value here does not take into account any potential costs.

Consequences of alternatives
To determine the gross value from crop yield resulting from each alternative, we digitally modified the landscapes based on the locations of each alternative (Figure 6.1) and converted the current land cover into a wildflower habitat. For example, the borders of farmers' fields may be early successional habitat that provides a moderate amount of floral and nesting resources. We can digitally replace this early successional habitat with a wildflower enhancement that we assume provides floral and nesting resources that are ideal for supporting wild pollinators. We then recalculate the yield in the focal field in response to the wild bees supported by the enhancement. We repeat this digital enhancement for

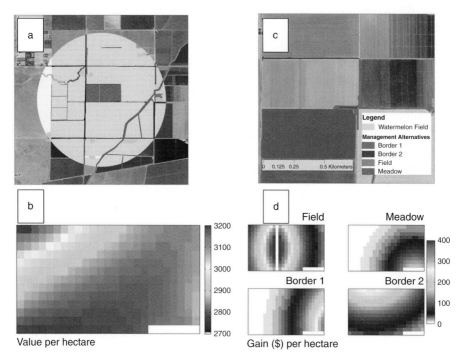

Figure 6.1 (a) Landscape map of Site A with different land-cover types classified within 1.5 km. Purple = focal watermelon field, green = annual row crop, yellow = managed grassland, pink = farmyards and homes, brown = unvegetated (including unpaved farm roads), black = paved roads. (b) Value map of the focal watermelon field based on modelled pollination with no habitat enhancement (status quo). Pixels are 30 × 30 m. (c) Landscape map showing locations of different enhancement alternatives. In this example the meadow surrounds a tail-water pond. (d) Marginal gain in value per hectare under four enhancement alternatives. (A black and white version of this figure will appear in some formats. For the colour version, please refer to the plate section.)

each potential alternative, treating the quality of enhancements equally for all alternatives. The difference in the effect on yield (measured in dollars gained or lost) is thus due to the relative difference in floral and nesting quality between the original cover and the enhancement as well as the location of the enhancement.

Economic analysis of net benefits

To determine which of the five alternatives is best, we choose the one with highest net benefit by also accounting for costs of implementation. Here we lay out the factors that affect our determination of benefits and costs, and illustrate how we use them to evaluate the overall decisions.

To determine net annual benefits, we first calculated the overall gross value of the watermelon field per year for each of the five alternatives as described above. Yield estimates were compared to published state-wide statistics (Census of AG, chapter 1, State Level/California). From the gross value, we subtracted the installation costs incurred only during the first year of restoration, plus maintenance costs that are incurred annually thereafter. Installation costs are incurred to prepare the site and include costs of removing any current vegetation, preparing site for planting, seeding and potentially irrigating if the site is outside of the watermelon field. Maintenance costs are incurred each year to remove weeds and to irrigate if needed. Additionally, if the wildflower plantings are directly adjacent to the fields, we assume that yield will be lost at the location of the planting and to allow farm operations in the field (e.g. tractor movement and spray buffers between enhancement and crop). We calculate this as 10 percent for every hectare of watermelon that bordered a flower strip.

Results and implications for decision-making
The yield in all but one field in our analysis would be strongly pollinator-limited without enhancement and if only pollination by wild bees was considered. Only Farm D, which has the most diverse landscape around the watermelon field, was predicted to have yields close to those observed under the status quo management action and without supplementing with managed pollinators, such as honey bees (Table 6.1). This is consistent with previous research at these sites that showed Farm D has greater abundance and diversity of native pollinators than the others included here (Kremen et al., 2002).

For an enhancement to be justified economically the benefit generated must outweigh the enhancements' associated costs. Although all types of enhancements caused yield and total gross value to increase, only enhancements in the largest field, Farm A (Figure 6.2), had net benefits that outweighed their cost. This benefit is the product of the marginal increase in value per area and the area planted. At Farm A the marginal increase per hectare was relatively large and the field was large. The marginal increases per hectare at Farm C and Farm D were comparable to Farm A, but their fields were too small to reap enough benefit from the enhancements. The field size of Farm B was large, but the locations of the enhancements were too far from the field to provide a high enough per hectare marginal gain and the resulting yield response was low. The results also reveal general locations where plants provide the greatest relative benefit within a farm. In three of four farms enhancing borders provided the greatest benefit. These locations place source habitats close to farm fields without removing much (or any) land from cultivation. Although enhancements placed within the crop field provide large benefits for the largest sized fields, they still suffer opportunity cost, the reduced yield through direct removal of the crop area, and can bring bees into areas where crop management activities may affect their survival.

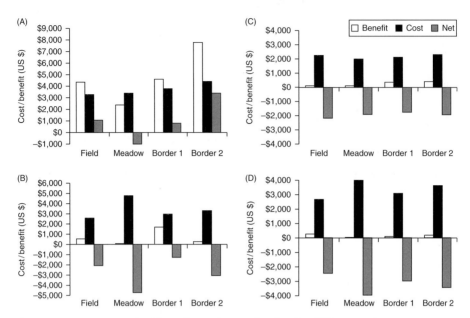

Figure 6.2 Costs, marginal benefits and net value for the different enhancement alternatives for Farms A–D, indicated by respective letters. In each panel benefits are shown by white bars and costs by black bars and net values by grey bars. The status quo, which is not plotted, would have zero costs and zero marginal benefits.

Our results are consistent with the findings of Ricketts and Lonsdorf (2013), who suggested that the most beneficial sites for enhancement are those areas in which the 'demand' for pollination is high and supply is currently low. Farm A, with the largest field, has the largest demand for pollinators and the landscape's supply (pollinator abundance index) is lowest. The demand for pollinators at Farms C and D is too small. From this general idea, it also follows that the pollinator habitat of greatest marginal value borders crop fields (Ricketts & Lonsdorf, 2013).

These economic analyses also provide new insights with respect to future planning of pollinator enhancements to promote resilient pollinator communities and stable economic returns in pollination-dependent agriculture. First, adding enhancements, even when fields are pollinator-limited, will not always make economic sense. Although we may have known this intuitively, it has never been formally explored. We find that there is interplay between the size of the fields benefitting from enhancements and the gain provided by the enhancement. For small fields even where pollination is limiting to yield, the costs of enhancements must be modest, because benefit is constrained by the total crop value. Reduced costs might be achieved by limiting the size of the enhancement, but this will also affect benefit. This also suggests that the crop value can affect the optimal enhancement strategy, with contrasting approaches likely needed in low- and high-value crops. Second, careful planning of enhancement location is necessary. According to our model the enhancements need to be placed as close to the target field as possible to maximise pollinator visitation. This placement will ultimately be influenced by practical considerations such as availability of area on the farm; however, when the area of pollinator-dependent crop is large, setting aside land to enhance pollination can be warranted even when it involves trading crop area for habitat (see Morandin & Winston, 2005). We note too that changing simple assumptions of pollinator movement may also increase the flexibility of which enhancement locations provide positive net benefits (Olsson et al., 2015), for example, if individual bees seek out and return to the most rewarding foraging patches (Cartar, 2004).

Our economic framework assumes average and constant values for habitat quality and bee behaviour. As such, it does not explicitly address certain resilience issues, such as temporal stability and the recovery from system shocks. The model can be extended, however, to address these aspects of resilience. For example, we could simulate the rapid and widespread extinction of a dominant pollinator (as was speculated to result from colony collapse disorder), or assign habitat values and foraging parameters stochastically to capture variations over space and time. Even in its current form, however, our approach can already help inform management actions to build resilience to changes in landscape configuration and management. In addition, the approach can already be extended to consider longer time stabilisation of profits by examining variation in pollination, yield, crop prices and longer-term costs of enhancements. Farm-wide implementation of pollinator enhancements can also be modelled over these larger spatial scales and over time to determine how this affects the stability of crop yields (biological resilience) and the revenue generated from different crops that rely on bees (economic resilience).

Conclusions

Diverse wild bee communities will need to be a key element of more integrated pollination strategies designed to maintain the resilience of agricultural pollination. There is increasing understanding of how to integrate these bees into crop pollination, yet also many areas where knowledge gaps exist. In this chapter, we reviewed the ecological relationships between bee diversity and pollination stability, discussed a variety of management practices intended to enhance wild bee abundance and diversity on or near farms, and presented a new framework and model or 'tool' to help decide among alternative actions to promote wild bees and their pollination services. The tool is grounded in a common currency of economic costs and returns from these management practices. With reasonable assumptions, the tool predicts substantial economic benefits from restoring habitat for bees, but these benefits vary with the spatial placement of the restored areas and with the management intensity of the surrounding landscape. Integrating gains and costs in a common economic currency also reveals the interplay of field size, total economic return and establishment cost in determining the viability of enhancement actions. These results indicate how important it is to understand the topics of the first two sections: the role of bee diversity in providing adequate and resilient pollination to crops, and the impact of common restoration efforts in enhancing that diversity and mitigating pollination deficits over space and time. Our review and modelling results reveal important gaps in understanding and opportunities for future research. These are as follows.

(1) Despite great progress on how to create habitats and forage resources for pollinators, little robust information exists on their ability to augment and stabilise pollinator populations and communities in the wider landscape. With few exceptions, studies are limited to monitoring of habitat plantings themselves.
(2) Longer-term studies are needed to understand the effects of landscape context and pollinator habitat enhancements on the stability and resilience of pollinator populations and communities.
(3) The importance of non-forage resources, such as nesting and overwintering habitat, relative to forage, has not been examined and is key to informing effective management decisions for pollinators and crop pollination.
(4) Perhaps most important, clear integration of economically grounded costs and benefits assessment of potential and existing actions is needed. Most challenging and most evidently lacking are empirical data linking different actions and the return on these investments for pollination and crop yield.

Advancing our understanding on these topics would allow more robust predictions about the ecological mechanisms and economic impacts of pollinator management efforts and can guide bee conservation efforts in farmland to improve resilience of crop pollination. In the end, we recognise that building resilience will need to involve more than the promotion of diverse pollinator communities within agricultural landscapes. Because agricultural systems vary widely in management, economics, spatial scale and geographic context, no single strategy will provide for resilient pollination into the future. We will need diversified approaches that involve combinations of managed honey bees, managed populations of native bee species and wild pollinator populations. Actions to promote robust wild bee communities within agricultural landscapes can also benefit managed bee populations, further enhancing the resilience of crop pollination in the face of various challenges facing bees and other pollinators.

References

Aizen, M.A., Garibaldi, L.A., Cunningham, S.A. & Klein, A. M. (2009). How much does agriculture depend on pollinators? Lessons from long-term trends in crop production. *Annals of Botany*, **103**, 1579–1588.

Artz, D.R. & Nault, B.A. (2011). Performance of *Apis mellifera, Bombus impatiens* and *Peponapis pruinosa* (Hymenoptera: Apidae) as pollinators of pumpkin. *Journal of Economic Entomology*, **104**, 1153–1161.

Bartomeus, I., Ascher, J.S., Wagner, D., et al. (2011). Climate-associated phenological advances in bee pollinators and bee-pollinated plants. *Proceedings of the National Academy of Sciences of the United States of America*, **108**, 20645–20649.

Bartomeus, I., Park, M.G., Gibbs, J., et al. (2013). Biodiversity ensures plant–pollinator phenological synchrony against climate change. *Ecology Letters*, **16**, 1331–1338.

Batáry, P., Báldi, A., Kleijn, D. & Tscharntke, T. (2011). Landscape-moderated biodiversity effects of agri-environmental management: a meta-analysis. *Proceedings of the Royal Society B: Biological Sciences*, **278**, 1894–1902.

Baumer, A., Hartz, T.K., Turini, T., et al. (2009). *Watermelon Production in California*. University of California, Division of Agriculture and Natural Resources, Publication 7213, ISBN-13: 978-1-60107-623-6.

Benjamin, F.E. & Winfree, R. (2014). Lack of pollinators limits fruit production in commercial blueberry (*Vaccinium corymbosum*). *Environmental Entomology*, **43**, 1574–1583.

Benjamin, F.E., Reilly, J.R. & Winfree, R. (2014). Pollinator body size mediates the scale at which land use drives crop pollination services. *Journal of Applied Ecology*, **51**, 440–449.

Biesmeijer, J.C., Roberts, S.P.M., Reemer, M., et al. (2006). Parallel declines in pollinators and insect-pollinated plants in Britain and the Netherlands. *Science*, **313**, 351–354.

Blaauw, B.R. & Isaacs, R. (2014). Flower plantings increase wild bee abundance and the pollination services provided to a pollination-dependent crop. *Journal of Applied Ecology*, **51**(4), 890–898.

Blüthgen, N. & Klein, A.-M. (2011). Functional complementarity and specialisation: the role of biodiversity in plant–pollinator interactions. *Basic and Applied Ecology*, **12**, 282–291.

Bosch, J., Kemp, W.P. & Trostle, G.E. (2006). Bee population returns and cherry yields in an orchard pollinated with *Osmia lignaria* (Hymenoptera: Megachilidae). *Journal of Economic Entomology*, **99**, 408–413.

Brittain, C., Kremen, C. & Klein, A.M. (2013). Biodiversity buffers pollination from changes in environmental conditions. *Global Change Biology*, **19**, 540–547.

Burkle, L.A., Marlin, J.C. & Knight, T.M. (2013). Plant–pollinator interactions over 120 years: loss of species, co-occurrence, and function. *Science*, **339**, 1611–1615.

Cameron, S.A., Lozier, J.D., Strange, J.P., et al. (2011). Patterns of widespread decline in North American bumble bees. *Proceedings of the National Academy of Sciences of the United States of America*, **108**, 662–667.

Cardinale, B.J., Duffy, J.E., Gonzalez, A., et al. (2012). Biodiversity loss and its impact on humanity. *Nature*, **486**, 59–67.

Cariveau, D.P., Williams, N.M., Benjamin, F.E. & Winfree, R. (2013). Response diversity to land use occurs but does not consistently stabilise ecosystem services provided by native pollinators. *Ecology Letters*, **16**, 903–911.

Cartar, R.V. (2004). Resource tracking by bumble bees: responses to plant-level differences in quality. *Ecology*, **85**, 2764–2771.

Carvalheiro, L.G., Seymour, C.L., Nicolson, S.W. & Veldtman, R. (2012). Creating patches of native flowers facilitates crop pollination in large agricultural fields: mango as a case study. *Journal of Applied Ecology*, **49**, 1373–1383.

Carvell, C., Westrich, P., Meek, W.R., Pywell, R. F. & Nowakowski, M. (2006). Assessing the value of annual and perennial forage mixtures for bumblebees by direct observation and pollen analysis. *Apidologie*, **37**, 326–340.

Chacoff, N.P., Aizen, M.A. & Aschero, V. (2008). Proximity to forest edge does not affect crop production despite pollen limitation. *Proceedings of the Royal Society B: Biological Sciences*, **275**, 907–913.

Chmielewski, F.-M. (2013). Phenology in agriculture and horticulture. In: *Phenology: an integrative environmental science*, edited by M.D. Schwartz. 2nd edition. Dordrecht: Springer, pp. 539–561.

Colla, S.R., Ascher, J.S., Arduser, M., et al. (2012). Documenting persistence of most eastern North American bee species (Hymenoptera: Apoidea: Anthophila) to 1990–2009. *Journal of the Kansas Entomological Society*, **85**, 14–22.

Corbet, S.A. (2006). A typology of pollination syndromes: implications for crop management and the conservation of wild plants. In: *Plant–Pollinator Interactions*, edited by N.M. Waser & J. Ollerton. Chicago, IL: University of Chicago Press, pp. 315–340.

Elmqvist, T., Folke, C., Nyström, M., et al. (2003). Response diversity, ecosystem change, and resilience. *Frontiers in Ecology and the Environment*, **1**, 488–494.

Folke, C., Carpenter, S., Walker, B., et al. (2004). Regime shifts, resilience, and biodiversity in ecosystem management. *Annual Review of Ecology, Evolution, and Systematics*, **35**, 557–581.

Forrest, J.R.K., Thorp, R.W., Kremen, C. & Williams, N.M. (2015). Contrasting patterns in species and functional-trait diversity of bees in an agricultural landscape. *Journal of Applied Ecology*, **52**, 706–715.

Garibaldi, L.A., Steffan-Dewenter, I., Kremen, C., et al. (2011). Stability of pollination services decreases with isolation from natural areas despite honey bee visits. *Ecology Letters*, **14**, 1062–1072.

Garibaldi, L.A., Steffan-Dewenter, I., Winfree, R., et al. (2013). Wild pollinators enhance fruit set of crops regardless of honey bee abundance. *Science*, **339**, 1608–1611.

Garibaldi, L.A., Carvalheiro, L.G., Leonhardt, S. D., et al. (2014). From research to action: enhancing crop yield through wild pollinators. *Frontiers in Ecology and the Environment*, **12**, 439–447.

Goulson, D., Lye, G.C. & Darvill, B. (2008). Decline and conservation of bumble bees. *Annual Review of Entomology*, **53**, 191–208.

Greenleaf, S.S. & Kremen, C. (2006). Wild bees enhance honey bees' pollination of hybrid sunflower. *Proceedings of the National Academy of Sciences*, **103**, 13890–13895.

Grundel, R., Jean, R.P., Frohnapple, K.J., et al. (2010). Floral and nesting resources, habitat structure, and fire influence bee distribution across an open-forest gradient. *Ecological Applications*, **20**, 1678–1692.

Haaland, C., Naisbit, R.E. & Bersier, L.F. (2011). Sown wildflower strips for insect conservation: a review. *Insect Conservation and Diversity*, **4**, 60–80.

Hannon, L.E. & Sisk, T.D. (2009). Hedgerows in an agri-natural landscape: potential habitat value for native bees. *Biological Conservation*, **142**, 2140–2154.

Heard, M.S., Carvell, C., Carreck, N.L., et al. (2007). Landscape context not patch size determines bumble-bee density on flower mixtures sown for agri-environment schemes. *Biology Letters*, **3**, 638–641.

Herrera, C.M. (2002). Censusing natural microgametophyte populations: variable spatial mosaics and extreme fine-graininess in winter-flowering *Helleborus foetidus* (Ranunculaceae). *American Journal of Botany*, **89**, 1570–1578.

Holling, C.S. (1992). Cross-scale morphology, geometry, and dynamics of ecosystems. *Ecological Monographs*, **62**, 447–502.

Holzschuh, A., Steffan-Dewenter, I. & Tscharntke, T. (2008). Agricultural landscapes with organic crops support higher pollinator diversity. *Oikos*, **117**, 354-361.

Houlahan, J., Currie, D., Cottenie, K., et al. (2007). Compensatory dynamics are rare in natural ecological communities. *Proceedings of the National Academy of Sciences*, **104**, 3273-3277.

Iler, A.M., Inouye, D.W., Høye, T.T., et al. (2013). Maintenance of temporal synchrony between syrphid flies and floral resources despite differential phenological responses to climate. *Global Change Biology*, **19**, 2348-2359.

Isaacs, R., Tuell, J., Fiedler, A., Gardiner, M. & Landis, D. (2009). Maximizing arthropod-mediated ecosystem services in agricultural landscapes: the role of native plants. *Frontiers in Ecology and the Environment*, **7**, 196-203.

Jauker, F., Peters, F., Wolters, V. & Diekötter, T. (2012). Early reproductive benefits of mass-flowering crops to the solitary bee *Osmia rufa* outbalance post-flowering disadvantages. *Basic and Applied Ecology*, **13**, 268-276.

Kennedy, C.M., Lonsdorf, E., Neel, M.C., et al. (2013). A global quantitative synthesis of local and landscape effects on wild bee pollinators in agroecosystems. *Ecology Letters*, **16**, 584-599.

Klein, A.M., Vaissiere, B.E., Cane, J.H., et al. (2007). Importance of pollinators in changing landscapes for world crops. *Proceedings of The Royal Society B: Biological Sciences*, **274**, 303-313.

Kleijn, D., Rundlöf, M., Scheper, J., Smith, H.G. & Tscharntke, T. (2011). Does conservation on farmland contribute to halting the biodiversity decline? *Trends in Ecology & Evolution*, **26**, 474-481.

Kleijn, D., Winfree, R., Bartomeus, I., et al. (2015). Delivery of crop pollination services is an insufficient argument for wild pollinator conservation. *Nature Communications*, **6**, 7414.

Knight, T.M., Steets, J.A., Vamosi, J.C., et al. (2005). Pollen limitation of plant reproduction: pattern and process. *Annual Review of Ecology Evolution and Systematics*, **36**, 467-497.

Kremen, C., Williams, N.M. & Thorp, R.W. (2002). Crop pollination from native bees at risk from agricultural intensification. *Proceedings of the National Academy of Sciences of the United States of America*, **99**, 16812-16816.

Laliberte, E., Wells, J.A., DeClerck, F., et al. (2010). Land-use intensification reduces functional redundancy and response diversity in plant communities. *Ecology Letters*, **13**, 76-86.

Lawton, J.H. & Brown, V.K. (1993). Redundancy in ecosystems. In: *Biodiversity and Ecosystem Function*, edited by E.D. Schulze & H.A. Mooney. Berlin: Springer-Verlag, pp. 255-270.

Lonsdorf, E., Kremen, C., Ricketts, T., et al. (2009). Modelling pollination services across agricultural landscapes. *Annals of Botany*, **103**, 1589-1600.

MacKenzie, D.I., Nichols, J.D., Hines, J.E., Knutson, M.G. & Franklin, A.B. (2003). Estimating site occupancy, colonization, and local extinction when a species is detected imperfectly. *Ecology*, **84**, 2200-2207.

Mandelik, Y., Winfree, R., Neeson, T. & Kremen, C. (2012). Complementary habitat use by wild bees in agro-natural landscapes. *Ecological Applications*, **22**, 1535-1546.

Matson, P.A., Parton, W.J., Power, A.J. & Swift, M.J. (1997). Agricultural intensification and ecosystem properties. *Science*, **277**, 504-509.

Mayfield, M., Bonser, S., Morgan, J., et al. (2010). What does species richness tell us about functional trait diversity? Predictions and evidence for responses of species and functional trait diversity to land-use change. *Global Ecology and Biogeography*, **19**, 423-431.

Meehan, T.D., Werling, B.P., Landis, D.A. & Gratton, C. (2011). Agricultural landscape simplification and insecticide use in the Midwestern United States. *Proceedings of the National Academy of Sciences*, **108**, 11500–11505.

Morandin, L. & Winston, M. (2005). Wild bee abundance and seed production in conventional, organic and genetically modified canola. *Ecological Applications*, **15**, 871–881.

Morandin, L.A. & Kremen, C. (2013). Hedgerow restoration promotes pollinator populations and exports native bees to adjacent fields. *Ecological Applications*, **23**, 829–839.

Morris, W.F., Vázquez, D.P. & Chacoff, N.P. (2010). Benefit and cost curves for typical pollination mutualisms. *Ecology*, **91**, 1276–1285.

Olsson, O., Bolin, A., Smith, H.G. & Lonsdorf, E.V. (2015). Modeling pollinating bee visitation rates in heterogeneous landscapes from foraging theory. *Ecological Modelling*, **316**, 133–143.

Osborne, J.L., Martin, A.P., Shortall, C.R., et al. (2008). Quantifying and comparing bumblebee nest densities in gardens and countryside habitats. *Journal of Applied Ecology*, **45**, 784–792.

Potts, S.G., Biesmeijer, J.C., Kremen, C., et al. (2010). Global pollinator declines: trends, impacts and drivers. *Trends in Ecology & Evolution*, **25**, 345–353.

Pywell, R.F., Meek, W.R., Loxton, R., et al. (2011). Ecological restoration on farmland can drive beneficial functional responses in plant and invertebrate communities. *Agriculture, Ecosystems & Environment*, **140**, 62–67.

Rader, R., Reilly, J., Bartomeus, I. & Winfree, R. (2013). Native bees buffer the negative impact of climate warming on honey bee pollination of watermelon crops. *Global Change Biology*, **19**, 3103–3110.

Redpath-Downing, N.A., Beaumont, D., Park, K. & Goulson, D. (2013). Restoration and management of machair grassland for the conservation of bumblebees. *Journal of Insect Conservation*, **17**, 491–502.

Richards, S.A., Williams, N.M. & Harder, L.D. (2009). Variation in pollination: causes and consequences for plant reproduction. *The American Naturalist*, **174**, 382–398.

Ricketts, T.H. & Lonsdorf, E. (2013). Mapping the margin: comparing marginal values of tropical forest remnants for pollination services. Ecological Applications, **23**, 1113–1123.

Ricketts, T.H., Regetz, J., Steffan-Dewenter, I., et al. (2008). Landscape effects on crop pollination services: are there general patterns? *Ecology Letters*, **11**, 499–515.

Robinson, R.A. & Sutherland, W.J. (2002). Post-war changes in arable farming and biodiversity in Great Britain. *Journal of Applied Ecology*, **39**, 157–176.

Roulston, T.H. & Goodell, K. (2011). The role of resources and risks in regulating wild bee populations. *Annual Review of Ecology and Systematics*, **56**, 293–312.

Rundlöf, M., Nilsson, H. & Smith, H.G. (2008). Interacting effects of farming practice and landscape context on bumblebees. *Biological Conservation*, **141**, 417–426.

Rundlöf, M., Persson, A.S., Smith, H.G. & Bommarco, R. (2014). Late-season mass-flowering red clover increases bumble bee queen and male densities. *Biological Conservation*, **172**, 138–145.

Russo, L., DeBarros, N., Yang, S., Shea, K. & Mortensen, D. (2013). Supporting crop pollinators with floral resources: network-based phenological matching. *Ecology and Evolution*, **3**, 3125–3140.

Stanghellini, M.S., Ambrose, J.T. & Schultheis, J.R. (1998). Seed production in watermelon: a comparison between two commercially available pollinators. *Hortscience*, **33**, 28–30.

Steffan-Dewenter, I. & Schiele, S. (2008). Do resources or natural enemies drive bee population dynamics in fragmented habitats? *Ecology*, **89**, 1375–1387.

Steffan-Dewenter, I., Munzenberg, U., Burger, C., Thies, C. & Tscharntke, T. (2002). Scale-dependent effects of landscape context on three pollinator guilds. *Ecology*, **83**, 1421–1432.

Steinhauer, N.A., Rennich, K., Wilson, M.E., et al. (2014). A national survey of managed honey bee 2012–2013 annual colony losses in the USA: results from the Bee Informed Partnership. *Journal of Apicultural Research*, **53**, 1–18.

Tepedino, V.J. (1981). The pollination efficiency of the squash bee (*Peponapis pruinosa*) and the honey bee (*Apis mellifera*) on summer squash (*Cucurbita pepo*). *Journal of the Kansas Entomological Society*, **54**(2), 359–377.

Tscharntke, T., Klein, A.M., Kruess, A., Steffan-Dewenter, I. & Thies, C. (2005). Landscape perspectives on agricultural intensification and biodiversity–ecosystem service management. *Ecology Letters*, **8**, 857–874.

Tylianakis, J.M., Didham, R.K., Bascompte, J. & Wardle, D.A. (2008). Global change and species interactions in terrestrial ecosystems. *Ecology Letters*, **11**, 1351–1363.

van der Zee, R., Brodschneider, R., Brusardis, V., et al. (2014). Results of international standardised beekeeper surveys of colony losses for winter 2012–2013: analysis of winter loss rates and mixed effects modelling of risk factors for winter loss. *Journal of Apicultural Research*, **53**, 19–34.

Vanbergen, A.J. and The Insect Pollinators Initiative. (2013). Threats to an ecosystem service: pressures on pollinators. *Frontiers in Ecology and the Environment*, **11**, 251–259.

vanEngelsdorp, D. & Meixner, M.D. (2010). A historical review of managed honey bee populations in Europe and the United States and the factors that may affect them. *Journal of Invertebrate Pathology*, **103**, S80–S95.

vanEngelsdorp, D., Speybroeck, N., Evans, J.D., et al. (2010). Weighing risk factors associated with bee colony collapse disorder by classification and regression tree analysis. *Journal of Economic Entomology*, **103**, 1517–1523.

Westphal, C., Steffan-Dewenter, I. & Tscharntke, T. (2003). Mass flowering crops enhance pollinator densities at a landscape scale. *Ecology Letters*, **6**, 961–965.

Westphal, C., Steffan-Dewenter, I. & Tscharntke T. (2009). Mass flowering oilseed rape improves early colony growth but not sexual reproduction of bumblebees. *Journal of Applied Ecology*, **46**, 187–193.

Williams, N.M. (2011). Restoration of nontarget species: bee communities and pollination function in riparian forests. *Restoration Ecology*, **19**, 450–459.

Williams, N.M. & Kremen, C. (2007). Resource distributions among habitats determine solitary bee offspring production in a mosaic landscape. *Ecological Applications*, **17**, 910–921.

Williams, N.M., Crone, E.E., Roulston, T.H., et al. (2010). Ecological and life-history traits predict bee species responses to environmental disturbances. *Biological Conservation*, **143**, 2280–2291.

Williams, N.M., Regetz, J. & Kremen, C. (2012). Landscape-scale resources promote colony growth but not reproductive performance of bumble bees. *Ecology*, **93**, 1049–1058.

Williams, N.M., Ward, K.L., Pope, N., et al. (2015). Native wildflower plantings support wild bee abundance and diversity in agricultural landscapes across the United States. *Ecological Applications*, **25**, 2119–2131.

Winfree, R. (2013). Global change, biodiversity, and ecosystem services: what can we learn from studies of pollination? *Basic and Applied Ecology*, **14**, 453–460.

Winfree, R. & Kremen, C. (2009). Are ecosystem services stabilized by differences among species? A test using crop pollination. *Proceedings of The Royal Society B: Biological Sciences*, **276**, 229–237.

Winfree, R., Williams, N.M., Dushoff, J. & Kremen, C. (2007). Native bees provide insurance against ongoing honey bee losses. *Ecology Letters*, **10**, 1105–1113.

Winfree, R., Bartomeus, I. & Cariveau, D.P. (2011). Native pollinators in anthropogenic habitats. *Annual Review of Ecology, Evolution, and Systematics*, **42**, 1–22.

Wright, C.K. & Wimberly, M.C. (2013). Recent land use change in the Western Corn Belt threatens grasslands and wetlands. *Proceedings of the National Academy of Sciences of the United States of America*, **110**, 4134–4139.

CHAPTER SEVEN

Conflicts and challenges to enhancing the resilience of small-scale farmers in developing economies

RICHARD EWBANK
Christian Aid

Introduction

Resilience, broadly defined as 'the capacity to buffer change, learn and develop' (Folke et al., 2002), has emerged as a significant priority in development practice comparatively recently. Three factors – the direct impact of climate change; the threat to small-scale farmers of some 'solutions' to climate change, such as agricultural biofuel production; and the contribution agriculture makes to climate change through greenhouse gas emissions – have brought the resilience debate with respect to small-scale agriculture sharply into focus. The importance to global food security and poverty reduction of supporting small-scale agriculture is well evidenced. Small-scale farmers comprise about 85 percent of the global total, manage about 60 percent of agricultural land and produce over half the world's food (IFPRI, 2005).

This provides 2.6 billion people with their livelihood. In many areas, their contribution is even higher, providing 90 percent of food produced in Africa. On the other hand, 87 percent of small-scale farmers, typically defined as those farming less than 2 hectares, are found in Asia. In many cases, small-scale farmers are more productive than large in terms of both productivity per unit area and per unit of energy (IAASTD, 2009). Conventional agriculture has driven declining energy productivity – the ratio of crop output to energy input – for example, the energy productivity of Bangladesh agriculture showed a decline from an output/input ratio of 3.9 to 3.0 from 1990 to 2005; in China, the ratio declined from 2 to 1.5 between 1978 and 2004 (Pelletier et al., 2011).

Most economies base their economic growth on a foundation of agricultural development. In many developing countries, agriculture is the main livelihood of the majority, delivering food security and typically contributing more to export earnings than any other sector. In these contexts, the majority of the poor rely on small-scale agriculture. Over 80 percent of the extreme poor (living on less than US$1.9 per day) and 76 percent of the poor (living on US$1.9–3.1 per day) live in rural areas with 64.7 and 52 percent of them, respectively, depending on the agricultural sector for their main livelihoods

(Castañeda et al., 2016). Investment focused on these farmers has tended to yield more poverty-reduction impact – a 1 percent increase in agricultural GDP has been shown to be 39 percent more effective than manufacturing and 104 percent more effective than the service sector in increasing the incomes of the poorest (Gallup et al., 1997).

However, the predominant paradigm has been a largely instrumental one – to increase short-term productivity on the assumption that this will sustainably raise income and thereby reduce poverty and vulnerability. Small-scale farmers have been widely viewed as risk-averse, which is often seen as a problem rather than a virtue. Their supposedly 'traditional' behaviour has prevented innovation with modern inputs and their concern to buffer their livelihoods against future uncertainty has created inefficiency as resources that should be used for increasing productivity were tied up in mitigating risk. When viewed from the narrow perspective of agricultural policymakers largely concerned with using small-scale agriculture as a lever of economic growth, these farmers are inefficient. However, when viewed from a wider resilience perspective, their priorities have often been more in tune with long-term sustainability and productivity across a diverse set of on-farm enterprises and ecosystems than those who have sought to guide them.

Advisory systems have been designed, and are largely still attempting, to overcome this by training farmers in the 'error of their ways', encouraging them to adhere to the top-down priorities of agricultural research and policy. The main vehicle for this transformation has been green revolution and post-green revolution dependency on increasing small-scale farmer use of so-called 'modern' inputs – chemical pesticides, herbicides and fertilisers and the high-yielding seed varieties designed to work with them – often attractively subsidised and allied to an expansion of cultivated area through ploughing or hoeing.

Although some of the negative environmental impacts were evident from the outset, this vehicle was developed prior to the greater understanding that now exists of climate change. It essentially assumed that climate would continue as it had before, within the bounds of normal variation. This intensive approach has also played a significant role in driving climate change. Agriculture contributes to decreased resilience through its direct generation of an estimated 25 percent of greenhouse gases (IPCC, 2014), including 50 percent of methane and most nitrous oxide (Park et al., 2012). More than half of all nitrogen fertiliser applied has been used in the last 30 years, disrupting the nitrogen cycle and contributing 6 percent of greenhouse gas emissions (IAASTD, 2009; IPCC, 2014).

Fossil fuels have driven mechanisation and are the feedstock for chemical fertilisers, pesticides and herbicides. With the exception of methane from rice production, the majority of these emissions are generated by larger-scale commercial agriculture, with land-use change accounting for more than half (IPCC, 2014). The threat of climate change alone suggests the need for a paradigm shift – moving large and small-scale agriculture onto a pathway that not only increases resilience to climate change but also reduces the capacity of agriculture to drive climate risks associated with the current high level of greenhouse gas emissions. For small-scale farmers, increasingly vulnerable to a climate they have had little part in creating, this would recognise their concern with risk as a positive feature and seek to work with rather than against it, strengthening their capacity to manage, mitigate and recover from shocks and stresses.

Perceptions of risk in small-scale agriculture

Risk profiles of small-scale farmers are always context-specific, but when asked about the shocks and stresses that contribute to their vulnerability, small-scale farmers in different agroecological circumstances with different holding sizes and social contexts do exhibit some overlap in the types of risk that concern them. For example, comparing farmers in Malawi and Burkina Faso (see Figure 7.1) shows that while the proportions change significantly, 96 percent of the risk categories have a similar focus on climatic risk, crop and livestock pests and disease and human health. Differences reflect the different environments – Burkina Faso's greater vulnerability to desertification and the prevalence in some areas of crop damage by wild animals in Malawi – but there is overlap in categories that preoccupy farmers' efforts to reduce their vulnerability in both places (Christian Aid Burkina Faso, 2011; Christian Aid Malawi, 2012).

Profiles of Dalit sharecroppers in rural Bihar, India, likewise illustrate the preoccupation with climatic risk, which covered three of the four main categories listed – increasingly intensive rainfall, mud slides and floods – with community conflict the only exception (CASA/Christian Aid India, 2012). These were all perceived to be the main threats to life, productive assets such as land and livestock and a driver of migration out of the area. Risk assessment by rural communities in Nicaragua and El Salvador showed similar concerns, with floods, mud slides and drought all highlighted. As in Burkina Faso, farmers in El Salvador expressed concern about land degradation, but some of the causative factors were different. Increased salination of groundwater in coastal land areas and agrochemical-contaminated run-off from adjacent commercial sugar plantations were both identified as affecting soil health and agricultural productivity, with the latter also identified as a driver of increased health vulnerability (Christian Aid Central America, 2012). A notable feature of these profiles is that the concern with recurrent and extensive risk and stress, such as repeated waterlogging or extended dry spells, tends to be as or more significant than concern about large-scale, intensive risks, such as cyclones and major droughts.

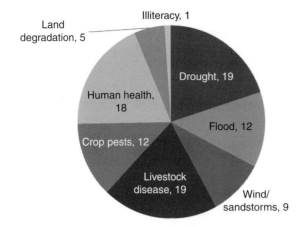

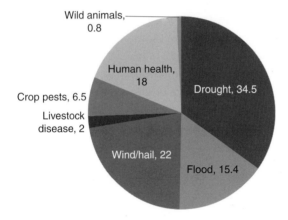

Figure 7.1 Farmer perceptions of risk (percent of responses) in Burkina Faso (top) and Malawi (bottom).

These analyses show two significant characteristics – first, while context and proportion may vary, there is considerable agreement in the risk perceptions of small-scale farmers across the tropics and subtropics; second, these perceptions encompass a range of shocks and stresses that informs the nature of resilience building that small-scale farmers are already implementing using their own resources. It clearly needs to more explicitly inform the support they may have access to from external agencies, such as local agricultural advisory services.

Based on how farmers themselves assess their vulnerability, a more evidence-based understanding of how various factors can either improve or reduce their capacity to buffer themselves and their key assets against further risk – so progressing towards an 'optimal level of redundancy', where buffers are absorbing and facilitating efficient recovery from as many shocks and stresses as possible, given the uncertainty about their future occurrence – can be developed. If (in Figure 7.2) the extra slices in each of the five asset

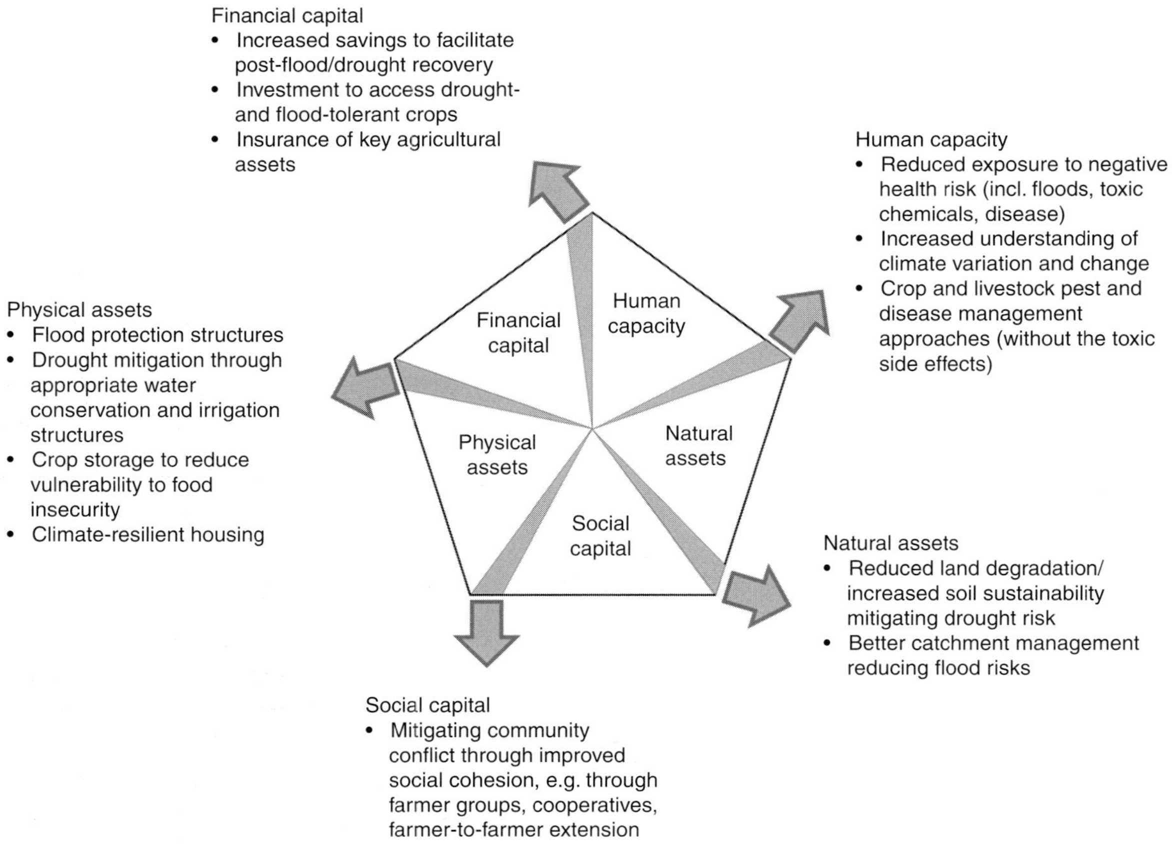

Financial capital
• Increased savings to facilitate post-flood/drought recovery
• Investment to access drought- and flood-tolerant crops
• Insurance of key agricultural assets
Human capacity
• Reduced exposure to negative health risk (incl. floods, toxic chemicals, disease)
• Increased understanding of climate variation and change
• Crop and livestock pest and disease management approaches (without the toxic side effects)
Physical assets
• Flood protection structures
• Drought mitigation through appropriate water conservation and irrigation structures
• Crop storage to reduce vulnerability to food insecurity
• Climate-resilient housing
Financial capital
Human capacity
Physical assets
Natural assets
Social capital
Natural assets
• Reduced land degradation/ increased soil sustainability mitigating drought risk
• Better catchment management reducing flood risks
Social capital
• Mitigating community conflict through improved social cohesion, e.g. through farmer groups, cooperatives, farmer-to-farmer extension

Figure 7.2 Options for strengthening farmer resilience based on their priorities (as expressed in Figure 7.1).

categories (adapted from DFID, 2001) represent the additional redundancy or buffers that are needed to improve resilience, measures to reduce agricultural risk based on farmers' own risk priorities and actions (above and in Figure 7.1) can be suggested.

Threats and challenges to the agricultural resilience of small-scale farmers

Climate variation and change

Of all livelihoods, small-scale agriculture exhibits one of the largest vulnerabilities to climate variation and change, both in terms of the inherent vulnerability of resource-poor households depending on it and the direct impacts on agroecologically based livelihoods. Climate change contributes to this vulnerability in two principal ways – through increased climate variation elevating the impact of shocks and hazards and through incremental change increasing stress on crops and livestock. Between 1980 and 2012, the number of extreme hydro-meteorological events more than doubled from about 350 to over 700 per annum (WMO, 2013). At the same time, incremental factors such as increasing temperatures over land areas, which interact with and amplify extreme events such as droughts, have risen by 1.5°C over the past 250 years and significantly more than this in many tropical and subtropical countries (BEST, 2012).

Extreme events have a profound impact on agricultural livelihoods, often leaving longer-term stresses behind that, combined with more incremental climate change, make it difficult for farmers to recover. For example, cyclone Aila struck Bangladesh and West Bengal in May 2009, inundating the Sundarbans with 2 m of sea water. In West Bengal this affected 2.2 million people, 2.89 million hectares of land and destroyed 61,000 houses (India Meteorological Department, 2009). Because the inundation occurred in May, before the summer monsoon, the soils were drier and more susceptible to saline infiltration by the flood. Two years later, farmers were still desalinating their soils and reporting yields of only a third pre-Aila levels (Ewbank, 2011).

While the attribution of individual cyclone events to man-made climate change remains uncertain, incremental sea-level rise has contributed to increased vulnerability from associated storm surges, which in this case was the cause of long-term damage. The same farmers also identified incremental risks, such as rising temperatures over the past 15 years, as a significant challenge to their livelihoods. Putting this into context more generally, each day spent above 30°C can reduce maize yields by 1.7 percent in drought conditions. Increased night-time temperatures also have profound impacts, reducing rice yields by up to 10 per cent for each 1°C increase in minimum temperature in the dry season (HLPE, 2012).

Rainfall intensity, which accelerates soil erosion and leaching of valuable soil nutrients, is forecast to increase even in subtropical areas where overall rainfall declines (IPCC, 2012). This concentrates rainfall into fewer rain days, increasing intraseasonal drought risks as the occurrence of longer dry spells stresses crops. If this occurs as crops are maturing, yield reductions can be particularly severe. So farmers in Mbeere District, central Kenya, may have experienced almost average total rainfall for the 2011 short rains but 35 percent of this fell on just four days (Ewbank, 2012). In the following long rains season in 2012, only 70 percent of anticipated rainfall occurred, it arrived late and 25 percent fell in a three-day period in April. Although the season technically ended at the anticipated time, the lack of rainfall for much of May resulted in farmers experiencing only 35 effective growing days once extended dry spells had been taken into account.

The East African region has experienced a drying trend, especially with respect to the main long rains season, and a potential climate change driver for increasingly dry long rains season conditions has been detected, as Indian Ocean sea surface temperatures affect atmospheric circulation patterns over the region, disrupting rainfall-bearing systems (Park Williams & Funk, 2011). As this increased temperature/reduced rainfall reliability/increased rainfall intensity scenario develops, it will have significant impacts. Depending on the region, substantial reductions in crop production are anticipated, including up to 30 percent declines in maize production in Southern Africa by 2030 and up to 10 percent declines in rice production in South Asia (PAR-FAO, 2010).

The enrichment effects of higher atmospheric carbon dioxide concentrations on crop production have been cited as a potential benefit that offsets the negative impacts of increasingly erratic climate, but evidence suggests that these have been overstated. Carbon fertilisation effects are substantially lower for the C4 crops that small-scale farmers in Africa rely on, such as maize, sorghum and millet, and have generally been found to be about half the level in the field compared to the enclosed experiments that first measured the effects (Leakey et al., 2009).

Higher atmospheric CO_2 has also been shown to increase susceptibility to crop pests (Zavala et al., 2008). Factors include the loss of natural resistance to pests by crops, increased longevity of and an increased preference for the consumption of crops over other food items by crop pest species. Added to this, higher CO_2 concentrations have been shown to inhibit the assimilation of nitrate, potentially leaving crops depleted of organic nitrogen compounds such as protein, and reducing food quality (Bloom et al., 2010). Given the more profound direct impacts of climate change, any marginal gains from carbon fertilisation are therefore unlikely to significantly mitigate the added levels of risk that climate change contributes to small-scale agriculture.

Land, soil and water degradation

Conventional agricultural systems rely on intensive soil cultivation, irrigation and high agrochemical use as the primary tools for increasing yields and productivity. Irrigation is frequently highlighted as a major priority in increasing small-scale farmer resilience, especially in Africa where only 5 percent of the cultivated area is irrigated (World Bank, 2007) and farmers rely on increasingly erratic rains for their moisture supply. Agriculture already uses 70 percent of freshwater withdrawals and is increasingly dependent on depleting finite groundwater resources at levels exceeding their recharge rates. Poorly managed use of water resources together with inappropriate cultivation methods and catchment deforestation are the main causes of land degradation. Globally, 34 million hectares or about 11 percent of the irrigated area is affected by some degree of salinity, with particularly acute problems in India, Sudan and Central Asia (Mateo-Sagasta & Burke, 2011).

South Asia as a whole accounts for about 30 percent of the total – by 2002, some 188 million hectares of India's cultivable land had become degraded by salinity, erosion and other factors largely as a result of the promotion of green revolution technologies and associated overuse of chemical inputs, a loss to national annual GDP of 1.1 percent but a loss to the GDP of the poor of 6.4 percent (Mani et al., 2012). India is currently withdrawing 260 cubic kilometres per year, over five times the 1970 level (Brown, 2012), a situation that has led to over 25 percent of the country's network of assessment units showing unsustainable, long-term declines in groundwater levels (Central Ground Water Board, 2009).

Accelerating overuse of mineral fertiliser has compounded this degradation through direct impacts on soil quality, climate change and through run-off into ground water and ultimately coastal marine environments. In the latter category, the number of documented cases of hypoxic coastal ecosystems, or 'dead zones', caused by fertiliser run-off of nitrogen and phosphate more than doubled to 400 in the 13 years to 2008, covering an area of 245,000 square kilometres (Rabalais et al., 2010), affecting the resilience of fishery resources that coastal communities rely on. Direct impacts on soils include deteriorating soil structure, especially the organic fraction, reducing vital nutrient and moisture retention capacity and the ability to absorb and sequester atmospheric carbon; acidification inhibiting crop growth through reductions in phosphate uptake; and subsurface salt pans restricting root growth and crop access to soil moisture (Kotschi, 2013). Nitrogen fertilisers, which make up three-quarters of all fertiliser used, are also directly responsible for the bulk of increased atmospheric concentrations of nitrous oxide, which comprise 6 percent of greenhouse gas emissions (Park et al., 2012).

Fertiliser loss through run-off and other factors is already higher in tropical and subtropical agricultural systems due to the prevalence of more intensive rainfall, causing over 50 percent loss of nitrogen and up to 90 percent loss of phosphorus (Baligar & Bennett, 1986). As a result, soil and water degradation by fertiliser and other chemical inputs used for mono-crop wheat and rice cultivation in the Ganges Plains of South Asia have cancelled the gains made by green revolution technologies (Ali & Byerlee, 2002). Studies in China on the level of nitrates in groundwater at 600 sites in 20 counties where fertiliser use in agriculture was high found that at nearly half of the sites, nitrate levels were above 50 milligrams per litre, considered the maximum safe level for human health in developed countries (Greenpeace China, 2010; FAO, 2013). As rainfall intensity increases through climate change, groundwater, river and marine pollution and hypoxic events are likely to spread and strengthen. Fertiliser subsidies in a variety of African countries typically consume 43–74 percent of the total agricultural budget (Kotschi, 2013), leaving little funding for the demand-led, resilience-building technical advisory support and other services that small-scale farmers value.

Cultivation by plough or hoe has also contributed through loss of fragile topsoil resources and increased evapotranspiration of soil moisture. Since 1950 one-third of soils have been profoundly altered from their natural ecosystem state because of moderate to severe soil degradation, with agricultural cultivation the main driver of this (IAASTD, 2009). Converting to conservation agriculture, which incorporates minimum tillage, reduces soil loss by 97 percent (Montgomery, 2007) and could play a key role in enhancing farmer resilience to increased drought and flood impacts by maintaining soil moisture – up to 2500 cubic kilometres of groundwater could be saved in Africa by 2050 (FAO, 2009a) – and improving soil structure. Where soil protection measures such as conservation agriculture, agroforestry, terracing and contour-aligned erosion-control embankments are combined with hydrological management measures at catchment level, such as protecting strategic catchment forests and wetlands, land degradation, drought and flood risks can be reduced in the most cost-effective way.

The challenge of agrochemical dominance

Increasing the use of chemical pest, disease and weed control has been integral to the green revolution model. For small-scale farmers, agrochemical promotion and sale has undermined resilience in a variety of ways. Chemical sales are often poorly regulated and information about safe use inadequately communicated, resulting in direct health impacts to both the farming household and the consumers of produce. For example, vegetable farmers in northern Tanzania were recorded using 41 different pesticides, eight of which were unregistered. They routinely suffered ill health after applying pesticide – 68 percent reported feeling sick, suffering skin problems and neurological symptoms (Ngowi et al., 2007). Several reasons were cited, including a lack of information provided with pesticides on their toxicity and environmental impact, the tendency of farmers to mix up to five chemicals together, the promotion of pesticides by government agricultural advisors without any training on their safe use and pressure from manufacturers and retailers to encourage farmers to use larger and larger doses to control pests and weeds.

This latter feature of the agrochemical industry has led to accelerating resistance by crop pests and weeds to chemicals which require increased applications to maintain their effect, capturing farmers on a treadmill of increased resistance requiring greater expenditure, often financed through credit with high interest rates, on more toxic products. For example, when the insecticide endosulfan, a chemical banned in 57 countries (Weinburg, 2009), was introduced to West African cotton farmers in 1999, it replaced less-toxic pyrethroid products without explanation of its hazards.

Pesticide poisoning is estimated to cause 355,000 deaths in developing countries annually (World Bank, 2008). It also results in pesticide residues becoming widespread in food products. In Ghana, for example, residues from six banned or restricted chemical pesticides – DDT, endosulfan, lindane, aldrin, dieldrin and endrin – have been found in food samples, and yet there is still no system of regular testing for food in domestic markets (Curtis, 2012). Research found widespread sale of banned pesticides in Ghanaian agro-dealer shops, a failure to observe the correct withdrawal period before harvest, almost total failure by farmers to use the right protective equipment when spraying, the storage of pesticides near to food stores contributing to widespread illness and fatalities and a lack of training from the Ministry of Food and Agriculture in the safe use of pesticides. Secondary pests also develop as the natural pest control balance is upset and biodiversity essential to the ecological functions that maintain agricultural productivity are damaged and destroyed, including the microfauna needed to maintain soil health and pollination. Despite these failings, large-scale development programmes have continued to support the agro-dealer sale of agrochemicals.

Few studies have actually attempted to quantify the damage of pesticide use in financial terms, but those that have indicate the costs are high. In Mali, the annual health costs associated with poisoning were up to $1.5 million, with an additional $8.5 million due to pesticide resistance and the destruction of natural pest control organisms. In Zimbabwe, acute health costs to each small-scale cotton farmer were equivalent to 45–83 percent of their annual pesticide expenditure (PAN UK, 2007).

In addition to the continuing use of banned chemicals, stockpiles of obsolete pesticides will require substantial resources to clear up. The African Stockpiles Programme has calculated that the 50,000 tonnes of obsolete pesticides in Africa will cost around US$150–175 million to clear up, with a further US$100 million on measures to prevent more accumulating in future (ASP, 2002). These figures do not include the cost of damage to human health, livestock or the environment from stockpiles leaking toxic products into soil, water or contaminating food or crops.

Neonicotinoid chemicals account for a third of global pesticide sales and are highly persistent in soil and water (Van der Sluijs et al., 2013). A variety of toxic effects on avian and fish species have been measured, from impaired immune function, reduced growth and reproductive success to mortality after consumption of even a few treated seeds (Gibbons et al., 2014). Studies of food stores in bee colonies have shown that routine exposure is repeated and chronic. Direct impacts include adverse effects on individual navigation, learning, food collection, longevity, resistance to disease and fecundity (Van der Sluijs et al., 2014). With pollination services worth an estimated $141 billion per annum (Kindemba, 2009), essential for over a third of global crop production and most of the vitamin and mineral-rich foods necessary to prevent nutritional deficiency, the lack of risk anticipation and regulation of corporate interests promoting these toxic chemicals represents a significant failure of conventional agricultural research and extension.

Loss of agricultural biodiversity

Maintaining and enhancing agricultural biodiversity – the wide variety of crop and livestock breeds that have been developed over millennia by farmers – is a key requirement in ensuring that current and future plant and livestock breeders have the genetic resources from which to draw in meeting changing environmental challenges and, in particular, changing climates. Unfortunately, agrobiodiversity has been substantially eroded by the widespread promotion of a limited selection of crop varieties, especially with respect to rice and wheat. Over the past 100 years, 90 percent of the biodiversity of the world's 20 major crops has been lost (IIED, 2012). This situation is replicated across small-scale farming areas. For example, 75 percent of India's rice varieties have been lost over the same period. Between 1975 and 1990 alone, Indonesia lost 1500 rice varieties (IAASTD, 2009). This has been driven by the increased commercialisation of seed production and supply focused on a narrow selection of traits designed to fit a chemical agricultural system that also requires sales of complementary chemical inputs.

This degradation affects the resilience of small-scale farmers in a number of ways – it reduces the genetic diversity from which they can draw on to cope with environmental change, both for direct use and as part of their own breeding and selection systems. However, it is also short-sighted from the perspective of commercial plant breeders, who benefit from wide genetic diversity from which to draw material for hybridisation and biotechnology, especially newer biotechnology techniques such as marker-assisted selection.

Increased genetic homogeneity increases susceptibility to crop pests and diseases, which are changing their range and intensity with climate change. The recent global spread of disease such as new, more virulent rust diseases in wheat (AGP-FAO, 2014) and banana xanthomonas wilt (Triparti et al., 2009) have demonstrated how disease threats will need to be addressed at least in part through plant breeding. Likewise, the traits needed to address climate change, such as drought and temperature resilience, depend on multiple gene associations and are therefore unlikely to be developed through breeding techniques aimed at modifying single or a few genes at a time (Gilbert, 2014). The options for breeding these new climate-resilient crop varieties are therefore compromised by continuing agrobiodiversity loss.

Exclusion from the research and extension agenda

The most extensive review of agricultural development to date, the International Assessment of Agricultural Knowledge, Science and Technology (AKST) for Development (IAASTD, 2009), concluded that despite being 'often ignored by formal AKST ... the combination of community-based innovation and local knowledge with science-based approaches in AKST holds the promise of best addressing the problems, needs and opportunities of the rural poor'. For this synergistic integration of local knowledge with appropriate technical and scientific support to generate results, an essential precondition is that external research and support respects local knowledge and is based on the expressed priorities of small-scale farmers. However, top-down approaches to rural extension have both persisted and often failed to generate sustainable changes in farming practices. The reasons are diverse, but typically include the use of pre-packaged extension messages that are not relevant to local conditions; farmers tend to distrust extension agents, who are not farmers themselves, and are often perceived as disrespectful, viewing small-scale farmers as passive service users in the process of 'modernising' agriculture.

A citizen jury of farmers from across West Africa (Pimbert et al., 2010) highlighted these typical problems – and while they acknowledged that the recurrent failure of research-generated crop varieties to find widespread uptake by farmers had motivated scientists to consult farmers more, they considered this relatively superficial. Various problems were identified, including a lack of meaningful involvement of farmers in on-farm research despite the importance in the process of the local crop varieties that farmers themselves had developed over generations.

Farmers were usually seen as little more than local labour and the involvement of women was limited despite their role in all production activities and their deeper knowledge of processing and nutritional factors. Research included the systematic use of inputs that were either not available or used very differently, such as chemical fertilisers, instead of those that farmers value, such as organic manure. The role of the private-sector input suppliers in driving policies that marginalise the involvement of small-scale farmers, such as legislation to ban the legal sale of local crop varieties that they have developed in order to artificially improve market conditions for their own products, was also highlighted.

The failure to address the research and extension priorities of women farmers has been viewed as especially acute, but is symptomatic of an inappropriately paternalistic paradigm. The basis of the problem has been a widespread bias in perceptions of the role of women in small-scale agriculture, who are often viewed, despite the evidence, as either not farmers at all or only involved in providing labour rather than making technical and managerial agricultural decisions (Madhvani & Pehu, 2010). Careers in agricultural research and extension are likewise biased against employment of women in most roles and especially those involved in direct advisory services. Only 15 percent of extension workers are women and only 5 percent of advisory services are actually received by women farmers, who make up 43 percent of the total workforce (FAO, 2008) and are the main activity implementers in many sectors, for example providing 90 percent of the labour for rice cultivation in South-East Asia and producing 80 percent of basic food items in sub-Saharan Africa (Meinzen-Dick et al., 2010).

More participatory, farmer group-based approaches can effectively overcome these failures and achieve significant impact through processes such as farmer field schools – which in East Africa achieved 50 percent participation of women, increasing their income by 189 percent (Davis et al., 2011) – and farmer participatory research, but these are typically implemented by the non-governmental sector and therefore tend to be highly localised in extent. For example, although India has a vibrant NGO sector spread widely across rural advisory services, it only covers about 1 percent of small-scale producers. Local government advisory extension systems generally do not fill the gap: farmers are either left unserved or look elsewhere. So 17 percent of farmers get their advice from other farmers and 13 percent from input suppliers (GFRAS, 2012).

Whole aspects of agricultural support that small-scale farmers prioritise, such as soil testing and post-harvest storage, are largely ignored. Despite being highlighted by the Food and Agriculture Organisation as a priority issue in 1981, by 2011, the World Bank declared that 'the world seems to have forgotten the importance of post-harvest food losses in the African grain sector' (World Bank, 2011). These can be as high as 40 percent, but the African

Postharvest Losses Information System (APHLIS) estimates generally range from 10 to 20 percent. The financial consequences of these losses can be severe, for example in Eastern and Southern Africa they amount to 13.5 percent of the total value of grain production, about US $1.6 billion per year. Relatively little attention is paid by research and extension to increase farmers' capacity to address this loss or assist with access to investment to upgrade farmer or farmer-group managed storage infrastructure that would increase their ability to mitigate crop loss and drought risks.

Developing country farmers rarely enjoy the climate services that their counterparts in temperate agriculture take for granted. Those that have had the opportunity to add meteorological forecasting and other climate services to their own, traditional forecasting systems have demonstrated the value of these to increased resilience. Increasing access to and understanding of seasonal forecasts with small-scale farmers in Zimbabwe resulted in yield increases of 9.4–18.7 percent compared to the typical range. Although smaller than the interannual variation in yield, this suggested that such forecasts were of positive value (Patt et al., 2004). Similar results from a longer-term process in Mali, supporting farmer use of rain gauges and short-term forecasts, increased yields across a range of staple crops by up to 48 percent (Helmuth et al., 2010).

More recent work in Kenya generally agreed with these yield improvement potentials, with farmers combining local bio-indicators such as the nature of acacia tree species' flowering patterns with seasonal and seven-day forecasts (Ewbank, 2012). In India, while farmers agreed with a 10–20 percent yield impact, they emphasised the cost savings achieved through using a tailored five-day agrometeorological forecast to better manage irrigation, biological pest control and organic soil management practices (Ewbank, 2014). Key to these approaches has been the participatory nature of learning, the interaction that has facilitated feedback from climate service users to meteorologist and agricultural advisors, the combination of weather and agricultural information and the importance of understanding how to manage the uncertainty inherent in forecast products.

Insecure land tenure

While the upsurge in international land acquisitions has gained prominent attention, it is only one manifestation of an increased level of tenure insecurity that undermines the capacity of small-scale farmers to make medium- to long-term resilience-building decisions and investments. Well-defined and secure property rights are fundamentally important in enabling farmers to invest – without them, few farmers will take the risk of improving land which they may soon lose (DFID, 2004). This security is especially important to resilience-building, where investments in measures such as terracing to reduce soil erosion, agroforestry to enhance resilience to increased temperatures, changing perennial crop enterprises such as coffee – 70 percent of which is small-scale farmer-produced (Panhuysen & Pierrot, 2014) – and water storage capacity to reduce vulnerability to drought loss typically require several years to generate a positive return.

Apart from post-apartheid land redistribution in South Africa and some land reform in Zimbabwe, however, few government-based initiatives currently recognise the need to improve the land tenure security of their small-scale farmers in order to facilitate this investment. The main direction of land reform is in the opposite direction – towards expropriating land managed by customary land tenure systems from small-scale farmers and pastoralists and allocating it to national and international commercial interests. In terms of foreign land deals, their closed-door nature has made determining exact amounts of land exchanged problematic. However, Land Matrix identifies 902 concluded land deals covering over 33 million hectares of land (Land Matrix, 2014), most of this in Africa.

Many of these have involved the arbitrary confiscation from small-scale farmers. Concerns have been raised over initiatives ranging from the purchase of small-scale farmer-operated land in Pakistan by Gulf interests to the allocation of 22,000 hectares of land in Tanzania to Swedish biofuel interests and 179,000 hectares of southern Sudan to a Norwegian forestry company. These have included the abolition of rural communities and their farms, the superficial extent and nature of local consultation, the lack of involvement of women in this process and the weak guarantees given for local development projects and employment opportunities (Centre for Human Rights and Global Justice, 2010).

In an assessment of three biofuel deals in Sierra Leone, farmers confirmed that they had lost access to most productive categories of land, the only exceptions being swamp and marginal river areas. The resources lost due to the favourable tax deals given to the foreign corporates involved, about US $18 million per annum, would have been sufficient to allow the Government to either increase the 2012 extension budget 13 times or the agricultural research budget by five times (Baxter, 2013).

Legislation designed to strengthen customary land rights can contain loopholes that allow land grabbing, such as the contradictions between the Tanzania Land Act 1999 and Village Land Act 1999, which enable the Government to claim lands unsettled or not farmed, including communal pasture and forest (Wily, 2010). Land acquisitions also include appropriation or diversion of other natural resources, especially water that small-scale farmers rely on. Many of these deals, however, fail to realise the benefits they anticipate. A World Bank study found many of the land-based investments reviewed had not only resulted in adverse local social impacts, such as local landholders not being consulted or paid adequate compensation, but also a lack of progress with implementation and commercial viability (Deininger & Byerlee, 2011).

The myth of low productivity

A recurring feature of the debate on the need for a climate 'smart' agriculture that can be intensified to meet the challenges of feeding an increasing global population is the assumed reduction in productivity that accompanies a switch from intensive use of chemical inputs to more sustainable, agro-ecological methods. Governments and their corporate partners assert that only more large-scale commercialisation with intensive use of fertilisers, pesticides and 'modern' crop varieties, often transgenetically modified, can deliver global food security in the face of an increasingly hostile climate. Experience suggests otherwise. For small-scale farmers in tropical and sub-tropical environments there is increasing evidence that the transition from conventional, chemical to sustainable or ecological agriculture is accompanied by increasing productivity, profitability and resilience and generates higher levels of soil carbon, thus mitigating agricultural emissions of greenhouse gases.

In a large-scale study of 360 reliable yield comparisons across 198 projects, an average yield increase of 79 percent was recorded after the adoption of sustainable agriculture. For the crops that small-scale farmers rely on, such as maize, millet, sorghum and a range of pulses, the average increase was over 100 percent (Pretty et al., 2006). A similar study examining the productivity improvements of farmers converting to organic or near-organic farming systems in Africa showed a 116 percent overall yield increase and a 128 percent increase for the projects in East Africa. These increases were typically accompanied by a greater diversity in food items that improved both nutritional quality and food security, increased farmer group activity that strengthened people's capacity to work together to solve common problems and improved access to markets (UNEP-UNCTAD, 2008).

Specific studies of sustainable techniques have likewise demonstrated their capacity to outperform conventional chemical systems. An assessment of Malawian farmers adopting conservation agriculture recorded average maize yields of 4450 kg per hectare compared with 1620 kg per hectare for those not adopting this method, an increase of 275 percent (Concern Universal, 2011). Farmer groups in Tanzania implementing farmer participatory research to evaluate crop varieties for adoption, together with enhanced soil fertility and pest management through use of manures, composts and locally produced bio-pesticides, demonstrated maize yield improvements of 124–162 percent and bean yield increases of 79 per cent. Terracing to reduce soil erosion control combined with manure use alone gave a 46 percent positive yield effect. This translated into substantial financial returns for the farmer groups involved.

Cost–benefit analysis demonstrated an internal rate of return of 56 percent (Ewbank et al., 2007a).

Farmers in Kiambu District of central Kenya achieved over four times the financial returns when switching from high-input conventional vegetable and flower production to cultivation of African indigenous green leaf vegetables, such as amaranthus and spider plant. This was associated with substantial reductions in pesticide expenditure, improved drought and flood resilience through soil quality enhanced by use of manure and diversified access to markets (Ewbank et al., 2007b). Before, farmers were completely reliant on expensive inputs and the arbitrary demands of agents for companies exporting horticultural products to Europe. After, farmers were able to free themselves to market indigenous vegetables through a variety of local channels, from Nairobi supermarket chains and independent retailers to local markets and traders, developing valuable group management and local marketing skills in the process.

The System of Rice Intensification (SRI), an approach that seeks to maintain or increase yields while reducing seed, agrochemical and water inputs and enhancing root growth and soil biodiversity, is now practiced by 600,000 farmers on about 1 million hectares in India. The average increase in income across eight countries using SRI (Bangladesh, Cambodia, China, India, Indonesia, Nepal, Sri Lanka and Vietnam) has been about 68 percent, with yield increases of 17–105 percent and decreases in water requirement of 24–50 percent (FAO, 2010). Production costs are typically 20 percent lower, largely due to the reduced need for chemical fertilisers (Uphoff, 2007), which in turn has the potential to reduce farmer indebtedness and the acute social problems that this has caused. Moving from intensive pesticide use to integrated pest management (IPM) has also generated positive results. With a supportive policy environment, a switch to IPM in Indonesia reduced expenditure on pesticides by 75 percent and increased yields by 25 percent from 1986 to 2001 (FAO, 2009b).

Enhanced resilience has been recognised with respect to cyclone occurrence, perhaps the most rigorous stress a farming system can be subjected to. In a post-Hurricane Mitch assessment of small-scale farmers in Central America, organic farmers suffered significantly less soil loss than conventional farms on all but the steepest slopes (where there was no change) and had a shorter recovery time. Their plots had 20–40 percent more topsoil, greater soil moisture, less erosion and experienced lower economic losses than their conventional neighbours (Holt-Gimenez, 2002).

The higher economic performance of sustainable agricultural approaches has been confirmed through wider combined survey studies. An FAO meta-study comparative analysis of organic and non-organic farming systems demonstrated higher economic profitability for organic systems in the 'overwhelming majority' of cases. Comparisons in developing countries were particularly significant, with a range of cases from 48 percent higher incomes for ecological rice farmers in the Philippines to 52–63 percent higher gross margins from organic cotton in India. The conclusions with respect to resilience were equally important as 'organic crop yields are higher in cases of bio-physical stress (e.g. drought)' (Nemes, 2009).

Priorities for local, national and international incentives to enhance agricultural resilience

Transforming support for small-scale farmers, women and men, to both respond to their priorities as the 'clients' of agricultural research and extension and to effectively build on these through collaborative approaches that focus on intensifying knowledge rather than just chemical inputs should not be the paradigm shift it seems. The discourses of *Farmer First* (Chambers, 1989) and its successors, and the approaches they developed such as farmer participatory research and farmer field schools, have been familiar agricultural development tools, used highly effectively for several decades. Unfortunately, the preoccupation with 'magic bullet' top-down solutions continue to dominate both public and private sector agricultural development, undermining environmental sustainability and farmer resilience.

Clearly, a substantial redesign and reorientation of agricultural research and extension is needed to a participatory, demand-led approach which explicitly addresses the needs of the most vulnerable, supporting the scale-up of ecologically sustainable agricultural technologies that show both better productivity and enhanced profitability over chemically intensive models. Inevitably this will require more focus on context-specific approaches, but small-scale farmers themselves highlight some shared priorities from region to region. These include increased access to climate services and early warning systems to improve their management of climate hazards and climate change. Seed production systems need to work with farmers' own selection and development methods, rewarding them for diversity conservation and compensating rather than abusing their genetic property rights. Other services that can enhance resilience, such as access to regular and reliable soil testing, integrated pest management, agroforestry techniques, restoring degraded land and water resources and seed and crop storage are likewise frequently raised as priorities but, as has been shown, often neglected.

A much higher profile needs to be given to both addressing the toxic environmental legacy of chemical agriculture and further scaling-up sustainable agriculture by global research networks such as the CGIAR, supporting the conclusions of the IAASTD and the call of the UN Special Rapporteur on the Right to Food (De Schutter, 2010) for the adoption of agroecological methods. Stronger regulation of pesticides and chemical fertilisers and the progressive replacement of input subsidies with resilience-based subsidies are urgent first steps. Some sustainable techniques, such as conservation agriculture, SRI and IPM have already demonstrated their effectiveness at scale and their potential for further scale up. Transferring further resources from chemical to poorly funded sustainable agricultural research would generate more.

Investment increases in these and other resilient agricultural approaches is clearly needed and should be a priority for agricultural development, climate change adaptation and loss and damage-focused funding schemes. Likewise, these funding sources need to recognise the interrelated nature of short-, medium- and long-term risk and develop the flexibility needed that ensures, for example, that arbitrary constraints are not imposed on agricultural development funding that restricts their reorientation to droughts, floods and other shocks when needed. Agricultural aid funding needs to be transformed from top-down, needs-based to bottom-up, risk-based approaches that empower small-scale farmers, and especially women farmers, to plan and support them to implement resilience-building based on their priorities.

Farmer groups, cooperatives and associations are growing in importance but need a transformative level of support so that they can play the role their members aspire for them. This should not only cover agricultural livelihoods but increasingly seek to use agriculture as a local platform to enable diversification into other resilient rural livelihoods, such as renewable energy provision that itself can provide, through reduced energy poverty, increased local livelihood options. These will help rural households to have more than a choice between a neglected, vulnerable agricultural livelihood and moving to an urban slum. Farmer organisations also need management as well as technical capacity support so that they can effectively manage their resources and accountably promote the interests of their members.

This means high-quality services in organisational and technological development, reversing the long-term disinvestment by most countries in their agricultural extension systems. However, these extension systems need to re-emerge in a new way that enables them to carry out this more demand-led, facilitating role for farmers and farmer groups, often acting as the connector between farmers and sources of resilience-enhancing support, such as national meteorological services and university expertise. This will also mean strengthening farmer-to-farmer linkages, maximising the known effectiveness of these connections in terms of knowledge and experience sharing. Connections with related functions, such as civil protection and disaster risk reduction, should also be better integrated so that all facets of resilience are strengthened effectively and efficiently.

The existing policy reticence that dictates land tenure reform as a 'non-negotiable issue' needs to be reversed. Instead, pragmatically strengthening land tenure, and especially common property mechanisms, in favour of small-scale farmers, pastoralists and their associations would provide the tenure security needed to facilitate longer-term resilience-building investments. Protecting small-scale farmers from non-transparent, large-scale land acquisition and setting and implementing rigorous environmental, consultative and labour use standards for any large-scale land purchase are therefore essential.

As well as improving small-scale farmers' own capacity to manage risks, this transformation ultimately requires political courage. Unsurprisingly, it is the policies and practices put in place by the short-term interests of corporate promoters of chemical inputs and large-scale land acquisition in pursuit of quick profits, coupled with their local political allies that remain some of the largest barriers to increasing small-scale farmer resilience. Sustainable, climate resilient agriculture uses less chemical input and thrives on small- to medium-scale farms with secure land tenure, both features that deliver less profit to traditional input suppliers and reduce the scope for land grabbing facilitated by the state.

The current global food system currently produces enough nutrition for 10 billion people (Holt-Gimenez et al., 2012), but at substantial cost in terms of human health, climate change and environmental degradation. Post-harvest losses in developing countries and consumption waste in developed countries conspire to generate twin problems of increasing malnutrition and food insecurity on the one hand and rising levels of obesity on the other. Global population is expected to reach 9 billion by 2050 and in some food insecure regions, such as Africa, is projected to double from just over 1 billion in 2010 to about 2 billion (HLPE, 2012). If the challenges of global food security, nutritional health, climate change and environmental degradation are to be addressed, policies and practices that promote sustainable, climate-resilient agriculture are essential to maintaining poverty reduction in the twenty-first century.

References

AGP-FAO. (2014). Wheat Rust Disease Global Programme. www.fao.org/agriculture/crops/core-themes/theme/pests/wrdgp/en/.

Ali, M. & Byerlee, D. (2002). Productivity growth and resource degradation in Pakistan's Punjab: a decomposition analysis. *Economic Development and Cultural Change*, **50**(4), 839–863.

ASP (2002). Programme overview. Africa Stockpiles Programme. www.africastockpiles.org

Baligar, V. & Bennett, O. (1986). NPK fertiliser efficiency: a situation analysis for the tropics. *Fertilizer Research*, **10**(2), 147–164.

Baxter, J. (2013). *Who is Benefitting? The social and economic impact of three large scale land Investments in Sierra Leone*. London: Action for Large-scale Land Acquisition Transparency (ALLAT)/Christian Aid.

BEST. (2012). Summary of findings. Berkeley Earth Surface Temperature Group. http://berkeleyearth.org/

Bloom, A., Burger, M., Rubio Asensio, J.S. & Cousins, A.B. (2010). Carbon dioxide enrichment inhibits nitrate assimilation in wheat and Arabidopsis. *Science*, **328**, 899–903.

Brown, L. (2012). *Full Planet, Empty Plates: the new geopolitics of food scarcity*. New York, NY: W.W. Norton & Company.

CASA/Christian Aid India. (2012). *Participatory Vulnerability and Capacity Assessment Consolidated Report*. London: Christian Aid.

Castaneda, A., Raul, A., Doan, D.T.T., et al. (2016). Who are the Poor in the Developing World? Policy Research working paper, no. WPS 7844. Washington, DC: World Bank Group.

Chambers, R. (1989). *Farmer First: farmer innovation and agricultural research*. Bradford: ITDG Publishing.

Christian Aid Burkina Faso/Serge Sedogo. (2011). *Programme Partnership Agreement Baseline Survey*. London: Christian Aid.

Christian Aid Central America/Jaime Guillen. (2012). *Participatory Vulnerability and Capacity Assessment Baseline Report*. London: Christian Aid.

Christian Aid Malawi. (2012). *Enhancing Community Resilience Programme Baseline Survey*. London: Christian Aid.

Central Ground Water Board. (2009). *Dynamic Ground Water Resources of India*. New Delhi: Ministry of Water Resources, Government of India.

Centre for Human Rights and Global Justice. (2010). *Foreign Land Deals and Human Rights: case studies on agricultural and biofuel investment*. New York, NY: NYU School of Law.

Concern Universal. (2011). *Conservation Agriculture Research Study*. Blantyre: Concern Universal Malawi.

Curtis, M. (2012). *Ghana's Pesticide Crisis*. Cambridge, MA: Northern Presbyterian Agricultural Services and Partners.

Davis, K., Nkonya, E., Kato, E., et al. (2011). Impact of farmer field schools on agricultural productivity and poverty in East Africa. *World Development*, **40**(2), 402–413.

De Schutter, O. (2010). *Report submitted by the Special Rapporteur on the Right to Food on Agro-ecology*. 3. Available from www2.ohchr.org/English/issues/food/docs/a-hrc-16-49.pdf

Deininger, K. & Byerlee, D. (2011). *Rising Global Interest in Farmland: can it yield sustainable and equitable benefits?* Washington, DC: The World Bank.

DFID. (2001). *The Sustainable Livelihoods Approach*. London: Department for International Development, UK Government.

DFID. (2004). *Agricultural Growth and Poverty Reduction*. London: Department for International Development, UK Government.

Ewbank, R. (2011). *Climate Change Review of India Partners*. London: Christian Aid.

Ewbank, R. (2012). *Developing Climate Services in Kenya*. London: Christian Aid.

Ewbank, R. (2014). *Developing Climate Services in India*. London: Christian Aid.

Ewbank, R., Kasindei, A., Kimaro, F. & Slaa, S. (2007a). Farmer participatory research in Northern Tanzania. FARM-Africa Working Paper 11. London: FARM Africa.

Ewbank, R., Nyang, M., Webo, C. & Roothaert, R. (2007b). Socio-economic assessment of 4 MATF-funded projects. FARM-Africa Working Paper 8. London: FARM Africa.

FAO. (2008). *Gender and Agriculture*. Rome: Food and Agriculture Organisation.

FAO. (2009a). *Scaling-up Conservation Agriculture in Africa: strategy and approaches*. Addis Ababa: FAO Sub-regional office for East Africa.

FAO. (2009b). *Increasing Crop Production Sustainably – the perspective of biological processes*. Rome: Food and Agriculture Organisation.

FAO. (2010). Biodiversity for Food and Agriculture. Contributing to food security and sustainability in a changing world. Outcomes of an Expert Workshop, Platform for Agribiodiversity Research and Food and Agriculture Organisation.

FAO (2013) *Guidelines to Control Water Pollution from Agriculture in China: decoupling water pollution from agricultural production* – FAO

Water Reports 40. Rome: Food and Agriculture Organisation.

Folke, C., Carpenter, S., Elmqvist, T., et al. (2002). *Resilience and Sustainable Development: building adaptive capacity in a world of transformations.* Scientific Background Paper on Resilience for the process of The World Summit on Sustainable Development on behalf of The Environmental Advisory Council to the Swedish Government. Stockholm: Environmental Advisory Council.

Gallup, J., Radelet, S. & Warner, A. (1997). Economic growth and the income of the poor. Harvard Institute for Economic Development, Discussion Paper No. 36.

GFRAS. (2012). Fact sheet on extension services – position paper. Lausanne: Global Forum for Rural Advisory Services.

Gibbons, D., Morrissey, C. & Mineau, P. (2015). A review of the direct and indirect effects of neonicotinoids and fipronil on vertebrate wildlife. *Environmental Science Pollution Research*, **22**, 103–118.

Gilbert, N. (2014). Cross-bred crops get fit faster – genetic engineering lags behind conventional breeding in efforts to create drought-resistant maize. *Nature*, **513**, 292.

Greenpeace China. (2010). *The Real Cost of Nitrogen Fertiliser in China.* Beijing: Greenpeace.

Helmuth, M., Diarra, D.Z., Vaughan, C. & Cousin, R. (2010). *Increasing Food Security with Agrometeorological Information: Mali's national meteorological service helps farmers manage climate risk.* Washington, DC: World Resources Report.

HLPE. (2012). Food Security and Climate Change. Rome: a report by the High Level Panel of Experts on Food Security and Nutrition of the Committee on World Food Security.

Holt-Gimenez, E. (2002). Measuring farmers' agroecological resistance after Hurricane Mitch in Nicaragua: a case study in participatory, sustainable land management impact monitoring.

Agriculture, Ecosystems and Environment, **93**, 87–105.

Holt-Gimenez, E., Shattuck, A., Altieri, M., Herren, H. & Gliessman, S. (2012). We already grow enough food for 10 billion people . . . and still can't end hunger. *Journal of Sustainable Agriculture*, **36**(6), 595–598.

IAASTD. (2009). *Agriculture at a Crossroads – the International Assessment of Agricultural Knowledge, Science and Technology for Development*, edited by B.D. McIntyre, H.R. Herren, J. Wakhungu & R.T. Watson. Washington, DC: Island Press.

IFPRI. (2005). *The Future of Small Farms: proceedings of a research workshop.* International Food Policy Research Institute, Wye, UK, 26–29 June 2005. Washington, DC: IFPRI.

IIED (2012). Strengthening biocultural innovation systems for food security in the face of climate change. Planning Workshop Report, International Institute for Environment and Development and Centre for Chinese Agricultural Policy.

India Meteorological Department. (2009). *Severe Cyclonic Storm Aila: a preliminary report.* New Delhi Regional Specialised Meteorological Centre – Tropical Cyclone.

IPCC. (2012). *Managing the Risks of Extreme Events and Disasters to Advance Climate Change Adaptation.* A Special Report of Working Groups I and II of the Intergovernmental Panel on Climate Change, edited by C.B. Field, V. Barros, T.F. Stocker, et al. New York, NY: Cambridge University Press.

IPCC. (2014). Summary for policymakers. In: *Climate Change 2014: mitigation of climate change. Contribution of Working Group III to the Fifth Assessment Report of the Intergovernmental Panel on Climate Change*, edited by O. Edenhofer, R. Pichs-Madruga, Y. Sokona, et al. New York, NY: Cambridge University Press.

Kindemba, V. (2009). *The Impact of Neonicotinoid Insectisides on Bumblebees, Honey Bees and Other Non-target Invertebrates* Peterborough: Buglife The Invertebrate Conservation Trust.

Kotschi, J. (2013). *A Soiled Reputation: adverse impacts of mineral fertilizers in tropical agriculture*. Berlin: Heinrich-Böll-Stiftung, WWF Germany.

Land Matrix. (2014). Land Matrix Online Database on Land Deals. www.landmatrix.org/

Leakey, A.D.B., Ainsworth, E.A., Bernacchi, C.J., et al. (2009). Elevated CO_2 effects on plant carbon, nitrogen and water relations – six important lessons from FACE. *Journal of Experimental Botany*, **60**(10), 2859–2876.

Madhvani, S. & Pehu, E. (2010). *Gender and Governance in Agricultural Extension Services: insights from India, Ghana, and Ethiopia*. Agricultural and Rural Development notes 53. Washington, DC: World Bank.

Mani, M., Markandya, A., Sagar, A. & Strukova, E. (2012). *An Analysis of Physical and Monetary Losses of Environmental Health and Natural Resources in India*. Policy Research Working Paper, 6219. World Bank, South-East Asia Region, Disaster Risk Management and Climate Change.

Mateo-Sagasta, J. & Burke, J. (2011). *Agriculture and Water Quality Interactions: a global overview*. SOLAW Background Thematic Report – TR08. Rome: FAO.

Meinzen-Dick, R., Quisumbing, A., Behrman, J., et al. (2010). *Engendering Agricultural Research*. International Food Policy Research Institute Discussion Paper 00973.

Montgomery, D.R. (2007). Soil erosion and agricultural sustainability. *Proceedings of the National Academy of Sciences*, **104**(33), 13268–13272.

Nemes, N. (2009). *Comparative Analysis of Organic and Non-Organic Farming Systems: a critical assessment of farm profitability*. Rome: FAO.

Ngowi, A.V.F., Mbise, T.J., Ijani, A.S.M., London, L. & Ajayi, O.C. (2007). Smallholder vegetable farmers in northern Tanzania: pesticides use, practices, perceptions, cost and health effects. *Crop Protection*, **26**(11), 1617–1624.

PAN UK. (2007). Hidden costs of pesticide use in Africa. Food and Fairness Briefing No. 2.

Panhuysen, S. & Pierrot, J. (2014). *Coffee Barometer 2014*. The Netherlands: Hivos.

PAR-FAO. (2010). *Biodiversity for Food and Agriculture. Contributing to food security and sustainability in a changing world*. Rome: FAO.

Park, S., Croteau, K.A., Boering, D.M., et al. (2012). Trends and seasonal cycles in the isotopic composition of nitrous oxide since 1940. *Nature Geoscience*, **5**, 261–265.

Park Williams, A. & Funk, C. (2011). A westward extension of the warm pool leads to a westward extension of the Walker circulation, drying eastern Africa. *Climate Dynamics*, **37**, 2417–2435.

Patt, A., Siarez, P. & Gwata, C. (2004). Effects of seasonal climate forecasts and participatory workshops among subsistence farmers in Zimbabwe. *Proceedings of the National Academy of Sciences*, **102**(35), 12623–12628.

Pelletier, N., Audsley, E., Brodt, S., et al. (2011). Energy intensity of agriculture. *Annual Review of Environment and Resources*, **36**, 223–246.

Pimbert, M., Boukary, B., Berson, A. & Tran-Thanh, K. (2010). *Democratising Agricultural Research for Food Sovereignty in West Africa*. Bamako and London: IIED, CNOP, Centre Djoliba, IRPAD, Kene Conseils, URTEL.

Pretty, J.N., Noble, A.D., Bossio, D., et al. (2006). Resource conserving agriculture increases yields. *Environmental Science and Technology*, **40**(4), 1114–1119.

Rabalais, N.N., Diaz, R.J., Levin, L.A., et al. (2010). Dynamics and distribution of natural and human-caused hypoxia. *Biogeosciences*, **7**, 585–619.

Triparti, L., Mwangi, M., Abele, S., et al. (2009). Xanthomonas wilt: a threat to agricultural production in East and Central Africa. *Plant Disease*, **93**(5), 440–451.

UNEP-UNCTAD. (2008). *Organic Agriculture and Food Security in Africa*. Geneva: United Nations Environment Programme/United Nations Conference on Trade and Development.

Uphoff, N. (2007). *The System of Rice Intensification: using alternative cultural practices to increase rice production and profitability from existing yield potentials*. International Rice Commission Newsletter, 55. Rome: FAO.

Van der Sluijs, J.P., Simon-Delso, N., Goulson, D., et al. (2013). Neonicotinoids, bee disorders and the sustainability of pollinator services. *Current Opinion in Environmental Sustainability*, **5**, 293–305.

Van der Sluijs, J.P., Amaral-Rogers, V., Belzunces, L.P., et al. (2014). Conclusions of the Worldwide Integrated Assessment on the risks of neonicotinoids and fipronil to biodiversity and ecosystem functioning. *Environmental Sciences and Pollution Research*, **22**(1), 148–154.

Weinburg, J. (2009). An NGO Guide to Hazardous Pesticides and SAICM. Berkeley, CA: IPEN.

Wily, L.A. (2010). Whose land are you giving away, Mr President? Paper presented to the Annual World Bank Land Policy & Administration Conference, Washington DC, 26–27 April 2010.

World Bank. (2007). Investment in Agricultural Water for Poverty Reduction and Economic Growth in Sub-Saharan Africa – a collaborative programme of ADB, FAO, IFAD, IWMI and the World Bank.

World Bank. (2008). *Agriculture for Development – World Development Report*. Washington, DC: The World Bank.

World Bank. (2011). Missing Food – the case of post-harvest grain losses in Sub Saharan Africa. Report No. 60371-AFR. Washington, DC: The World Bank.

WMO. (2013). *The Global Climate 2001–10. A decade of climate extremes*. Geneva: World Meteorological Organisation.

Zavala, J.A., Casteel, C.L., DeLucia, E.H. & Berenbaum, M.R. (2008). Anthropogenic increase in CO_2 compromises plant defence. *Proceedings of the National Academy of Sciences*, **105**(13), 5129–5133.

CHAPTER EIGHT

Modern biotechnology and sustainable intensification: chances and limitations

ROLF MEYER
Institute for Technology Assessment and Systems Analysis,
Karlsruhe Institute of Technology

Introduction

Sustainable intensification is an evolving concept which aims to produce more food from the same area of land while reducing the environmental impacts, under socially and economically beneficial conditions (Royal Society, 2009; Garnett et al., 2013). The connection to resilience in agriculture is that sustainable intensification tries to answer how environmental (e.g. climate change) and economic shocks (e.g. input and commodity price volatility) can be handled and how agricultural production (e.g. increased food production) and other ecosystem services can be reorganised and improved (see Hodge, Chapter 10).

Resilience is an approach for understanding the dynamics of social–ecological systems, emphasising non-linear dynamics. The resilience approach tries to understand how periods of gradual change interplay with periods of rapid change (Folke, 2006) or, in other words, how path dependencies are interlinked with path creation. Resilience includes three different components as exemplified for farm management (Darnhofer, 2014):

- buffer capability is the ability to assimilate a perturbation without a change in structure or function;
- adaptive capability describes the ability to adjust by incremental changes, building on established structures and functions, marked by path dependencies;
- transformative capability relates to the ability to implement radical changes which imply a transition to a new system where a different suite of factors becomes important in the design and implementation of new strategies.

Arrangements of knowledge, technologies and management approaches influence the resilience of farms and the agricultural sector. The objectives of this chapter are to place possible contributions of modern biotechnologies in plant breeding in the context of overall approaches for sustainable intensification, to discuss the appropriateness for different European farming systems and to assess their impacts on resilience.

The context of modern biotechnology applications

As a baseline, this chapter gives short introductions to approaches for sustainable intensification, European farming systems and modern biotechnologies as part of plant breeding.

Approaches for sustainable intensification

Multiple pathways for agricultural research and technology development which may contribute to sustainable intensification are under discussion (Royal Society, 2009). Overall, improved crop production under changing environmental conditions can be achieved by plant breeding on the one hand and by improved technologies and management systems of crop production on the other hand (Meyer et al., 2013). In this context, general objectives of plant breeding are to increase yield potentials of crops and to improve safeguarding of yields. Improved technologies and management systems of crop production include different objectives:

- reducing yield gaps,[1]
- improving input use efficiency, and
- increasing the site-specific yield potential.

In general (and simplified), two types of approaches can be distinguished (Meyer et al., 2013).

- High-tech approaches focus on improving input use efficiency and are aimed at specialised crop production with high external inputs. An important example is precision agriculture; this approach seeks to apply the right treatment in the right place at the right time by taking into account in-field variations of soil and crop. Various advanced technologies such as satellite-supported positioning systems, yield mapping, remote sensing, sensor technologies, geo-information systems, as well as variable rate application techniques and decision support systems are used.
- Agroecological approaches aim to reduce yield gaps and to improve the site-specific yield potentials. In this context, maintenance and enhancement of soil fertility is an important issue. Diversified farming systems at farm and landscape levels are sought to intentionally include functional biodiversity at multiple spatial and/or temporal scales in order to maintain ecosystem services that provide critical inputs to agriculture (Kremen et al., 2012). Different crop production systems such as conservation agriculture,

---

[1] Yield gap is the difference between the average farmer's yield and the yield potential which can be achieved with optimal management of all yield-restricting production factors such as seed date, plant population, nutrition supply, protection against pests, disease and weed competition. Yield potential is assessed by model simulations, field experiments or maximum farmer yields.

agroforestry systems, integrated crop–livestock systems, and organic farming include agroecological approaches.

European farming systems

Farming systems are characterised – *inter alia* – by farm size, production intensity, specialisation and integration in food chains. They apply different crop management practices and new technologies are often not equally suitable for each farming system. Thus, they also differ in their environmental impacts. A simplified scheme of farming systems in the EU comprises the following (Meyer et al., 2013):

- *Extensive small-scale, semi-subsistence farming*: over 40 percent of all holdings in the EU-27 produce food for family and relatives, with only surplus going to the market. This farming system is only of any importance in the new Member States and Mediterranean countries, with Romania being the most important one. The small-scale farms apply extensive production methods, partly without external inputs.
- *Extensive farming in less-favoured areas*: 54 percent of all farms in the EU-27 are located in less-favoured areas. High coverage of less-favoured areas (over 50 percent of the total agricultural area) is designated in 12 Member States. Farming in less-favoured areas is characterised by extensive production systems and traditional land-use systems, often based on grazing livestock. However, cereal production is also important in less-favoured areas.
- *Medium intensive, mixed farming systems*: mixed farming systems combine crop and livestock production in different patterns and have a relatively low level of specialisation. Around 13 percent of all farms in the EU-27 are mixed farms. Mixed farming systems occupy over 10 percent of the total utilised agricultural area in Belgium, Czech Republic, Denmark, Germany, France, Latvia, Lithuania, Hungary, Poland, Portugal, Romania, Slovenia and Slovakia.
- *Intensive, larger-scale crop farming*: the regions with concentrated cereal and specialised crop production are at the same time the areas with a high degree of larger-scale farms. Larger-scale farming, based on high external inputs, is associated with lowland areas with high productivity. High-input farm types are predominant in the Netherlands, Belgium, south-eastern England, northern France, north-western Germany, northern Italy and northern Greece.
- *Large-scale corporate farming*: large-scale corporate farming comprises production cooperatives and various types of farming companies. They are the result of the transition process in Central and Eastern Europe since 1990. Corporate farms hold over 50 percent of the total agricultural area in Bulgaria, Czech Republic and Slovakia. Large corporate farms tend to specialise in cereals and oil crops.

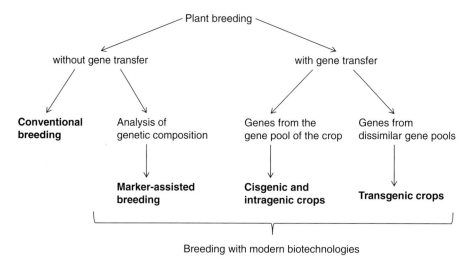

Figure 8.1 Current approaches of modern biotechnology used in plant breeding.

These very different settings of farming demand well-adapted approaches to sustainable intensification.

Modern biotechnology and plant breeding

In recent decades, modern biotechnology has opened new possibilities in plant breeding. From the modern biotechnologies used in plant breeding (Figure 8.1), two approaches in particular have potential relevance for agricultural resilience:

- breeding with genetic modification to introduce new genetic variation and
- marker-assisted breeding to select favourable genotypes.

In addition, modern biotechnology includes different tissue culture methods, which will not be discussed here.

Breeding with genetic modification
With genetic engineering, it has become possible to transfer single genes of interest from any genome. Since the 1990s, varieties created by genetic modification, so-called genetically modified crops (GM crops), have become available. Worldwide, the area cultivated with GM crops has increased continuously, to around 190 million ha in 2017. They are used mainly in North and South America, China and India. Currently, two traits (herbicide tolerance, insect resistance) and four major cash crops (cotton, maize, rapeseed, soybean) are dominating the area cultivated with GM crops (ISAAA, 2017). Only one GM crop (Bt maize MON810) is authorised in the EU for cultivation; the area cultivated with GM crops in the EU is very restricted.

After the introduction of a new trait by genetic engineering, conventional breeding methodologies remain indispensable for further breeding steps (Meyer et al., 2013).

Recently, a set of new GM technologies has emerged, also for the creation of single genetic variations. The breeding of cisgenic and intragenic crops belongs to these new approaches in development. Both follow the principle that the gene of interest originates from the species' own gene pool (same plant species or a closely related species). This means that, in principle, the gene transfer could also be arranged using classical breeding methods, although this would take much more time (Müller-Röber et al., 2013). These new plant breeding techniques aim partly at more targeted modifications and are associated with the hope of higher consumer acceptance, but have an uncertain regulatory status and are 'tiptoeing' around transgenic and GM regulation (Lusser et al., 2011; Waltz, 2012).

The discussion in Europe about the use of GM crops for food and feed is highly controversial and the future introduction of GM crops in European agriculture is very uncertain. Forecasting future regulation and use of GM crops in the EU is therefore associated with high uncertainties and is not part of this contribution. The aim here is to identify possible impacts of GM crops on agricultural resilience in general. The authorisation requirements for GM crops and nearly 20 years of practical experience have led to a broad spectrum of publications on the agronomic application and impacts of GM crops.

Marker-assisted breeding
The best individuals of an initial variation must be identified and selected in the second major step of any breeding process. Marker-assisted breeding opens new possibilities to shift from phenotypic screening to more genotype-based methods, such as marker-assisted selection (MAS). Their general approach is to analyse the DNA composition of plants and identify individuals with the best genetic characteristics for particular traits. At the moment, MAS is widely applied by major breeding companies for different crop plants. MAS is mainly used for breeding targets such as biotic and abiotic stress resistance, classification of gene pools or quality assurance in seed production (Meyer et al., 2013). In marker-assisted pyramiding approaches, molecular markers are used to incorporate not only one but several genes in one genotype (Collard & Mackill, 2008).

Gene-based selection methods are gaining more and more importance due to the rapid progress in the gene sequencing and identification sector. Gene-based selection methods allow in most cases a much more precise and effective selection in breeding programmes and will increase the accuracy and success of breeding, especially in combination with improved modern phenotyping methods (Meyer et al., 2013). Varieties produced with marker-assisted breeding technologies are not labelled as such. Hence, information on agronomic impacts is very restricted and must be derived by inductive conclusions.

Farming systems, crop production approaches and breeding with modern biotechnology: how does it all fit together?

In the following, the relevance of different breeding approaches with modern biotechnology is compared with crop production approaches, and their resilience outcomes are assessed. Then the appropriateness of crop varieties based on modern biotechnology is discussed for the European farming systems considered earlier.

Breeding with modern biotechnology and crop production approaches

Herbicide tolerance and weed management

Herbicide tolerance is the dominant trait in GM crops, but non-GM breeding approaches for herbicide tolerance are also available. The main attractiveness for farmers is that herbicide tolerance makes weed management easier and saves time. Additionally, it is an element in reduced or no-tillage systems. The range of weed management tools employed in GM crops with herbicide tolerance is markedly reduced (Powles, 2008).

The development of herbicide resistance in weeds is in principle a risk of weed management strategies with herbicides. Heavy reliance on a single herbicide – glyphosate – in different GM crops has placed weed populations under increasingly intense selection pressure (Owen, 2008; Duke & Powles, 2009). Glyphosate-resistant weeds have now been found in 18 countries worldwide, with significant impacts in Brazil, Australia, Argentina and Paraguay. For example, glyphosate-resistant Palmer amaranth (*Amaranthus palmeri*) has become a severe problem in cotton cultivation of the southeastern USA since the mid-2000s (Gilbert, 2013).

In response to resistance problems, adaptive measures are being undertaken by farmers and the seed industry. Farmers respond by increasing herbicide application rates, making multiple applications of herbicides and mixing glyphosate with additional herbicide active ingredients (Benbrook, 2012). The impacts on herbicide use and thus on the environment are contested. It is generally believed that the overall toxicity of herbicides has declined because glyphosate is more environmentally benign than the herbicides it replaced (Ervin et al., 2010). Regarding the total amount of herbicide used, an assessment estimates that herbicide-tolerant GM crops led to a considerable increase in herbicide use in the USA between 1996 and 2011 (Benbrook, 2012). In plant breeding with genetic engineering, Monsanto and other biotechnology companies are developing new herbicide-resistant GM crops that work with different herbicides (stacked traits of herbicide tolerance), which they expect to commercialise within a few years (Gilbert, 2013). Criticisms of this strategy are that GM crops with stacked herbicide tolerance are likely to increase the severity of resistant weeds and will facilitate a significant increase in herbicide use (Mortensen et al., 2012).

From the above, we can see that in the case of weed resistance, farmers have buffer capability by short-term changes of herbicide application regimes and biotechnology companies have adaptive capability by mid-term development of combined herbicide tolerance. However, addressing the herbicide-resistant weed problem only with herbicides could hinder transformation and produce a 'lock-in' on high external inputs (Ervin et al., 2010). This may lead to critical situations because the increase in the number of multiple resistant weeds is confronted with a dramatic slowdown in discovery and development of new herbicide active ingredients in recent decades. Finally, the short-term fix will continue to discourage public research and extension in integrated weed management (Mortensen et al., 2012).

Transformation of weed management is difficult due to very different innovation systems: development of herbicides and herbicide-tolerant crops is highly centralised and done by the private sector, representing a common solution for different farming situations. In contrast, integrated weed management based on agroecological approaches includes different measures such as crop rotation, cover crops, competitive crop cultivars, mechanical weed control and targeted herbicide applications. Such approaches need on-farm adaptation and are knowledge-intensive practices, not based on saleable products. They need ongoing public research, combined with effective extension (Mortensen et al., 2012).

Insect resistance and pest management
GM crops which produce insecticidal toxins from the bacterium *Bacillus thuringiensis* (Bt) are the only currently available GM crop with insect resistance. The refuge strategy has been the primary approach used worldwide to delay pest resistance to genetically modified Bt crops and has been mandated in the USA, Australia and elsewhere (Tabashnik et al., 2013). Typically, the refuge is an area of conventional crop not expressing the Bt trait planted within a certain distance of the Bt field, in order that the rare resistant pest surviving on Bt crops will mate with the relatively abundant susceptible pest nearby, slowing down the rate of development of insect resistance (Head & Greenplate, 2012).

Field-evolved resistance associated with reduced efficacy of Bt crops has been reported for five major target pests in eight countries around the world, parallel to the increased area planted to Bt crops. Resistance can evolve in as few as two years under the worst circumstances; however, under the best circumstances, efficacy can be sustained for 15 years and more (Tabashnik et al., 2013). A common pattern for resistance development is continuous maize cultivation and continued use of one specific Bt trait.

First-generation Bt crops were based on a single gene producing one Bt toxin. As an adaptive measure to resistance problems, they are increasingly replaced by second-generation Bt crops, named pyramids, that produce two or more distinct Bt toxins active against a particular pest. This is based on the assumption that selection for resistance to one Bt toxin does not cause cross-resistance to other toxins in the pyramid, so that insects resistant to all toxins are extremely rare. However, when a pyramid of two toxins is adopted only after resistance is no longer rare to one of the toxins, the benefits of this approach seem to be greatly reduced (Tabashnik et al., 2013). As an adaptive measure at the farm level, it has been proposed that a greater diversity of practices be used, such as crop rotation, cultivation of different Bt events and seed mixtures, and alternating cultivation of non-Bt crops with insecticides (Gassmann et al., 2012).

Many studies show decreases in the amount of insecticide and/or number of insecticide applications used on Bt crops compared with conventional crops in Argentina, Australia, China, India and the USA (Carpenter, 2010). These positive effects could be challenged by the emergence of secondary pests because the toxin of Bt crops is specifically effective against a restricted number of insects and the insecticide treatments (with a broad spectrum of targeted insects) are reduced with the introduction of Bt crops. The emergence of minor pests becoming major pests is reported for Bt cotton in China, India, and the USA (Bergé & Ricroch, 2010). The underlying mechanism, well known in crop protection, is that when a primary pest is successfully targeted, other species are likely to move into its ecological niche. The impact of farmers' buffer measures (as one of the resilience components in farm management, see introduction) on insecticide use is variable: a case study in China reports increasing pesticide use (Pemsl et al., 2011). A longer-term survey of cotton farmers in India indicates that the pesticide-reducing effect of Bt cotton is stronger than the increasing effect through secondary pests (Krishna & Qaim, 2012).

Marker-assisted selection is an alternative to transgenic approaches in breeding. The most important objective of MAS applications in plant breeding is disease and pest resistance. Long-term resistances are quantitative (polygenic) traits that require the combination of several or many genes. Marker-assisted pyramiding is the process of combining several genes together into a single genotype in order to develop durable, broad-spectrum resistance. A precondition is that genetic variation for the desired traits can be found in the breeding material from crossable species. Such pyramiding can be much more complex than pyramiding approaches in GM crops.

Pyramiding may be possible through conventional breeding, but phenotypic testing of individual plants for all traits can be time-consuming and sometimes very difficult (Collard & Mackill, 2008). Marker-assisted resistance breeding therefore has the potential to increase resilience of crop production systems, e.g. in the face of increasing pest and disease pressure due to climate change.

There are far more publications on the development of markers than on successful use of MAS in breeding (Brumlop & Finckh, 2011). Reasons are seen in different technical application obstacles, the high cost of MAS and the non-disclosure of applied MAS methodologies for new registered varieties (Collard & Mackill, 2008). Nonetheless, an increasing relevance and application of MAS is expected for the coming years, but MAS will probably never replace phenotypic selection entirely (Brumlop & Finckh, 2011).

The current importance of breeding for resistance is controversial. While resistance is often mentioned as a priority in breeding programmes, some private breeders recognise that resistance is not as important an objective in their work as it is in public breeding institutions. During the last decades, the existence of pesticides and the focus on high-input systems has led in Europe to a greater focus on yield than on resistance (Vanloqueren & Baret, 2008). Incentives from public authorities and policies are needed to improve the adaptive capacity of the breeding sector to move towards polygenic, multi-resistant cultivars.

A case study for winter wheat in the cereal-growing part of Wallonia (Belgium) shows that the adoption of existing multi-resistant cultivars is low. Factors impeding a broader use of multi-resistant wheat cultivars were identified at different levels, from input suppliers over farmers and extension services to current regulations. Vanloqueren and Baret (2008) draw the conclusion that modern wheat cropping systems are now 'locked-in' to a fungicide-dependency situation. This probably also applies to other high-input cropping systems.

Diversification strategies are another approach in plant breeding to achieve pest control. In the last hundred years, the trend in modern plant breeding has been towards genetic uniformity within crops. This has greatly enhanced possibilities for mechanisation and intensification. However, the resilience of agricultural ecosystems is based on complex interactions among species and genotypes at all levels of the system, and lack of diversity renders crops and cropping systems vulnerable to pest and disease attacks. Diversity can be achieved at the level of species (intercropping), varieties (variety mixtures), or genes within species (multi-lines). Important impediments are legal problems arising from existing seed regulation (Finckh, 2008).

The use of multi-resistant cultivars can be a component of integrated pest management (IPM). Integrated pest management aims to maintain pest infestations below economically acceptable levels by encouraging natural control of pests. Different appropriate techniques should be used, such as enhancing natural enemies, planting pest-resistant crops, adapting crop management and using pesticides judiciously (Oerke & Dehne, 2004). In high external input agriculture of industrialised countries, IPM in the past paid special attention to field-based management and restricting pesticide use by economic treatment thresholds (Brewer & Goodell, 2012; Ehler, 2006). In contrast, the adoption of agronomic options of IPM, such as cultivar mixtures, changed seed regimes, reduced nitrogen fertilising, maintaining natural habitats, or the introduction of biological control methods, is low in arable crop production systems with high external inputs.

Future opportunities for IPM could be opened up by the EU pesticide regulation (Directive No. 2009/128/EC on sustainable use of pesticides) which requires all farmers to use pesticides within an Integrated Pest Management framework by 2014. Member States have to adopt National Action Plans setting out the objectives for reducing hazards, risks and dependence on chemical control for plant protection. Low pesticide-input farming is to be promoted, and all farms in the EU must implement principles of IPM for their crop protection activities (Hillocks, 2012). Future dissemination of advanced IPM strategies to support resilience is dependent on the implementation of the new EU regulation in the Member States.

Finally, how do IPM and GM crops fit together? Bt crops – like resistant crops in general – can be regarded as a preventive measure with high specificity, which reduces pesticide use and fits well in the IPM approach. However, the insecticidal Bt protein is produced season-long in relatively high concentrations and independently of the actual pest pressure, which is in conflict with IPM principles. The growing importance of GM crops with stacked genes will increase the situations where target pests of some expressed Bt proteins are not present or not expected to reach damaging levels (Meissle et al., 2011).

A key advantage of Bt maize is that it simplifies pest control operations. Bt maize makes redundant scouting for pest manifestation, forecasting and appropriate timing of chemical or biological control systems. At the same time, growing Bt maize in Europe is associated with more administrative work due to notifications, detailed bookkeeping and resistance management (Meissle et al., 2011). These facts make it less probable that IPM measures such as changed cultivation, e.g. against secondary pests, will be introduced because they are more work-intensive and demand a rethinking of pest control. Therefore, GM crop systems with insect resistance have buffer capability by chemical pest control, but the transformation capability could be low.

Drought tolerance and management of abiotic stress
Drought means a reduction in water availability that reduces crop yields below what farmers could produce with adequate water supply. Drought events, like other abiotic stresses, are complex with variations in timing, duration and

intensity of stress interacting with different stages of plant development (Varshney et al., 2011). In the field, many crops routinely encounter a combination of drought and other stresses, such as heat or salinity (Mittler, 2006). Additionally, agronomic factors such as soil fertility, nutrition supply, and pest and disease pressure influence how plants cope with water scarcity (Passioura, 2006).

Water scarcity is one of the major causes of poor plant performance and limited crop yields worldwide (Cominelli & Tonelli, 2010). In the EU, the percentage of river basin area under water stress is estimated to be around 10 percent over the year and 23 percent over the summer period (Strosser et al., 2012). In the future, global climate change will be associated with increasing temperatures and changing rainfall patterns, exacerbating the negative impacts of water deficiencies (Morison et al., 2008).

The increasing identification of signalling pathways and regulatory genes and networks controlling complex traits related to environmental stresses is based on the upsurge of genomic information and the use of associated computational biology tools (Varshney et al., 2011). Genetic engineering approaches to obtain drought-tolerant plants include (Cominelli & Tonelli, 2010):

- single-action genes responsible for the modification of a single metabolite or protein that would confer increased tolerance (e.g. targeting at osmo-regulation or heat regulation);
- overexpression of genes encoding stress-regulating transcription factors which typically regulate several genes;
- stress-inducible (rather than constitutive) promoters that allow expression of a transgene only when it is required.

There are hundreds of patents claiming inventions that may improve drought tolerance (Passioura, 2006). Many genes for different stresses have been cloned and characterised in model plants as well as in some crop plants and in some cases, successful developments of transgenics have also been reported (Varshney et al., 2011). However, only one trait, Monsanto's so-called DroughtGard corn, has received regulatory approval (Gurian-Sherman, 2012).

Molecular marker technologies can break down quantitative traits such as drought tolerance into individual components, known as quantitative trait loci (QTLs), enabling marker-assisted selection of desired traits. Several reports are available on the identification or validation of QTLs or markers for abiotic stress tolerance. After identifying the markers associated with QTLs or genes, the candidate QTLs or genes can be crossbred in elite lines, also by pyramiding several in the same genetic background (Varshney et al., 2011). MAS for drought tolerance is not an easy task because dozens of QTLs for drought-related traits have been identified. The contribution of genomics-assisted breeding to the development of drought-tolerant cultivars has so far been restricted, and only a few examples of MAS for traits associated with drought tolerance have been reported (Cattivelli et al., 2008).

A widening gap between the rate of development of new biotechnologies and their deployment in applied breeding programmes for abiotic stress tolerance has been stated (Varshney et al., 2011). Important reasons which make the timeframe and success of modern biotechnology approaches uncertain are as follows (Cattivelli et al., 2008; Mittler, 2006; Sinclair, 2011; Varshney et al., 2011).

- Due to the high variation of drought events and the complex interactions between physiological plant processes and the abiotic environment in the field, it is unlikely that small changes in a trait will result in a uniform increase in yield across different environments. Therefore, approaches involving the introgression of one gene are usually not sufficient to develop drought-tolerant lines.
- The effect of a gene can be influenced by the genetic background. The effectiveness of genes for drought tolerance can therefore vary among varieties.
- Abiotic stress tolerance is often measured by indicators such as yield under stress, integrating over time many processes. Accurately assessing drought tolerance of different varieties requires extensive field trials over a number of growing seasons in varied environments.
- The response of plants to a combination of two different abiotic stresses is unique and cannot be directly extrapolated from the response of plants to each of the different stresses applied individually. Tolerance to a combination of different stresses is a complex trait and mapping genes essential for tolerance to a combination of different stresses could be challenging.

Participatory Plant Breeding (PPB) is an alternative approach developed for marginal and often drought-prone regions in developing countries which are dominated by small-scale farming systems. PPB describes the approach to involve collaborations between plant breeders and farmers in breeding programmes (Vernooy et al., 2009). The aim is to involve the individual demands and experiences of the farmers in the breeding process with a view to creating varieties that are well adapted, especially to marginal regions. It is assumed that seed supply in marginal regions is improved by strengthening local seed systems, rather than by replacing local varieties with seeds from the formal sector. By 2009, there were around 100 PPB programmes worldwide, implemented by diverse institutions (PRGA, 2009).

Beside plant breeding, agronomic approaches are available to increase the amount of plant available water by improved soil and residue management that increase infiltration, improve water storage and reduce run-off. No-till and reduced tillage in combination with permanent organic soil cover are examples of such practice (Cassman, 1999). In dry Mediterranean climates, yield differences resulting from improved soil moisture and nutrient availability have been reported in the range of 20–120 percent and more between conservation agriculture systems and tillage systems. Beside improved soil fertility, conservation agriculture reduces the risk of soil erosion and landslides (Kassam et al., 2012). Therefore, the impacts of such agronomic changes include more than only increased drought tolerance. Some assessments see for these approaches greater impact on yield and yield stability than can be expected from genetic improvements (Cassman, 1999). Instead of single technologies or fixed technology packages, agroecological approaches in crop management are diversified in terms of local conditions. The principles of conservation agriculture need to be adapted to specific farming situations. Consequently, the undertaking of adaptive research and chances of learning from farmer to farmer are important (Meyer et al., 2013).

Increasing yield potential and reducing yield gaps
In past decades, the yield potential of major crops in naturally favourable or irrigated cropping regions has increased steadily through breeding, with gains of 1 percent or less per annum. Average farm yields followed this progress closely (Fischer & Edmeades, 2010), but there are indications that breeders have been less successful in achieving higher yield potentials than in improving the biotic and abiotic stress resistance of the major cereals (Cassman, 1999). For the main cereals, increase in yields by breeding has mostly been achieved through improved harvest index (HI). Maize is an exception because increased yields are mainly based on larger biomass production and an extended period of photosynthesis (Hall & Richards, 2013). Switching from inbred lines to hybrids provides a one-time boost to yield potential on the order of 10 percent (Cassman, 1999); others estimate an advantage of up to 20 percent (Khush, 2013).

Yield potential, like drought tolerance, is under complex genetic control involving thousands of genes (Cassman, 1999). Yield is determined by source (overall biomass production) and sink (reproductive part of the plant to be harvested) capacities. In principle, higher yield potential can be achieved by:

- increasing photosynthetic capacity and/or radiation use efficiency,
- more efficient root systems and/or better root architecture,
- modifying plant growth cycle and/or plant architecture.

The main target of improving yield potential by genetic engineering is photosynthesis. Improving the photosynthetic performance of crops through bioengineering aims at basic processes of carbon fixation by changing biochemical and photochemical processes of photosynthesis or by introducing key biochemical components of C_4 pathways of photosynthesis into C_3 plants. The development of transgenic varieties in the form of farmer-ready cultivars with C_4 traits is a very great challenge which will likely need 20–30 years for significant improvements in photosynthetic performance of C_3 crops, provided that realisation of these goals does not bring into play unconsidered trade-offs that often complicate the transition in scales from biochemical process to crop. So far, there is no commercial release of GM crops that directly target yield (Hall & Richards, 2013).

So-called 'yield hierarchy' implies that a great part of the impact of improvements in the molecular capacity of photosynthesis and enhanced photosynthetic capacity at the leaf level diminish when scaled up to harvestable yield (Sinclair et al., 2004). Additionally, translation in enhanced yield is dependent on changes in other plant characteristics. For example, wheat yield is mainly limited by sink strength during grain filling, and further increases in yield depend upon increases in the sink strength of the grains (Araus et al., 2008).

Marker-assisted selection has the potential to assist breeders in revealing the best allele combinations, even for polygenic yield-related traits (Stamp & Visser, 2012). For rice, for example, such traits include tiller number, number of grains per panicle, grain weight, and grain filling. Additionally, a wide range of germplasm, e.g. from wild species, can be scanned to identify genes for yield components (Khush, 2013). Some optimistic molecular biologists already predict phenotyping as the new bottleneck in breeding progress (Stamp & Visser, 2012). As for other quantitative traits such as drought tolerance, as we have argued, successful application of MAS in breeding for increased yield potential is challenging and depends on further scientific and technological advances.

Although complex traits are the most difficult to handle during a breeding programme, they are responsible for most breeding progress in features such as yield, yield stability and adaptation (Araus et al., 2008). Conventional breeding methods will remain the cornerstone of crop improvement because they effectively and efficiently combine multiple traits together into a single new variety. Many traits contribute to yield and adaptation, and these are most effectively selected in the target environment where large populations can be grown. Therefore, conventional breeding will remain the mainstay for yield potential improvement in the next decades (Hall & Richards, 2013).

Besides breeding for increased yield potential, greater agricultural production can be achieved by better exploring the current yield potential; in other words, by reducing yield gaps. Yield gaps are low for major crops and intensive production in Western Europe, but are quite high in Eastern and Southern Europe. With conventional crop management practices, bringing crop yields up to their climatic and genetic potential would probably require more external inputs such as nutrients, pesticides, and water (Licker et al., 2010). This conventional pathway of intensification has a high probability of negatively affecting ecosystem goods and services. Diverse agroecological approaches are proposed as an alternative for reducing yield gaps. In this context, maintenance and enhancement of soil fertility is a key issue (Meyer, 2010).

Intellectual property rights and seed industry concentration
The economic and regulatory framework of breeding and modern biotechnologies influences the direction of breeding and the resilience effects. Major players in the international seed business are private breeders. The 10 largest companies have a global share in seed sales of more than 60 percent. They only invest in the leading cash crops for economic reasons. Presently, only hybrid seeds provide a solid return on investment (Stamp & Visser, 2012). In consequence, breeding activities and biotechnology applications are concentrated on a small number of major crops. Minor crops are more or less disregarded in plant breeding, and neglected and underutilised crops are often not present in formal seed systems. Breeding progress for minor crops, increasing their economic viability, is an important element of more diversified cropping systems, which is seen as important for higher resilience of farming.

The introduction of patent rights, in addition to plant breeder rights, in the breeding sector resulted from the development of modern biotechnologies. Breeders' exemptions and farmers' privilege are lost with the patentability of plant-related innovations. GM patent applications are dominated by the private sector, with a high share in the top 10 companies (Louwaars et al., 2009). Patents are also of high relevance for other biotechnology applications in plant breeding, such as MAS. Biological technologies are interdependent technologies requiring multiple key components to function, and denial of access to any component prevents the use of the technology. Therefore, patent rights hold possibilities for strategic use which is impossible under plant breeders' rights. Patent rights, together with the way these are granted and exerted, contribute to a decreasing diversity in breeding companies and threaten innovation in plant breeding (Louwaars et al., 2009). Future design of intellectual property rights will influence the potential contribution of plant breeding to sustainable intensification.

Farming systems and modern biotechnology
New varieties (with or without modern biotechnology) are in principle available for all farmers. However, they differ in their adaptability in specific European farming systems, because varieties are linked to the method of crop production management.

Extensive small-scale, semi-subsistence farming
This farming system is characterised by low or no use of external inputs. In contrast, varieties bred with the help of modern biotechnology (breeding

with genetic engineering as well as with marker-assisted selection) are aligned to intensive farming with high external inputs. In the case of GM crops, administrative burden and coexistence regulations would impede the use of GM crops by small-scale farmers. Therefore, semi-subsistence farming and modern biotechnology do not fit together.

So far, semi-subsistence farming has been ignored by EU agricultural policies. In the face of the tenuous situation of many semi-subsistence farmers especially in Eastern Europe, adapted support measures are needed to improve the resilience of this farming system. In this context, participatory plant breeding could be an approach to address European semi-subsistence farming which would need public support. Additionally, participatory plant breeding could contribute to the on-farm preservation of landraces. Overall, more important than improved varieties are agroecological approaches to sustainable intensification.

Extensive farming in less-favoured areas
Less-favoured areas are characterised by relatively low land productivity, and farming in these areas is dominated by extensive production systems. Available GM crops with herbicide tolerance and/or insect resistance do not fit well with extensive crop production. Organic farming is concentrated on less-favoured areas and not allowing GM crops. Potential future GM crops with abiotic and/or biotic stress resistance will only be helpful if they are adapted to extensive crop production.

Pest- and/or disease-resistant crops achieved with MAS could be an interesting option for extensive farming in less-favoured areas and would supplement well agroecological approaches in crop management. A precondition is that these resistance traits are integrated into locally adapted varieties. Diversification strategies in plant breeding (multi-lines and variety mixtures) could increase resilience in less-favoured areas. Appropriate research and breeding activities and adjustments of the seed legislation are required to bring this into practice.

The presence of traditional agroforestry systems, integrated crop–livestock farming and organic farming is an important baseline for agroecological approaches to sustainable intensification. In the face of relatively high yield gaps, improvements of crop management could contribute here more to resilience than higher-yielding varieties.

Medium intensive, mixed farming systems
Mixed farming is mostly located in intermediate areas and by definition is characterised by relatively low specialisation. Improvements of fodder crops are of particular interest, but breeding activities with modern biotechnology

in this area are restricted. The exception is GM maize. However, smaller fields in this farming system often make compliance with coexistence regulations complicated. Once again, improved biotic and abiotic stress resistance achieved by modern biotechnology is only valuable when integrated into locally adapted varieties. In contrast, the overall chances for agroecological approaches to crop management are good. For example, mixed farming is a key element of many organic farms so that the conversion potential is in many cases high.

Intensive, larger-scale crop farming
Yield gaps in intensive crop farming are low. Broad introduction of GM crops with available traits (herbicide tolerance, insect resistance) in Europe would mainly make weed and pest management easier, with restricted effects on yields and pesticide use. The main agronomic risk is the development of resistances which could offset the positive impacts. In the case of increasing future drought risk, GM crops with drought tolerance could increase resilience, given sufficient time and if they work well in different crops and for varying drought situations. Advances in marker-assisted breeding for drought tolerance and disease and/or pest resistance would be of great interest. A major uncertainty is the progress in successful breeding with MAS for stress tolerance. Increasing yield potential through modern biotechnology is very uncertain, so conventional breeding will probably remain the cornerstone for this breeding goal.

In crop management, precision agriculture has high potential to enhance input use efficiency and to reduce production costs. Agroecological approaches to sustainable intensification demand major changes in crop management and farm organisation and will only take place with substantial incentives.

Large-scale corporate farming
Corporate farms specialise in capital-intensive production and in products with low labour requirements, such as cereals and oil crops. Therefore, the feature of simplified crop management of current GM crops is attractive. Compliance with the technical and administrative burdens of GM regulation is most convenient in this farming system. Crop management approaches for sustainable intensification – be it high-tech or agroecological approaches – show a mixed picture, potentially restricted by missing management skills and the associated workload.

Conclusions
Modern biotechnologies for plant breeding are strongly supported by European and national research policies, are an increasingly important research field of the private sector, are protected by new intellectual property

rights and are under continuing development. However, their potential contribution to sustainable intensification depends on their appropriateness for different farming systems. Varieties bred with modern biotechnologies are most suitable for intensive and specialised cropping systems with high external inputs, e.g. larger-scale crop farming systems.

For the GM crops with available traits (monogenic herbicide tolerance and/or insect resistance), the development of resistance is the main agronomic risk which could endanger potential economic and environmental benefits. Possible reactions (increased pesticide use, pyramiding of traits) tend to reinforce the pathway to high external inputs. These GM crops have the tendency to hinder transitions to agroecological approaches to crop management for sustainable intensification because they do not encourage farms to develop transformation capabilities.

Genetic engineering and marker-assisted selection approaches for quantitative (polygenic) traits such as biotic and abiotic stress resistance and yield potential are still in development and much more difficult to achieve, with uncertain time horizons. In contrast to monogenic traits, they are not simply transferable to different crops and locations. Additionally, there is a widening gap between the development of new, basic approaches and their deployment in applied breeding programmes. Therefore, potential contributions to resilience of farming are uncertain.

Modern biotechnology is associated with changes in intellectual property right regimes (introduction of patents) and concentration processes in the seed industry. In consequence, economic incentives lead to a concentration of breeding activities and biotechnology applications on a small number of major crops and broadly applicable traits. This is in conflict with desirable higher diversity (at genetic and species level) as an important element of improved resilience.

Alternative breeding approaches such as breeding for diversity (e.g. multi-lines), participatory plant breeding and breeding for organic farming can address the needs of extensive farming systems, especially with agroecological approaches in crop management, and support their resilience. These are still niche activities that need public support due to low economic incentives. With mainstreaming of agroecological approaches and more local differentiation of crop management, an overall closer collaboration of plant breeders and farmers would become more important.

Technologies and management systems for improved crop production are supplements or alternatives to breeding approaches. High-tech approaches such as precision agriculture aim first of all to improve input use efficiency, without fundamentally changing intensive production systems. A shift to agroecological approaches to sustainable intensification implies deeper transformations in crop production and farm organisation. Such changes in dominant technological trajectory and break out of lock-in situations (Vanloqueren & Baret, 2009) require the promotion of niche innovations, the development of new forms of cooperation and knowledge exchange between scientists and practitioners, differentiated research agendas specifically addressing European farming systems, revitalisation of publicly funded extension services and enabling Common Agricultural Policy reforms (Meyer et al., 2013).

In the case of major economic and/or environmental disturbances, the transformative capability will become most important for the resilience of farming. For strengthening the transformative capability, technological approaches – modern biotechnology as well as single-crop production technologies – can make only restricted contributions because only interacting changes in all system components make transformations possible.

References

Araus, J.L., Slafer, G.A., Royo, C. & Serret, M.D. (2008). Breeding for yield potential and stress adaptation in cereals. *Critical Reviews in Plant Science*, **27**, 377–412.

Benbrook, C.M. (2012). Impacts of genetically engineered crops on pesticide use in the U.S. – the first sixteen years. *Environmental Science Europe*, **24**, 24.

Bergé, J.B. & Ricroch, A.E. (2010). Emergence of minor pests becoming major pests in GE cotton in China. What are the reasons? What are the alternative practices to this change of status? *GM Crops*, **1**, 214–219.

Brewer, M.J. & Goodell, P.B. (2012). Approaches and incentives to implement Integrated Pest Management that addresses regional and environmental issues. *Annual Review of Entomology*, **57**, 41–59.

Brumlop, S. & Finckh, M.R. (2011). *Application and Potentials of Marker Assisted Selection (MAS) in Plant Breeding*. BfN-Skripten 298. Bonn: Federal Agency for Nature Conservation.

Carpenter, J.E. (2010). Peer-reviewed surveys indicate positive impact of commercialized GM crops. *Nature Biotechnology*, **28**, 319–321.

Cassman, K.G. (1999). Ecological intensification of cereal production systems: yield potential, soil quality, and precision agriculture. *Proceedings of the National Academy of Sciences*, **96**, 5952–5959.

Cattivelli, L., Rizza, F., Badeck, F.-W., et al. (2008). Drought tolerance improvement in crop plants: an integrated view from breeding to genomics. *Field Crops Research*, **105**, 1–14.

Collard, B.C.Y. & Mackill, D.J. (2008). Marker-assisted selection: an approach for precision plant breeding in the twenty-first century. *Philosophical Transactions of the Royal Society B*, **363**, 557–572.

Cominelli, E. & Tonelli, C. (2010). Transgenic crops coping with water scarcity. *New Biotechnology*, **27**(5), 473–477.

Darnhofer, I. (2014). Resilience and why it matters for farm resilience. *European Review of Agricultural Economics*, **41**(3), 461–484.

Duke, S.O. & Powles, S.B. (2009). Glyphosate-resistant crops and weeds: now and in future. *AgBioForum*, **12**, 346–357.

Ehler, L.E. (2006). Integrated pest management (IPM): definition, historical development and implementation, and the other IPM. *Pest Management Science*, **62**, 787–789.

Ervin, D.E., Glenna, L.L. & Jussaume Jr, R.A. (2010). Are biotechnology and sustainable agriculture compatible? *Renewable Agriculture and Food Systems*, **25**(2), 143–157.

Finckh, M.R. (2008). Integration of breeding and technology into diversification strategies for disease control in modern agriculture. *European Journal of Plant Pathology*, **121**, 399–409.

Fischer, R.A. & Edmeades, G.O. (2010). Breeding and cereal yield progress. *Crop Science*, **50**, S85–S98.

Folke, C. (2006). Resilience: the emergence of a perspective for social–ecological systems analyses. *Global Environmental Change*, **16**, 253–267.

Garnett, T., Appleby, M.C., Balmford, A., et al. (2013). Sustainable intensification in agriculture: premises and policies. *Science*, **341**, 33–34.

Gassmann, A.J., Petzold-Maxwell, J.L., Keweshan, R.S. & Dunbar, M.W. (2012). Western corn rootworm and Bt maize. Challenges of pest resistance in the field. *GM Crops and Food*, **3**(3), 235–244.

Gilbert, N. (2013). A hard look at GM crops. Superweeds? Suicides? Stealthy genes? The true, the false and the still unknown about transgenic crops. *Nature*, **497**, 24–26.

Gurian-Sherman, D. (2012). *High and Dry. Why genetic engineering is not solving agriculture's drought problem in a thirsty world*. Cambridge: Union of Concerned Scientists.

Hall, A.J. & Richards, R.A. (2013). Prognosis for genetic improvement of yield potential and water-limited yield of major grain crops. *Field Crops Research*, **143**, 18–33.

Head, G.P. & Greenplate, J. (2012). The design and implementation of insect resistance management programs for Bt crops. *GM Crops and Food*, **3**(3), 144–153.

Hillocks, R.J. (2012). Farming with fewer pesticides: EU pesticide review and resulting challenges for UK agriculture. *Crop Protection*, **31**, 85–93.

ISAAA (International Service for the Acquisition of Agri-biotech Applications). (2017). Global status of commercialised Biotech/GM crops in 2017: biotech crop adoption surges as economic benefits accumulate in 22 years. *ISAAA Brief No. 53*. Ithaca, NY: ISAAA.

Kassam, A., Friedrich, T., Derpsch, R., et al. (2012). Conservation agriculture in the dry Mediterranean climate. *Field Crops Research*, **132**, 7–17.

Khush, G.S. (2013). Strategies for increasing the yield potential of cereals: case of rice as an example. *Plant Breeding*, **132**, 433–436.

Kremen, C., Iles, A. & Bacon, C. (2012). Diversified farming systems: an agroecological, system-based alternative to modern industrial agriculture. *Ecology and Society*, **17**(4), 44.

Krishna, V.V. & Qaim, M. (2012). Bt cotton and sustainability of pesticide reductions in India. *Agricultural Systems*, **107**, 47–55.

Licker, R., Johnston, M., Foley, J.A., et al. (2010). Mind the gap: how do climate and agricultural management explain the "yield gap" of croplands around the world? *Global Ecology and Biogeography*, **19**, 769–782.

Louwaars, N., Dons, H., van Overwalle, G., et al. (2009). *The Future of Plant Breeding in the Light of Developments in Patent Rights and Plant Breeder's Rights*. CGN Report 2009-14. Wageningen: Centre for Genetic Resources, Wageningen University and Research Centre.

Lusser, M., Parisi, C., Plan, D. & Rodríguez-Cerezo, E. (2011). *New Plant Breeding Techniques. State-of-the-art and prospects for commercial development*. EUR 24760 EN. Seville: Institute for Prospective Technological Studies (IPTS), Joint Research Centre.

Meissle, M., Romeis, J. & Bigler, F. (2011). Bt maize and integrated pest management – a European perspective. *Pest Management Science*, **67**, 1049–1058.

Meyer, R. (2010). Low-input intensification in agriculture. Chances for small-scale farmers in developing countries. *GAIA*, **19**, 263–268.

Meyer, R., Ratinger, T. & Voss-Fels, K.P. (2013). *Options for feeding 10 billion people – Plant breeding and innovative agriculture*. Report prepared for STOA, the European Parliament Science and Technology Options Assessment Panel, under contract IP/A/STOA/FWC/2008-096/LOT3/C1/SC3.

Institute for Technology Assessment and Systems Analysis (ITAS), Karlsruhe Institute of Technology, member of ETAG, the European Technology Assessment Group.

Mittler, R. (2006). Abiotic stress, the field environment and stress combination. *Trends in Plant Science*, **11**(1), 15–19.

Morison, J.I.L., Baker, N.R., Mullineaux, P.M. & Davies, W.J. (2008). Improving water use in crop production. *Philosophical Transactions of the Royal Society B*, **363**, 639–658.

Mortensen, D.A., Egan, J.F., Maxwell, B.D., Ryan, M.R. & Smith, R.G. (2012). Navigating a critical juncture for sustainable weed management. *BioSience*, **62**(1), 75–84.

Müller-Röber, B., Boysen, M., Marx-Stölting, L., Osterheider, A. (Eds.) (2013). *Grüne Gentechnologie – Aktuelle wissenschaftliche, wirtschaftliche und gesellschaftliche Entwicklungen*. 3rd Edition. Berlin: Berlin-Brandenburgische Akademie der Wissenschaften (BBAW).

Oerke, E.-C. & Dehne, H.-W. (2004). Safeguarding production – losses in major crops and the role of crop protection. *Crop Protection*, **23**, 275–285.

Owen, M.D.K. (2008). Weed species shifts in glyphosate-resistant crops. *Pest Management Science*, **64**, 377–387.

Passioura, J. (2006). Increasing crop productivity when water is scarce – from breeding to field management. *Agricultural Water Management*, **80**, 176–196.

Pemsl, D.E., Volker, M., Wu, L. & Waibel, H. (2011). Long-term impact of Bt cotton: findings from a case study in China using panel data. *International Journal of Agricultural Sustainability*, **9**(4), 508–521.

Powles, S.B. (2008). Evolved glyphosate-resistant weeds around the world: lessons to be learnt. *Pest Management Science*, **64**, 360–365.

PRGA (Participatory Research and Gender Analysis) Program. (2009). *Participatory Plant Breeding*. PRGA Program Thematic Brief No. 2. Cali, Colombia: CGIAR Systemwide Program on Participatory Research and Gender Analysis.

Royal Society. (2009). *Reaping the Benefits. Science and the sustainable intensification of global agriculture*. London: The Royal Society.

Sinclair, T.R. (2011). Challenges in breeding for yield increase for drought. *Trends in Plant Science*, **16**(6), 289–293.

Sinclair, T.R., Purcell, L.C. & Sneller, C.H. (2004). Crop transformation and the challenge to increase yield potential. *Trends in Plant Science*, **9**, 70–75.

Stamp, P. & Visser, R. (2012). The twenty-first century, the century of plant breeding. *Euphytica*, **186**, 585–591.

Strosser, P., Dworak, T., Andrés, P., et al. (2012). *Gap analysis of the water scarcity and droughts policy in the EU*. Final report. Study for the European Commission.

Tabashnik, B.E., Brévault, T. & Carrière, Y. (2013). Insect resistance to Bt crops: lessons from the first billion acres. *Nature Biotechnology*, **31**(6), 510–521.

Vanloqueren, G. & Baret, P.V. (2008). Why are ecological, low-input, multi-resistant wheat cultivars slow to develop commercially? A Belgian agricultural "lock-in" study. *Ecological Economics*, **66**, 436–446.

Vanloqueren, G. & Baret, P.V. (2009). How agricultural research systems shape a technological regime that develops genetic engineering but locks out agroecological innovations. *Research Policy*, **38**, 971–983.

Varshney, R.K., Bansal, K.C., Aggarwal, P.K., Datta, S.K. & Craufurd, P.Q. (2011). Agricultural biotechnology for crop improvement in a variable climate: hope or hype? *Trends in Plant Science*, **16**(7), 363–371.

Vernooy, R., Shrestha, P., Ceccarelli, S., et al. (2009). Towards new roles, responsibilities and rules: the case of participatory plant breeding. In: *Plant Breeding and Farmer Participation*, edited by S. Ceccarelli, E.P. Guimarães & E. Weltzien. Rome: FAO, pp. 617–626.

Waltz, E. (2012) Tiptoeing around transgenics. *Nature Biotechnology*, **30**(3), 215–217.

CHAPTER NINE

Pastoralism, conservation and resilience: causes and consequences of pastoralist household decision-making

KATHERINE HOMEWOOD
University College London
MARCUS ROWCLIFFE
Intitute of Zoology, London
JAN DE LEEUW
ISRIC World Soil Information, Wageningen
MOHAMED Y. SAID
Institute for Climate Change and Adaptation (ICCA), University of Nairobi and Norwegian University of Science and Technology
and
AIDAN KEANE
University of Edinburgh

Introduction: pastoralism and conservation in sub-Saharan Africa's arid and semi-arid lands

Recent estimates put global pastoralist populations at around 120 million people,[1] those of sub-Saharan Africa at around 50 million (plus 200 million agro-pastoralists: Inter-Réseaux Développement Rural, 2012) and those of the Horn of Africa (including East Africa) at between 12 million and 22 million people. These are debatable figures. There are different definitions of what constitutes a pastoralist, ranging from 'someone who depends primarily on mobile livestock, extensively reared on open range' to 'someone who self-identifies socially and culturally as a pastoralist, even though (s)he may own no animals and may rely primarily on other livelihood activities'. Many pastoralists and most observers of pastoralism agree that both definitions are valid. This is because the vagaries of pastoralist lives are such that any one individual, household or community may move to and fro between the two extremes once or more in the course of their life cycle. As well as the fluidity of definitions, there are complexities around population estimates because of pastoralists' spatial mobility (whether managing animals, or temporarily

---

[1] www.worldbank.org/en/news/press-release/2014/03/18/world-bank-pastoralists-horn-africa (accessed 27 March 2014).

otherwise engaged) and their often remote locations. Few sub-Saharan nations have good census data on mobile people or on mobile production systems.

Two-thirds of Africa south of the Sahara is arid or semi-arid. In these areas rain-fed cultivation is risky at best, but livestock can move to exploit key grazing, water and mineral resources, and to avoid seasonally high populations of disease vectors, or areas that are temporarily unsafe because of conflict or raiding. By the same count, enterprises based on relatively mobile wildlife may be very profitable. So, for example, wildlife-based tourism is regularly listed among the top three contributors to national GDP in both Kenya and Tanzania (Homewood et al., 2012). These official figures usually overlook the very significant contribution of pastoral livestock production to national GDP. Although largely invisible to the official gaze, that contribution, where it has been fully evaluated, is estimated to match agricultural GDP from crops (Behnke & Muthami, 2011).

Because both wildlife and pastoralist livestock exploit open rangelands, there are potential synergies between the two enterprises (du Toit et al., 2010). Those synergies hinge particularly on mobility across unfenced, non-fragmented, extensive rangelands, allowing domestic and wild herbivores to seek out the best conditions of pasture and water, and to minimise challenge from biting flies and environmental diseases. Through their grazing, burning, nutrient redistribution and waterhole maintenance, pastoralists and their herds exert a positive influence in shaping grazing lands for wildlife as well as for themselves (Mershall et al., 2018). A large literature explores the rich biodiversity of current and former pastoralist rangelands (Sinclair et al., 2008); the density and health trade-offs between wildlife and livestock where these compete for key resources of forage and water (Prins, 2000; Odadi et al., 2011; Butt & Turner, 2012); where they prevent (Sinclair & Arcese, 1995) and/or transmit infectious diseases between themselves; and the thresholds at which optimum combinations of wild and domestic herbivores can be maintained (Sinclair et al., 2008; Niamir-Fuller et al., 2012; Western et al., 2009a, 2009b). Pastoralist herds thus make a very major economic contribution to the wildlife sector, for example in Tanzania (Nelson, 2012).

East African rangelands in particular are famous both for their pastoralist and their wildlife populations, but their social and environmental sustainability is in doubt. Pastoralist livestock production is important both to local livelihoods (Homewood et al., 2009) and to national economies (Behnke & Muthami, 2011), yet most of these peoples are chronically poor. The same rangelands are renowned for their biodiversity, with wildlife tourism income a major component of national GDP, but the wildlife populations are in drastic decline (Craigie et al., 2010; Western et al., 2009a, 2009b; Ogutu et al., 2012), threatened by changes in land use (Norton Griffiths & Said, 2010; Homewood et al., 2001), possibly exacerbated by climate trends.

Pastoralism thus presents conservation enterprises both with complementarities (Nelson, 2012) and also with possible conflicts. Many national or international conservation agencies prefer to exclude pastoralists and livestock from protected areas set aside for conservation. This is often justified on the basis of environmental degradation narratives, even where demonstrably far more damaging extractive endeavours are favoured in their place (Walsh, 2008; IWGIA, 2013).

At the same time, many pastoralists complement their own livestock-based livelihoods with farming, however low and risky the returns, converting rangeland in ways that may exclude most large mammal wildlife and other valued components of biodiversity. While fragmentation due to the spread of small-scale settlement and farming affects wildlife (Ogutu et al., 2012), large-scale farming for cash crops may displace them completely, driving major declines (Homewood et al., 2001). As the global human population continues to grow, pressure from national and international investors intensifies to convert sub-Saharan rangelands to large-scale cultivation of food, fuel or fibres, and to capture already limited water resources with which to irrigate these otherwise unpredictably rain-fed arid and semi-arid lands (ASALs) (e.g. Catley et al., 2012; Fratkin, 2014; Galaty, 2014). Although pastoralism, wildlife and farming have coexisted for millennia across sub-Saharan Africa, these rangelands are shrinking under the combined pressures of land conversion for large-scale cultivation, and land set-aside to meet international and national conservation targets (Lambin & Meyfroidt, 2011; Fairhead et al., 2012; Galvin et al., 2018). Pastoralist populations and their livelihoods are being squeezed into ever more constrained areas with ever dwindling prospects for mobility. Where mobility has allowed pastoralists across sub-Saharan Africa to survive through extreme events of drought, epidemic and conflict over the last 4000–5000 years, this central coping strategy is now threatened (Galaty, 1999; Niamir-Fuller, 1999; Reid et al., 2014). At the same time, wildlife populations are in sharp decline across pastoralist rangelands, despite the large and increasing areas set aside for them and the high earnings they bring to governments and entrepreneurs, because land conversion, and land fragmentation with privatisation, limit mobility – and therefore survival – for wildlife as much as for pastoralists and livestock.

Past and current work on pastoralist household decision-making, land use and livelihoods is central to the debate over the resilience of pastoralist social–ecological systems (SESs) and over resilience thinking more generally. Pastoralist strategies have long been recognised as managing risk in ways that absorb the shocks and stresses of sub-Saharan Africa's ASALs' unpredictable, fluctuating biophysical and socioeconomic environments (Davies & Nori, 2008). They have also long been recognised as displaying ongoing dynamic innovation and adaptation (Galvin, 2009; Catley et al., 2012). The emergence and survival of pastoralism in sub Saharan Africa over the last 4000–5000 years, and the survival of the African savanna megafauna (including large predators) into the twenty-first century are measures of the social and environmental sustainability of the system. However, for all its past and continuing internal resilience as a production system within ASALs, the continued survival of pastoralism and its continued coexistence with wildlife through the twenty-first century are now both under threat.

This chapter starts by considering multi-scale resilience of pastoralist SESs and associated biodiversity, the complementarities and conflicts of biodiversity conservation with these SESs, and the extent to which trends in pastoralist economics and ecology can be understood through and informed by resilience thinking. The chapter goes on to outline a conceptual model of the main biophysical and socioeconomic processes these SESs encompass, and to set out the main dimensions of mobile pastoralism in sub-Saharan African (sSA) rangelands.

Shocks, stresses and opportunities engendered by unpredictable, fluctuating biophysical and socioeconomic conditions are integrated at the level of individual households. In making decisions about whether (and to what extent) to herd livestock or pursue wildlife-based enterprises on open rangeland, or to convert land to small- or large-scale farming, households are evaluating these multi-dimensional trade-offs. Individual household-level decisions aggregate to shape landscape-level trends affecting the viability of pastoralist systems and of wildlife populations. We introduce our work on household decision-making by first outlining what constitutes the pastoralist household and what this unit of analysis entails.

We go on to describe current research on pastoralist household decisions, their determinants and outcomes. Building on empirical ethnographic and livelihoods data from earlier multi-site studies across Kenya and Tanzania Maasailand (Homewood et al., 2009), current research uses economic games, choice experiments, stochastic dynamic programming and other modelling approaches, alongside empirical work, to explore the way trade-offs (and decisions) vary between locations and households of different wealth, shifting with rates of return from different enterprises and with policy incentives or constraints. To date, this work has focused on household decisions in and around Kenya's proliferating conservancies, particularly around the Maasai Mara National Reserve (MMNR), its highest-earning wildlife area. Conservancies are defined by Kenya's Wildlife and Conservation Management act of 2013 as 'Land set aside by an individual landowner, body corporate, group of owners or a community for purposes of wildlife conservation'. They now number some 160 across Kenya (over 100 being wildlife- and others forest-related), covering several thousand square kilometres of rangelands, and are hailed as an effective answer to poverty and conservation issues, although more nuanced perspectives are starting to emerge (Osano, 2013; Bedelian, 2014). Further work now in progress explores the contrasting situations engendered by Tanzania's Wildlife Management Areas, originally conceived as a national programme for community-based natural resource management initiatives and as a key component of the national Poverty Reduction Strategy. We expect that our emerging findings will be generalisable to other sub-Saharan rangelands across the Horn of Africa and beyond, particularly in terms of their underlying approaches to understanding people's choices and to scoping policy scenarios.

Resilience of pastoralist SESs

Studies of pastoralist household strategies tend to use resilience thinking in either more applied or more theoretical ways. Applied work focuses on resilience thinking as a framework for understanding risk coping strategies and disaster management, and as a potential tool in prioritising need for humanitarian and development aid (Oxfam, 2010; Robinson & Berkes, 2010; DFID, 2011). These applied approaches build to a significant extent on earlier understandings of risk and vulnerability (Wisner et al., 2004). For example, Robinson and Berkes (2010) look at the potential of resilience approaches to identify surrogate indicators of stress and adaptation among Gabra pastoralists in northern Kenya. Building on the long-established understanding of pastoralist mobility, herd management and social risk spreading systems as adaptive strategies buffering shocks and stresses, they identify a set of indicators (species-specific herd numbers, specific dimensions of pasture condition, etc.) which can be tracked to reveal trends in vulnerable pastoralist populations potentially in need of humanitarian intervention.

Although not explicitly phrased in terms of resilience thinking, risk management and household outcomes have been studied in these same Gabra populations by McPeak et al. (2011), who have developed a useful classification of household trajectories. In their schema, households are categorised according to their economic position on two axes: livestock assets and income versus other assets and income. Households are categorised as being below or above average for each of these axes, giving rise to four household categories. These are:

(a) households with above-average assets and income from both livestock and other activities. These households are successfully combining pastoralism with diversification;
(b) households with above-average livestock assets but which are below average for other assets and income. These households are seen as staying on as pastoralists;
(c) households with below-average livestock assets and income but having above-average assets and income from other activities. These are seen as moving out of pastoralism; and

(d) households that are below-average both for livestock and for other assets and income. These households are effectively being left behind in poverty.

While this approach gives a good rule of thumb, it is not necessarily the case that the average represents a realistic threshold: for example, average values in pastoralist SES are driven upwards by the few very wealthy households. However, the scheme brings home the way different strategies and outcomes lead to different possible futures, including the decline of pastoralist livelihoods, and thus of pastoralism, for households with decreasing livestock numbers.

More theoretical approaches (building on Folke, 2006) have analysed pastoralist strategies as a basis for developing resilience theory per se. Leslie and McCabe (2013) analyse individual-, household- and village-level responses to varying circumstances (particularly around differing interactions with national park and conservation easement initiatives) as forms of response diversity, in itself a dimension of functional diversity, as a central component of the resilience of pastoralist SESs. They describe the different strategies adopted by different individuals, households and villages in their interactions with the vigorous spread of state and private conservation set-aside across formerly communal rangelands in northern Tanzania. For example, at village level, these strategies include:

- opting to cultivate to forestall boundary extensions by the National Park;
- resisting creation of supposedly community-based natural resource management (CBNRM) Wildlife Management Areas, because of their failure to replace lost resource access and associated income with commensurate revenues, due to state and elite capture of revenues including payments for environmental services (PES) or payments for wildlife conservation (PWC);
- negotiating conservation easements with private tour operators, allowing villagers to graze their livestock in the set-aside, while agreeing not to cultivate or build there.

These variants, and comparable variation observed at individual and household level, can be related to the specific social and historical circumstances of each case. Interpreting them within the context of resilience thinking, Leslie and McCabe (2013) take the wider view that these variants are a manifestation of response diversity, forming the raw material underpinning the potential for adaptation and resilience.

Resilience thinking does not only seek to analyse and predict the possibilities of adaptation to and survival through shocks and stresses, it also considers the possibility of systems collapsing and reorganising as a new, different state, which is characterised by quite different components, processes and linkages

(Folke, 2006). Across sub-Saharan Africa, especially in East Africa, the competition for land has become so intense and land grab by the state, by investors and elites so overwhelming (Zoomers, 2010), that rangelands available for pastoralism (and wildlife dispersal outside protected areas) are dwindling fast (Lambin & Meyfroidt, 2011; Land Matrix, www.landmatrix.org/en/). In Tanzania, there have been high-profile forced displacements of many thousands of pastoralists and their herds in recent years (Loliondo: BBC, 2013; Usangu: Walsh, 2008; Morogoro: IWGIA, 2013). Similar events are taking place in Kenya (Laikipia: Letai & Lind, 2012; Galaty, 2014; Tana River: Duvail et al., 2012); Ethiopia (HRBDF, 2013; Fratkin, 2014) and elsewhere. These evictions and associated land grabs are made all the more easy by land laws which generally do not recognise customary ownership of pastoralist grazing lands and official views holding that such lands are not in productive use. By contrast, farmland ploughed and sown with crops is recognised as being under private or customary tenure warranting at the very least compensation for any displacement.

Pastoralist SESs may therefore be highly resilient on all indicators of internal functioning, with livestock production contributing half or more of pastoral people's household incomes (Homewood et al., 2009, 2012) and of the national agricultural GDP (Behnke & Muthami, 2011); displaying considerable response diversity (Leslie & McCabe, 2013), innovation and adaptive capacity to deal with climatic and economic shocks and stresses (Davies & Nori, 2008; Galvin, 2009; Catley et al., 2012) and showing finely calculated, rapid and effective responses to policy and economic changes (Keane et al., 2016). At the same time, though, despite this internal resilience, they are being pushed towards collapse by external drivers, with the state, and state-networked investors, progressively capturing control of key resources of water and land and converting or removing these from the pastoralist SES, whether for cultivation or for conservation purposes.

Resilience thinking emphasises the need to look at multiple scales (Folke, 2006). Certainly in this case, internal resilience is negated by political and economic vulnerability on the national and international scale, in the context of twenty-first-century global population growth and globalisation processes. This raises the question as to quite what theoretical resilience thinking adds to our understanding of pastoralist SESs. It is perfectly possible to phrase pastoralist strategies and systems in resilience terms, but it is not clear that the exercise brings any newer insights or greater depth of understanding than do, say, risk management and political ecology frameworks.

Given the marginalised status of many pastoralist groups, the highly political use of poorly substantiated pastoralist degradation narratives to justify either displacement of pastoralist households from a coveted area, or privatisation of resources (which subsequently become available to global capital, whether for large-scale cultivation or for commercial conservation-based enterprises), and the universal nature of these processes (e.g. Sternberg & Chatty, 2012), political ecology might be seen as the more incisive analytical tool in understanding trends in twenty-first-century pastoralist SES (Robbins, 2012). Political ecology, the political economy of natural resources control, exposes the interplay of vested interests, the impact of neo-liberalisation and of potential profits to state, external investors and local elites in capturing rangelands formerly under pastoralist common property management. It also makes clear the strategic use of narratives of pastoralism's low productivity and environmental degradation (Robbins, 2012). Pastoralist production is poorly captured by official statistics, and official policies commonly dismiss now widely available evidence of the ecological and economic efficacy and sustainability of pastoralist production. Narratives portraying pastoralism in a negative light are commonly used as tools in displacing pastoralists from rangelands targeted for large scale cultivation or conservation set-aside (Benjaminsen & Bryceson, 2012; Galaty, 2014).

As the human population nears 9 billion, approaching or exceeding planetary boundaries on biogeochemical processes that have maintained a favourable, stable environment for the last 10,000 years (Rockström et al., 2009), African rangelands increasingly constitute the main remaining land area available for conversion to cultivation – but at great social and economic cost to their long-resident pastoralist inhabitants (Lambin & Meyfroidt, 2011). Pastoralist livelihoods are squeezed between competing land uses, driven by global capital, for large-scale cultivation on the one hand and for national and international conservation set-aside on the other. In principle these competing forces are meant to be integrated at the local level through CBNRM, allowing communities and households to develop socially, economically and environmentally sustainable compromises. This is the picture presented in the Poverty Reduction Strategies of Tanzania and Ethiopia (e.g. URT, 2005, 2010) and on the websites and literature of many conservation initiatives. In practice, CBNRM outcomes for both environment and for poverty reduction have been mixed, and often disappointing (e.g. Blaikie, 2006; Nelson & Agrawal, 2008; Dressler et al., 2010; Bedelian, 2014), or simply left undocumented. For example, recent work (Osano, 2013) explores the trade-offs between pastoralism and conservation in a series of papers around the potential of different types of PES schemes for poverty alleviation and adaptation to climate change in Kenya's rangelands, including conservancy (around Maasai Mara) and conservation easement schemes (Kitengela). While meticulous in flagging up theoretically possible negatives, Osano (2013) focuses on measuring the benefits of PES and PWC schemes among participants: the costs, particularly to those unable to participate, remain unexplored in his work. Other studies attempt a more balanced approach (Sulle et al., 2011; Bedelian, 2014; Homewood et al., 2015). Together with the behavioural economics approaches outlined in this chapter, ongoing empirical studies deliver a robust statistical approach using before/after, control/impact design and matched comparisons to evaluate the social and environmental impacts of conservation interventions in East African rangelands.[2]

Conceptual model: factors shaping pastoralist household decisions

In order to unpack further these ideas of resilience or the lack of it in pastoralist SESs, we now focus on pastoralist livelihoods, based loosely around the sustainable livelihoods framework (Scoones, 2009). This is informed by complementary approaches based on pastoralist perceptions of risk (Oxfam, 2010; McPeak et al., 2011), on understandings of pastoralist poverty (Anderson & Broch-Due, 1999; Barrett et al., 2011); on pastoralist management of climate and environmental change (Davies & Nori, 2008; Galvin, 2009) and on our understanding of diversification and change in African pastoralist livelihoods more generally (Fratkin & Roth, 2005; Catley et al., 2012) and across African rural societies (Ellis, 2000). Our work is also informed by research on the poverty and livelihoods implications of conservation interventions in East African rangelands, from the economic (Emerton, 2001) to the more political (Jones, 2006; Adams & Hutton, 2007; Igoe & Croucher, 2007; Benjaminsen & Bryceson, 2012), particularly around ostensibly community-based initiatives (Dressler et al., 2010; Nelson, 2010). This chapter goes on to look at the implications for our emerging understanding of pastoralist livelihoods and their interplay with biodiversity conservation, drawing on the rich tradition of scientific study of ecological processes and wildlife populations in East African rangelands (e.g. Sinclair & Norton Griffiths, 1979; Sinclair & Arcese, 1995; Sinclair et al., 2008).

In the course of earlier work,[3] Homewood et al. (2001) established a conceptual framework bringing together the main biophysical and socio-economic factors feeding into pastoralist household decision-making (Figure 9.1). The full model is explained in detail elsewhere[4] (Homewood et al., 2009, chapter 1). In brief, building on past work, our conceptual model centres on the increasing competition between alternative uses (1) for land area overall and (2) for net primary production within the grasslands, between competing wildlife and livestock. Cash crop prices and policies drive the extent of large-scale commercial farms and of their associated ecological impacts. Small-scale farming is driven by population growth, migration and education – all factors affecting aspirations and economic trade-offs between

---

[2] ESPA-funded Poverty and ecosystem impacts of payment for wildlife conservation initiatives in Africa: Tanzania's Wildlife Management Areas (PIMA: NERC grant NE/L00139X/1) UCL PI
[3] EU-DGXII-funded Savanna Land Use Policy Outcomes, UCL with Prof Eric Lambin and Suzy Serneels, then of the Catholic University of Louvain.
[4] See also short video at www.ucl.ac.uk/best

alternative livelihoods. Biophysical factors (rainfall, fire, ground water, grazing and browsing populations, etc.) influence the proportional extent of woodland and grassland in land that has not been converted to cultivation. These processes lead to active trade-offs by individual Maasai deciding whether to herd livestock or to run wildlife-related enterprises on open rangelands. They also involve trade-offs in whether to convert land to small- or large-scale farming. Trade-offs around land-use and labour investment decisions are calculated at the level of the individual household. Those calculations are influenced by social hierarchies and institutions, which affect the distribution of access to and benefits from resources and shape the different types of revenues available (Figure 9.1).

Thus, household-level decisions driving major changes in the study region are made on the basis of:

- revenue from different enterprises (herding, farming, wildlife, tourism),
- distribution of revenue between households,
- how people manage their assets, with livestock remaining a major potential wealth store and investment, and
- intangible, non-monetary values attached to livestock-related activities rather than other occupations. Owning livestock and successfully managing a herd carries 'social capital' in pastoralist societies. People make

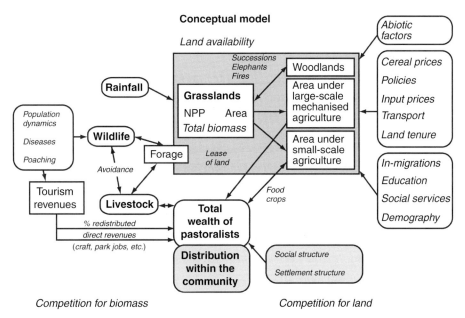

Figure 9.1 Conceptual model of the main drivers of pastoralist household decision making: see Homewood et al. (2001) and Homewood et al. (2009) for full exposition.

choices that safeguard their standing within the social group and hence their ability to call on that group in possible future times of need. This is tied up with the concept of respect, and self-respect within one's social group (*enkanyit*, in Maasai culture; '*pulaaku*' among Peul or Fulani pastoralists); concepts central to people's well-being (Gough & McGregor, 2007; Agarwala et al., 2014; Milner-Gulland et al., 2014).

Finally, irrespective of limited yields, farming may be effective in staking a claim to land in a way herding does not, because governments often do not recognise customary communal tenure of pasture land as official ownership. Therefore, decisions to farm may represent a tenure strategy as much as an economic choice, especially when there is the prospect of formerly communal lands being subdivided and allocated as private plots. Similarly, it may represent a pre-emptive bid for subsequent compensation if local people suspect that formerly communal lands may be expropriated by the state and leased to investors for large-scale cultivation or wildlife-based enterprises.

These decisions, taken on the basis of site-specific trade-offs evaluated at the level of the individual household, aggregate to shape landscape-level changes in the amount and distribution of land converted to cultivation, land set-aside for wildlife conservation and land available for pastoralist livestock production, and for the mobility essential for both livestock and wildlife populations needing to access shifting key resources.

Methodological note: the pastoralist household

As we focus on the pastoralist household as the main unit of economic and ecological analysis, it is important to flag up an underlying conceptual issue around pastoralist households.

The standard definition of a household might read: 'the basic residential unit in which economic production, consumption, inheritance, child rearing, and shelter are organised and carried out'; [the household] 'may or may not be synonymous with family' (Haviland, 2003). It is an often unspoken assumption that the concept of the household is readily transferable across places and cultures, but this general definition may not apply in very different socio-cultural contexts of pastoralist societies (Randall et al., 2011). This is particularly the case where people are extremely mobile and where individuals of different ages, genders or wealth may move through different patterns of residence, subsistence and migration both through the year and in the longer term. Also, where men are often polygamous, wives may not be co-resident, and men may divide their time, and their assets, equally or otherwise between spatially distant homesteads. In surveys and censuses, this may lead to one unit being recorded as male-headed and as endowed with the male household head's assets (radio, bicycle, mobile phone, etc.), and the other as a female-headed household without assets, where these superficially separate units would be better understood as a single bi-local, polygamous household under a single male head, with both units sharing his occasional presence and his assets to some extent (Randall et al., 2011). Alternatively, households may be multi-local, with related units in complementary locations (pastoral rangeland homestead with livestock; upland- or wetland-based homestead with farmed land; urban base hosting wage-earners, children in school, relatives attending hospital: see for example Homewood et al., 2009, pp. 45–47, 151–158).

As much of the present chapter focuses around Maasai pastoralist systems in East Africa, we illustrate the potential complexity of the household concept in pastoral societies by outlining the different structural levels contributing to the concept of the household among Maasai. Here, the 'household' refers to the Olmarei (pl. Ilmareita), which may be best understood as the family of one household head with his or her dependants. These may include, in the case of male-headed households, more than one wife and her children and grandchildren, as well as the household head's parents and dependent siblings, together with non-related individuals who reside with the family and depend on them for food in return for assistance with household chores (most commonly herding).

The Olmarei may or may not be spatially and physically congruent with the homestead (Maa: enkang Pl. Inkang'itie), as the wives of a polygamous male Olmarei head may or may not be co-resident. Customarily they do often co-reside within a single homestead, with each wife building a small house (aji) for herself, her children and the occasional presence of her husband. The physical structure, positioning and occupancy of such houses have been described elsewhere (Spencer, 1988; Homewood & Rodgers, 1991; Coast, 2002), but broadly speaking, the enkang customarily comprises a number of these enkaji (Pl. inkajijik) built around one or more linked livestock corrals.

Men traditionally lived in their wife's house or moved between several wives' houses. Increasingly, men invest – if they can – in a house of their own, built to a 'modern', non-traditional design (rectangular plan, mud and wattle walls, if possible with plaster, cement floor and corrugated iron roof). Such houses may be built at the rural homestead (and often used as a store and site of more formal meetings). In some cases men invest in modern houses located in trading centres or urban sites as a business or property investment, and may place one of their wives there to manage it, generating a multi-local household. Other multi-local households with two or more related bases might be in other complementary locations (upland- or swamp-based farm as well as pastoral rangeland and urban settlement).

Ecological and economic approaches to analysing drivers and outcomes of pastoralist household decisions

The conceptual model and issues described above have informed several subsequent studies. The conceptual model formed the basis for a multi-site study of the ways land use and livelihood choices vary in different parts of Kenya and Tanzania Maasailand (Homewood et al., 2009, 2012).[5] Together with the site-specific relevance and idiom conferred by local knowledge, the model has also informed the design of economic games, choice experiments, stochastic dynamic programming and other models exploring the implications of different climate events and policy/economic incentives for household decisions.[6] The model also underpins research currently in progress on poverty and ecological impacts of Tanzania's Wildlife Management Areas, both in pastoralist rangelands and in miombo woodland areas (Bluwstein et al., 2018).

The multi-site study of land use and livelihoods (Homewood et al., 2009, 2012) showed that across Maasai rangelands in Kenya and Tanzania, livestock remain the central pillar of people's livelihoods, accounting on average for around half or more of household income in all sites (Figure 9.2). In four of five sites, off-farm wage or salaried work, including remittances, constituted the second and crops the third most important source of income. In four of five sites, wildlife-based income was the least important, accounting for less than 5 percent of household income on average, and contributing to fewer than 15 percent of households overall in any one site. However, in households around the Maasai Mara National Reserve, Kenya's highest-earning wildlife area, wildlife-based income contributed around 20 percent of household income on average across all wealth quintiles, and flowed to over 50 percent of households (Thompson et al., 2009; Homewood et al., 2009, 2012). Although these figures dropped off with declines in tourism during periods of election-related disruption, with a progressive concentration of revenues in the hands of the wealthiest quintile, land-owning households have subsequently showed a consolidation and on average a doubling of conservation-related income (Bedelian, 2014).

Building on both the conceptual model and on these initial empirical data on livelihood choices and household incomes, recent work designed economic games, choice experiments and policy scenario models (Figure 9.3a,b), looking at the following questions.

- How does conservation set-aside affect pastoralist household decisions over land use and livelihoods?

---

[5] Belgian DGIC-funded *Reto-e-Reto* research collaboration, led by Robin Reid, then of ILRI Nairobi, UCL Co-I.

[6] ESPA-funded project Biodiversity, Ecosystem Services, Social Sustainability and Tipping Points in East African Rangelands (ESPA-BEST: NERC grant NE/1003673), UCL PI.

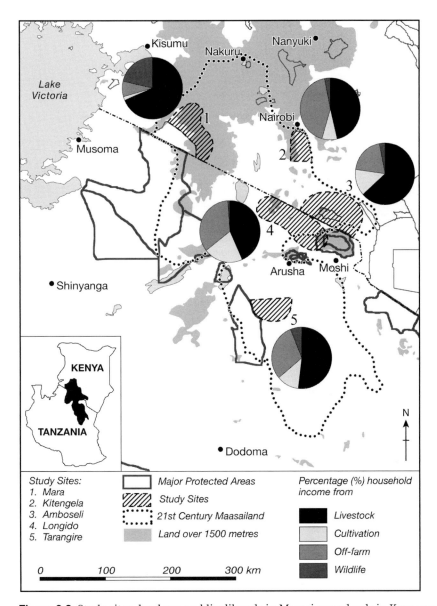

Figure 9.2 Study sites, land use and livelihoods in Maasai rangelands in Kenya and Tanzania.

- What are the economic and ecological outcomes of these decisions and the trade-offs they involve?
- How do impacts differ by age, gender and wealth and between households in different circumstances (e.g. conservancy members versus non-members)?
- What are the unanticipated outcomes of policy and economic incentives?

(a)

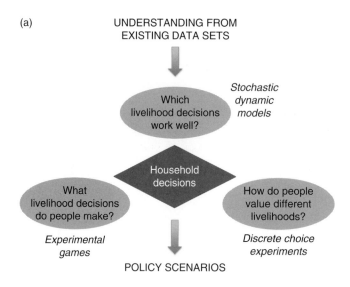

(b)

Figure 9.3 (a) Questions and tools used to explore the decision-making process in pastoralist households. (b) An economic game where the blue counters represent wealth held either as cash or livestock. The blue square on the board represents the bank: this player chooses to hold only a small part of his wealth there. The green square represents his own land, with more of his wealth held as livestock on this private grazing. The largest share of his wealth is held as livestock grazing illegally in the Maasai Mara National Reserve, represented by the red square on the board. These games and choice experiments revealed the extent to which pastoralists' strategies (and theoretically optimal choices) shift according to household wealth and land tenure context, as well as individual characteristics (age, gender, education ...), and the unanticipated outcomes of poorly-conceived conservation policies. These context-sensitive strategies represent an important dimension of response diversity crucial to the internal resilience of the pastoralist SES. (A black and white version of this figure will appear in some formats. For the colour version, please refer to the plate section.)

- Where do they result in adverse ecological impacts through, for example, illegal grazing versus intended ecological benefits?
- Where do they lead to adverse socioeconomic outcomes through elite or state capture (whether of CBNRM resource access, or of PES and/or PWC) versus intended poverty reduction outcomes?

We focused initially on households around the Mara – of our five in-depth study sites, the only one where conservation income is of demonstrably high importance to many households, and cited by many policymakers and practitioners as leading the way, despite Mara's distinctive (and arguably non-replicable) context. Common property resource games based on Maasai Mara grazing resources have explored the ways conservancies prompt changes in livelihood decisions. These games represent the closest we can come to controlled experiments studying household decision-makers' behaviour, presenting alternative investment opportunities – livestock, conservancy revenue or cash-earning activities – under different permutations of land tenure scenarios. Successive rounds may or may not include a drought year, in which pay-offs to the different investment choices change dramatically. The games explore situations where individual interests can conflict with those of the group: where individual decisions can mean that too many animals are set to graze on the same land unit, conditions deteriorate and all herds suffer.

Alongside economic games, stochastic dynamic programming (SDP) models were also used to explore decisions made by single households over a 15-year timeline (the period for which land must commonly be covenanted to set-aside by households wishing to join a conservancy). SDP allows calculation of the theoretically optimal decision sets for households of differing wealth and in different policy scenarios.

Choice experiments are described here in more detail to illustrate our approach to exploring conservancy effects on well-being. These field experiments offer choices between pairs of alternative livelihood futures, each with contrasting levels of herd size, crop area, wage, land set aside, etc. On the basis of the choices made, the experimenter estimates threshold values at which decisions change, and thus values representing equivalence between alternative options (Keane et al., 2016). The results of these choice experiments, carried out with 388 respondents (189 men, 199 women) around Maasai Mara National reserve in Kenya in 2013 are set out below. Full details of the sample and of the procedures are given in Keane et al. (2016).

Figure 9.4 shows marginal rates of substitution for each of the choice attributes in terms of cattle, based on a model which differentiates between men/women and members/non-members, but not between other respondent characteristics. For example, for male non-members, a monthly wage of 10,000 KSh (Kenyan shillings) is approximately equivalent to 76 (95 percent

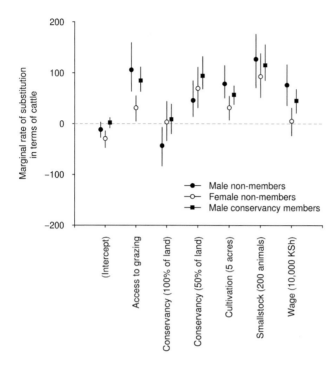

Figure 9.4 Marginal rates of substitution (in terms of cattle numbers) for different land-use and livelihood decision choices.

Credible Interval [CI]: 36, 116) head of cattle. For male conservancy members, it is worth 45 (95 percent CI: 21, 68) head of cattle, but for female non-members this monthly wage only rates five head of cattle (95 percent CI: 23, 31). Figure 9.4 also shows marginal rates of substitution for access to grazing, for covenanting half or all of a household's land to conservancy, for small-scale cultivation and for the value placed on a flock of 200 small stock, in terms of the numbers of cattle seen as approximately equivalent to each of these options. The individual results are interesting. For example, it is striking that Maasai – customarily known as 'people of cattle' – all place such a high value on small stock. In each case respondents are divided into male conservancy members, male and female non-members. For example, women non-members place lower values on access to grazing, and on a man's regular cash wage, than do male non-members – meshing with other studies suggesting women have little knowledge of or access to cash earned by their husbands (Bedelian, 2014). The differences these groups display in their valuing of different components of livelihoods are further explored in Figures 9.5–9.7 and below.

Figure 9.5 shows the proportion of the respondents captured in our sample who would choose to covenant their land to conservancy membership under three different scenarios: (1) when grazing is allowed on conservancy land during times of drought, (2) when grazing is not allowed on conservancy land

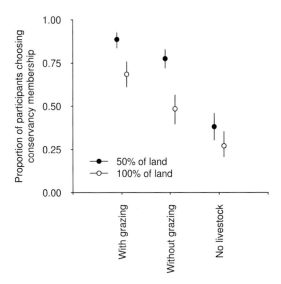

Figure 9.5 The proportion of respondents choosing to covenant their land to conservancy membership under different grazing allowance scenarios.

and (3) when grazing is not allowed and as a consequence no livestock can be kept, alongside two levels of conservancy membership (50 percent or 100 percent of household land placed under conservancy management). In each case, the alternative scenario is assumed to be one in which the household owns 40 head of cattle and 80 sheep or goats, cultivates 5 acres of land, and earns 6000 KSh per month in income from wages. Cultivation and wage earnings are assumed to be unaffected by conservancy membership. Figure 9.5 is based on a model including all of the recorded participant characteristics as covariates. While the great majority of participants chose conservancy membership when grazing is possible, Figure 9.5 makes clear people's preferences to set aside part rather than all of their land, and to do so preferably where dry season grazing is allowed. It also shows the powerful disincentive effect of conservancy management precluding livestock keeping.

Figure 9.6 shows the same data as Figure 9.5 above, but splitting the respondents into men ($N = 189$) and women ($N = 199$). Both show rather similar patterns, and the main gender difference to emerge here is that access for grazing is less important to female non-members. Women participants show marginally more interest in conservancy membership than do male non-members and are less put off by restrictions on dry season grazing.

Figure 9.7 again shows the same data as Figure 9.6, but splitting the respondents into those who were not conservancy members at the time of the survey ($N = 149$) and those who were ($N = 239$). Interestingly, there is a suggestion of unmet demand for more conservancies here – many non-members choose the scenario involving conservancy membership, provided this is compatible with maintaining their livestock. Conversely, even among existing conservancy

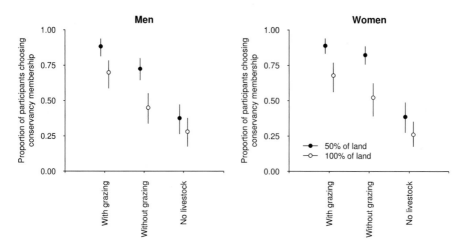

Figure 9.6 Responses of men and women to the option of conservancy membership under different grazing allowance scenarios.

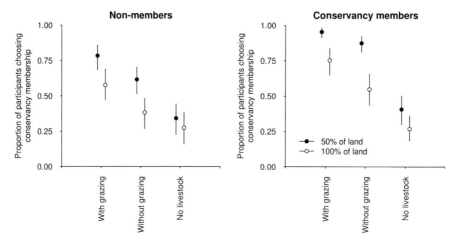

Figure 9.7 Responses of non-members and conservancy members to the option of covenanting land under different grazing allowance scenarios.

members, management scenarios or household choices meaning they would not be able to have grazing land, or that would undermine livestock keeping, are a clear disincentive to participating in conservancies.

Discussion

The results of this work help elucidate pastoralist decision-making and its implications both for the interplay of pastoralism and conservation, and for resilience of both the pastoralist production system and wildlife populations in East African rangelands. We set out to explore how conservation set-aside affects pastoralist household decisions over land use and livelihoods. As predicted by our overarching conceptual model, changes both in land availability (captured in our conservancy scenarios which set aside 50 percent or 100 percent land), and in the level of PES/PWC payments (built into our experimental choices), lead to changes in people's evaluation of alternative options, and consequently in their land-use and livelihood decisions.

Outcomes include changes in the extent to which people invest in cattle as land tenure changes, and choices to maintain grazing access in the face of expanding conservancy set-aside. The outcomes are sensitive to context, with strategies (and theoretically optimal choices) shifting according to household wealth and land tenure context. The experiments make it possible to disentangle variables relating to the scenario under exploration (land tenure system; household wealth; proportion of wealth invested in livestock) and individual characteristics of participant players, which can also be significant predictors of outcomes (Keane et al., 2016). For example, people with more cattle in real life allocate more resources to cattle when making decisions within our experimental scenarios.

We wanted to explore the economic and ecological outcomes of these decisions, and the trade-offs and sometimes unanticipated outcomes they involve. Conservancies maintain open rangeland both for wildlife and also potentially for livestock, as in Kenya's Northern Rangeland Trust conservancies. However, around the Mara, as well as in most Tanzanian Wildlife Management Areas (WMAs), herds are usually only allowed under severe restrictions, or not at all. Conservancy managers explain this as necessary because tourists in high-end conservancy camps do not wish to see Maasai with cattle. Elsewhere, such as in the Northern Rangelands Trust, exclusion or tightly controlled grazing patterns for pastoralist cattle stems from adherence to a particular (much debated) theory of 'holistic management' (Savory, 1988). Denial of grazing access is at the root of much of the conflict, resentment and illegal behaviour associated with rangeland conservancies in Kenya and WMAs in northern Tanzania. It is continually contested and in some cases access has been successfully negotiated, both in Mara conservancies and some Tanzanian WMAs. Our findings make clear that monetary returns do not compensate for loss of grazing access; conservation and tourism-based livelihoods are seen as complementary, not as alternative to livestock and livestock-based activities, nor as fully compensating for their loss. These ecological and economic trade-offs are highlighted in the way most participants in our experiments made land-use choices, with diversification common but with the almost universally preferred choices being those guaranteeing access to dry season grazing within the conservancy, over alternative benefits.

Our choice experiments illuminate household decisions that underpin trends reported by independent lines of evidence. For example, the universally high value placed on small stock in choice experiments confirms and helps explain existing aerial survey data showing increasing numbers of small stock in the area, and the increasing proportion they make up of all livestock there (cf. Ogutu et al., 2011). Local informants cite the relative ease of management of Dorper sheep within reduced grazing areas, and high market prices, as factors driving this trend.

We sought to explore how impacts differ by wealth, gender and age, and between households in different circumstances (e.g. conservancy members versus non-members).

Empirical observation and theoretical prediction have long seen households as starting with small stock and progressively, with increasing wealth, switching to large stock (with or without cash investment). Our findings suggest that changing land tenure scenarios can shift optimal household strategies away from this classic progression with increasing wealth. The choice experiments highlight a divergence between men and women's interests at the level of the household over land-use and livelihood strategies, particularly with respect to cash payments from conservancy and from wage employment, respectively. Although the data are not detailed here, choice experiments also show significant age effects, for example in preferences for cattle (Keane et al., 2016).

These findings offer potential for fine-tuning policy and economic incentives to encourage more economically and ecologically sustainable livelihood choices in the face of rapid pace of change and shifting baselines, not least through recognising the major effect of (externally set) PES thresholds on decisions. Around the Mara, Kenya's highest-earning wildlife tourism destination, the value Maasai place upon access to land and mobility for grazing animals means that setting aside 50 percent of one's land in a conservancy scheme is seen as more acceptable than covenanting 100 percent of one's land. In other study areas, where the income from wildlife is less important to the household, the incentives and the will to set land aside for conservation are far lower.

The choice experiments reveal underlying decision rules and unanticipated outcomes. For example, our findings show the potential of PES/PWC for divergent effects on the well-being of men and women within the household. This is thought to stem from men and women's different degrees of access to and control of different income streams. Maasai men earning a cash wage are very unlikely to share it with their womenfolk, nor to divulge the amounts involved (Bedelian, 2014). While conservancy payments are made to a bank account established in the name of, and controlled by, the (usually male) household head, they involve standard sums paid at known points in time (often coinciding with the dates when school fees fall due). Even though they cannot access or control these monies directly, women are in a better position to press for these funds to be used to the benefit of the wider household than is the case with men's wages. The findings help make policymakers aware of the perspectives and responses of grassroots users, hopefully informing more effective policy and practice.

Ongoing work is currently exploring the detail of these issues. For example, economic games not reported here explore the way increasing conservancy set-aside drives choices to graze illegally in formally gazetted areas (particularly by those benefiting most from conservancy revenue, and thus most able to pay cash fines if caught – c.f. Butt, 2011). Further work is needed to investigate emerging questions around who is or is not able to engage with conservancies; how measuring broader dimensions of well-being might change conclusions; and on the social, economic and ecological leakage effects on surrounding areas (Bedelian, 2014). Modelling could explore overall effects at the community level of multiple, heterogeneous households' decision-making, identifying winners and losers at the household level.

Our choice experiments were conducted with people responding to the expansion of conservancies in Kenya's highest-earning wildlife tourism site. However, the insights they reveal are generalisable to the many initiatives attempting to use PES or PWC incentives to bring about socially and environmentally sustainable conservation across East African rangelands. As outlined in our introduction, such initiatives span a wide range, from those covenanting privately owned land to conservancies (as in, but also beyond the high-PES Mara area), to Tanzania's WMAs (where communal, rather than private, land is pooled and set aside to form multi-village WMAs, and the state taxes PES revenue heavily: Homewood et al., 2015), to local conservation easements negotiated with conservation NGOs (e.g. in Kitengela, Kenya: Homewood et al., 2009) or with tour operators (e.g. in Simanjiro, Tanzania: Leslie & McCabe, 2013). Households' and individuals' responses are contingent not only on wealth, gender and occupation (this chapter; Keane et al., 2016) but also on perceived well-being, including non-economic dimensions such as autonomy, security, identity and aspirations (Woodhouse et al., 2015). Conservation initiatives that fail to balance benefits, broadly defined, with the costs that people experience, or fail to distribute those benefits sufficiently equitably, ultimately undermine both social and ecological resilience. Supposedly community-based conservation initiatives may then elicit passive resistance, subversive or illegal behaviours, or outright conflict (Greiner, 2012; Mariki et al., 2015; Homewood et al., 2015; Homewood, 2017; Galvin et al., 2018).

This work thus has wider applicability in exploring whether policies promoting and governing PES can enhance financial sustainability in less-favourable areas, for whom and how. Our findings apply to conditions for a wider set of conservancies in Kenya, throughout Tanzania, across East Africa, the Horn and beyond. This research could also be extended to explore the implications of changing climate conditions. It develops a better understanding of household-level processes driving landscape-level changes, and starts to unpick the potential for rangeland policy to generate unexpected consequences.

Beyond its immediate applied use, this work contributes a new layer to our understanding of the resilience or otherwise of pastoralist social ecological systems. Findings to date are consistent with the broader conclusion that while mobile pastoralism can be highly resilient to external biophysical drivers, and shows a fine-tuned and rapid response to changing socio-political and economic factors, the interaction of external biophysical changes and new external socioeconomic drivers (political marginalisation of pastoralist peoples in modern nation states, global human population growth, globalisation, resource grab for large-scale cultivation and for conservation) are creating trends of rangeland loss and conversion that pastoralist SESs may not be equipped to survive. The same drivers, alongside the ratchet effects of increasingly frequent extreme climatic events, are undermining the resilience and ultimately the survival of savanna biodiversity. Work on pastoralist household decision-making aims to inform resource users, conservation workers, development planners, policymakers and donor agencies as to the ways that people's behaviour may respond to opportunities and constraints in these shifting contexts.

In wider perspective, the current downward spiral of pastoral rangelands, their people and wildlife, is thus perhaps best understood by looking beyond internal and local effects to the overwhelming pressures of external market forces and global capital currently converting pastoral rangelands, constraining local choices and driving adverse local social and environmental outcomes. While that overarching context offers little cause for optimism, the internal resilience of pastoralist social–ecological systems can be better understood and managed by taking into account the way local behaviours represent fine-tuned strategic responses to local political and economic drivers, including the perverse incentives these may set up.

Acknowledgements

This work builds on research projects *Biodiversity, Ecosystem Services, Social Sustainability and Tipping Points in East African rangelands (BEST)*, NE/I003673 and *Poverty and Ecosystem Impacts of Tanzania's Wildlife management Areas (PIMA)*, NE/L00139X/1, both funded with support from the Ecosystem Services for

Poverty Alleviation (ESPA) programme. The ESPA programme is funded by the Department for International Development (DFID), the Economic and Social Research Council (ESRC) and the Natural Environment Research Council (NERC). We would like to thank the Governments of Kenya and of Tanzania, respectively, for their permission to carry out field work associated with these research projects, the Kenyan and Tanzanian communities who participated in the research, and our field research colleagues Heather Gurd, Dickson ole Kaelo and Stephen Sankeni, as well as other colleagues too numerous to mention from UCL, Imperial, ILRI, ATPS, TAWIRI, TNRF and Copenhagen University. We are grateful to an anonymous reviewer for constructive suggestions on an earlier draft.

References

Adams, W.M. & Hutton, J. (2007). People, parks and poverty: political ecology and biodiversity conservation. *Conservation and Society*, **5**, 147–183.

Agarwala, M., Atkinson, G., Palmer Fry, B. et al. (2014). Assessing the relationship between human wellbeing and ecosystem services: a review of frameworks. *Conservation and Society*, **12**(4), 437–449.

Anderson, D. & Broch-Due, V. (eds.). (1999). *The Poor Are Not Us*. Oxford: James Currey.

Barrett, C., Travis, A.J. & Dasgupta, P. (2011). On biodiversity conservation and poverty traps. *Proceedings of the National Academy of Sciences*, **108**, 13907–13912.

BBC. (2013). Tanzania's Maasai battle game hunters for grazing land. www.bbc.co.uk/news/world-africa-22155538 (accessed 30 August 2013).

Bedelian, C. (2014). *Conservation, tourism and pastoral livelihoods: wildlife conservancies in the Maasai Mara, Kenya*. PhD thesis, University College London.

Behnke, R. & Muthami, D. (2011). The contribution of livestock to the Kenyan economy. IGAD LPI Working Paper 03-11. Addis Ababa, Ethiopia.

Benjaminsen, T. & Bryceson, I. (2012) Conservation, green/blue grabbing and accumulation by dispossession in Tanzania. *Journal of Peasant Studies*, **39**(2), 335–355.

Blaikie, P. (2006). Is small really beautiful? Community-based natural resource management in Malawi and Botswana. *World Development*, **34**, 1942–1957.

Bluwstein, J., Homewood, K., Lund, J.F., et al. (2018). A quasi-experimental study of impacts of Tanzania's wildlife management areas on rural livelihoods and wealth. *Scientific Data*, **5** [180087] on https://doi.org/10.1038/sdata.2018.87

Butt, B. (2011). Coping with uncertainty and variability: the influence of protected areas on pastoral herding strategies in East Africa. *Human Ecology*, **39**(3), 289–308.

Butt, B. & Turner, M.D. (2012). Clarifying competition: the case of wildlife and pastoral livestock in East Africa. *Pastoralism: Research, Policy and Practice*, **2**(9), doi:10.1186/2041-7136-2-9.

Catley, A., Lind, J. & Scoones, I. (eds.). (2012). *Pastoralism and Development in Africa: dynamic change at the margins*. Abingdon: Earthscan/Routledge.

Coast, E. (2002). Maasai socio-economic conditions: cross-border comparison. *Human Ecology*, **30**(1), 79–105.

Craigie, I.D., Baillie J.E.M., Balmford, A., et al. (2010). Large mammal population declines in Africa's protected areas. *Biological Conservation*, **143**, 2221–2228.

Davies, J. & Nori, M. (2008). Managing and mitigating climate change through pastoralism. In *Policy Matters*, 16, Commission on Environmental, Economic and Social Policy, IUCN-CEESP, pp. 127–141.

DFID (2011). *Defining Disaster Resilience: a DFID approach paper*. London: Department for International Development.

Dressler, W., Buscher, B., Schoon, M., et al. (2010). From hope to crisis and back again? A critical history of the global CBNRM narrative. *Environmental Conservation*, 37, 5–15.

du Toit, J.T., Kock, R. & Deutsch, J.C. (eds) (2010). *Wild Rangelands: conserving wildlife while maintaining livestock in semi-arid ecosystems*. Chichester: Wiley-Blackwell.

Duvail, S., Médard, C., Hamerlynck, O. & Nyingi, D.W. (2012). Land and water grabbing in an East African coastal wetland: the case of the Tana delta. *Water Alternatives*, **5**(2), 322–343.

Ellis, F. (2000). *Rural Livelihoods and Diversity in Developing Countries*. Oxford: Oxford University Press.

Emerton, L. (2001). The nature of benefits and the benefits of nature. In: *African Wildlife and Livelihoods: the promise and performance of community conservation*, edited by D. Hulme & M. Murphree. Oxford: James Currey. For a more detailed treatment see: www.sed.manchester.ac.uk/idpm/research/publications/archive/cc/cc_wp05.htm.

Fairhead, J., Leach M. & Scoones I. (2012). Green grabbing: a new appropriation of nature? *Journal of Peasant Studies*, **39**(2), 237–261.

Folke, C. (2006). Resilience: the emergence of a perspective for social–ecological systems analyses. *Global Environmental Change*, **16**, 253–267.

Fratkin, E. (2014). Ethiopia's pastoralist policies: development, displacement and resettlement. *Nomadic Peoples*, **18**, 94–114.

Fratkin, E. & Roth, E. (2005). *As Pastoralists Settle: social, health, and economic consequences of the pastoral sedentarisation in Marsabit District, Kenya*. Dordrecht: Kluwer Academic Press.

Galaty, J. (1999). Grounding pastoralists: law, politics and dispossession in East Africa. *Nomadic Peoples* **3**(2), 56–73.

Galaty, J. (2014). "Unused" land and unfulfilled promises: justifications for displacing communities in East Africa. *Nomadic Peoples*, **18**, 80–93.

Galvin, K.A. (2009). Transitions: pastoralists living with change. *Annual Review of Anthropology*, **38**, 185–198.

Galvin, K.A., Beeton, T.A. & Luizza, M.W. (2018). African community-based conservation: a systematic review of social and ecological outcomes. *Ecology and Society*, **23**(3), 39.

Gough, I. & McGregor, J.A. (eds.). (2007). *Wellbeing in Developing Countries: new approaches and research strategies*. Cambridge: Cambridge University Press.

Greiner, C. (2012). Unexpected consequences: wildlife conservation and territorial conflict in Northern Kenya. *Human Ecology*, **40**, 415–425.

Haviland, W.A. (2003). *Anthropology*. Belmont, CA: Wadsworth.

Homewood, K. (2017). "They call it Shangri-La": sustainable conservation, or African enclosures? In: *The Anthropology of Sustainability: beyond development and progress*, edited by M. Brightman & J. Lewis. New York, NY: Palgrave Macmillan.

Homewood, K. & Rodgers, W.A. (1991). *Maasailand Ecology: pastoralist development and wildlife conservation in Ngorongoro, Tanzania*. Cambridge: Cambridge University Press, reprinted 2004.

Homewood, K., Lambin, E.F., Coast, E., et al. (2001). Long-term changes in Serengeti–Mara wildebeest and land cover: pastoralism, population or policies? *Proceedings of the National Academy of Science*, **98**(22), 12544–12549.

Homewood, K., Kristjanson, P. & Chenevix Trench, P. (2009). *Staying Maasai? Livelihoods, conservation and development in East African rangelands*. New York, NY: Springer.

Homewood K., Chenevix Trench, P. & Brockington, D. (2012). Pastoralist livelihoods and wildlife revenues in East Africa: a case for coexistence? *Pastoralism: Research, Policy, Practice*, **2**(19). www.pastoralismjournal.com/content/2/1/19

Homewood, K., Bluwstein, J., Lund, J., et al. (2015). The economic and social viability of Tanzanian wildlife management areas. Policy Briefs No. 4, Copenhagen Centre For Development Research. http://ccdr.ku.dk/policy_briefs/

HRBDF. (2013). Ethiopian sugar alleged to destroy pastoral communities of Lower Omo, Ethiopia. Human Rights and Business Dilemmas Forum, http://human-rights-forum.maplecroft.com/showthread.php?15012-Ethiopian-Sugar-Alleged-to-Destroy-Pastoral-Communities-of-Lower-Omo-Ethiopia&s=6a203e6c83c733e2c18220adfdea0c2c (accessed 30 August 2013).

Igoe, J. & Croucher, B. (2007). Conservation, commerce and communities. *Conservation and Society*, **5**, 534–561.

Inter-Réseaux Développement Rural. (2012). Pastoralism in Sub-Saharan Africa: Know its Advantages, Understand its Challenges, Act for its Sustainability. Food Sovereignty Brief, no. 5 (7 pp). Available at www.fao.org/fileadmin/templates/agphome/documents/rangelands/BDS_pastoralism_EN.pdf

IWGIA. (2013). Forced evictions of pastoralists in Kilombero and Ulanga Districts in Morogoro Region in Tanzania. Briefing Note. IWGIA (International Work Group For Indigenous Affairs) Classensgade 11 E, DK 2100 – Copenhagen, Denmark. 5 pp.

Jones, S. (2006). Political ecology of wildlife conservation in Africa. *Review of African Political Economy*, **33**(109), 483–495.

Keane, A., Gurd, H., Kaelo, D., et al. (2016). Gender differentiated preferences for a community-based conservation initiative. *PLoS ONE*, **11**(3), e0152432. doi:10.1371/journal.pone.0152432

Lambin, E.F. & Meyfroidt, P. (2011). Global land use change, economic globalisation and the looming land scarcity. *Proceedings of the National Academy of Sciences of the USA*, **108**, 3465–3472.

Leslie, P. & McCabe, J.T. (2013). Response diversity and resilience in social–ecological systems. *Current Anthropology*, **54**(2), 114–143.

Letai, J. & Lind, J. (2012). In: *Pastoralism and Development in Africa: dynamic change at the margins*, edited by A. Catley, J. Lind, & I. Scoones. Abingdon: Earthscan/Routledge.

Mariki, S., Svarstad, H. & Benjaminsen, T. (2015). Elephants over a cliff: explaining wildlife killings in Tanzania. *Land Use Policy*, **44**, 19–30.

Marshall, F., Reid, R., Goldstein, S., et al. (2018). Ancient herders enriched and restructured African grasslands. *Nature*, **561**, 387–391.

McPeak, J.G., Little, P.D. & Doss, C.R. (2011). *Risk and Social Change in an African Rural Economy: livelihoods in pastoralist communities*. London: Routledge

Milner-Gulland, E.J., Mcgregor, J., Agarwala, M., et al. (2014). Accounting for the impact of conservation on human well-being. *Conservation Biology*, **28**, 1160–1166.

Nelson, F. (ed.). (2010). *Community Rights, Conservation & Contested Land: the politics of natural resource governance in Africa*. London: Earthscan.

Nelson, F. (2012). Natural conservationists? Evaluating the impact of pastoralist land use practices on Tanzania's wildlife economy. *Pastoralism: Research, Policy and Practice*, **2**(15). www.pastoralismjournal.com/content/2/1/15

Nelson, F. & Agrawal, A. (2008). Patronage or participation: community-based natural resource management reform in Sub-Saharan Africa. *Development and Change*, **39**(4), 557–585.

Niamir-Fuller, M. (ed.). (1999). *Managing Mobility in African Rangelands: the legitimation of transhumance*. London: Intermediate Technology Publications.

Niamir-Fuller, M., Kerven, C., Reid, R. & Milner-Gulland, E.J. (2012). Co-existence of wildlife and pastoralism on extensive rangelands: competition or compatibility? *Pastoralism: Research, Policy and Practice*, **2**(8). doi:10.1186/2041-7136-2-8

Norton-Griffiths, M. & Said, M. (2010). The future for wildlife on Kenya's rangelands: an economic perspective. In *Wild Rangelands: conserving wildlife while maintaining livestock in semi-arid ecosystems*, edited by J. Deutsch, J. Du Toit & R. Kock. London: Zoological Society of London Symposium volume with WCS, pp. 367–392.

Odadi, W.O., Karachi, M., Abdulrazak, S. & Young, T.P. (2011). African wild ungulates compete with or facilitate cattle depending on season. *Science*, **333**, 1753–1755.

Ogutu, J., Owen-Smith, N., Piepho, H.P. & Said, M. (2011). Continuing wildlife population declines and range contraction in the Mara region of Kenya during 1977–2009. *Journal of Zoology*, **285**, 99–109.

Ogutu, J., Owen-Smith, N., Piepho, H.P., Kuloba, B. & Edebe, J. (2012). Dynamics of ungulates in relation to climatic and land use changes in an insularised African savanna ecosystem. *Biodiversity and Conservation*, **21**(4), 1033–1053.

Osano, P. (2013). *Direct payments to promote biodiversity conservation and the implication for poverty reduction among pastoral communities in east African arid and semi-arid lands*. PhD thesis, McGill University, Canada.

OXFAM. (2010) *Disaster Risk Reduction in Drought Cycle Management: a learning companion*. London: Oxfam Disaster Risk Reduction and Climate Change Adaptation Resources.

Prins, H.H. (2000). Competition between wildlife and livestock in Africa. In: *Wildlife Conservation by Sustainable Use*, edited by H.H.T. Prins, T. Geu Grootenhuis & T. Dolan. Boston, MA: Kluwer Academic Publishers, pp. 53–80.

Randall, S., Coast, E. & Leone, T. (2011). Cultural constructions of the concept of household in sample surveys. *Population Studies – A Journal of Demography*, **65**(2), 217–229. doi:10.1080/00324728.2011.576768.

Reid, R., Fernández-Giménez, M. & Galvin, K. (2014). Dynamics and resilience of rangelands and pastoral peoples around the globe. *Annual Review of Environment and Resources*, **39**(1), 217–242.

Robbins, P. (2012) *Political Ecology: a critical introduction*. 2nd edn. Chichester: Wiley.

Robinson, L. & Berkes, F. (2010). Applying resilience thinking to questions of policy for pastoralist systems: lessons from the Gabra of northern Kenya. *Human Ecology*, **38**(3), 335–350.

Rockström, J., Steffen, W., Noone, K., et al. (2009). A safe operating space for humanity. *Nature*, **461**(7263), 472–475.

Savory, A. (1988). *Holistic Resource Management*. Washington, DC: Island Press.

Scoones, I. (2009). Livelihoods perspectives and rural development. *Journal of Peasant Studies*, **36**(1), 171–196.

Sinclair, A.R.E. & Arcese, P. (1995). *Serengeti II: dynamics, management and conservation of an ecosystem*. Chicago, IL: Chicago University Press.

Sinclair, A.R.E. & Norton Griffiths, M. (1979). *Serengeti I: dynamics of an ecosystem*. Chicago, IL: Chicago University Press.

Sinclair, A.R.E., Packer, C., Mduma, S. & Fryxell, J. (2008). *Serengeti III: human impacts on ecosystem dynamics*. Chicago, IL: Chicago University Press.

Spencer, P. (1988). *The Maasai of Matapato: a study of rituals of rebellion*. Manchester: Manchester University Press for the International African Institute.

Sternberg, T. & Chatty, D. (2012). *Modern Pastoralism and Conservation: old problems, new challenges*. Whitstable: White Horse Press.

Sulle, E., Lekaita E. & Nelson, F. (2011). From Promise to Performance? Wildlife Management Areas in Northern Tanzania. www.maliasili.org/publications/WMA%20TNRF%20Brief%20Nov%202011.pdf

Thompson, D.M., Serneels, S., Ole Kaelo, D. & Chenevix Trench, P. (2009). Maasai Mara – land privatisation and wildlife decline: can conservation pay its way? In: *Staying Maasai? Livelihoods, conservation and development in East African rangelands*, edited by K. Homewood, P. Kristjanson & P. Chenevix Trench. New York, NY: Springer.

URT. (2005). *Mkukuta: Tanzania's national strategy for growth and reduction of poverty*. Dar es Salaam: United Republic of Tanzania: Vice-President's Office.

URT. (2010). *Mkukuta II: national strategy for growth and reduction of poverty II*. Dar es Salaam: Ministry of Finance and Economic Affairs, United Republic of Tanzania.

Walsh, M. (2008). The not-so-Great Ruaha and hidden histories of an environmental panic in Tanzania. *Journal of Eastern African Studies*, **6**(2), 303–335.

Western, D., Groom, R. & Worden, J. (2009a). The impact of subdivision and sedentarisation of pastoral lands on wildlife in an African savanna ecosystem. *Biological Conservation*, **142**, 2538–2546.

Western, D., Russell, S. & Cuthill, I. (2009b). The status of wildlife in protected areas compared to non-protected areas of Kenya. *PLoS ONE*, **4**(7), e6140.

Wisner, B., Blaikie, P., Cannon, T. & Davis, I. (2004). *At Risk: natural hazards, people's vulnerability and disasters*. London: Routledge Publication.

Woodhouse, E., Homewood, K.M., Beauchamp, E., et al. (2015). Guiding principles for evaluating the impacts of conservation interventions on human well-being. *Philosophical Transactions of the Royal Society B: Biological Sciences*, **370**, 1681.

Zoomers, A. (2010). Globalisation and the foreignisation of space: seven processes driving the current global land grab. *Journal of Peasant Studies*, **37**(2), 429–447.

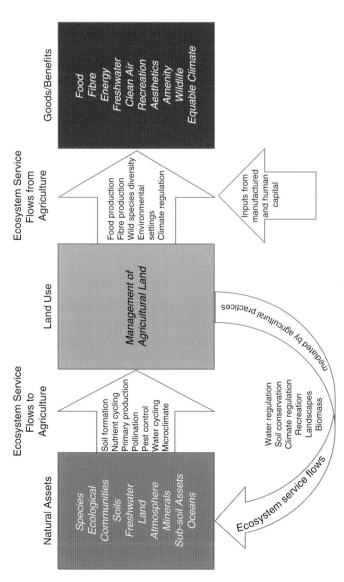

Figure 4.1 The influence of agriculture on the flow of ecosystem services. (A black and white version of this figure will appear in some formats.)

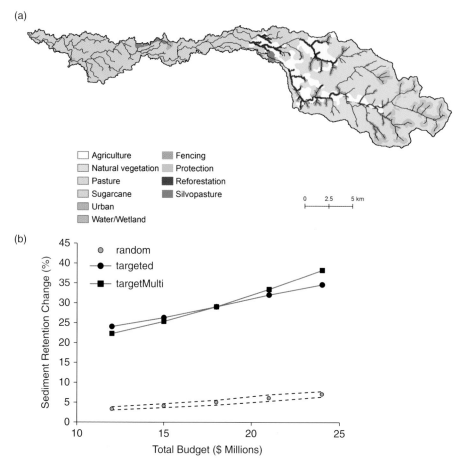

Figure 4.2 (a) Investment portfolio for where to prioritise different conservation activities in the Cauca Valley of Colombia. (b) Estimated improvement in sediment retention resulting from conservation activities that were either targeted to sediment specifically (targeted), targeted to multiple ecosystem services (targetMulti), or randomly allocated across the watershed. Figure credit: S. Wolny and A. Vogl. (A black and white version of this figure will appear in some formats.)

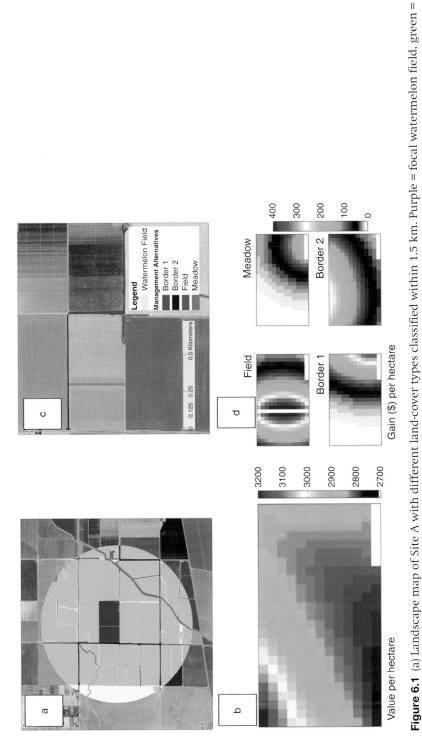

Figure 6.1 (a) Landscape map of Site A with different land-cover types classified within 1.5 km. Purple = focal watermelon field, green = annual row crop, yellow = managed grassland, pink = farmyards and homes, brown = unvegetated (including unpaved farm roads), black = paved roads. (b) Value map of the focal watermelon field based on modelled pollination with no habitat enhancement (status quo). Pixels are 30 × 30 m. (c) Landscape map showing locations of different enhancement alternatives. In this example the meadow surrounds a tail-water pond. (d) Marginal gain in value per hectare under four enhancement alternatives. (A black and white version of this figure will appear in some formats.)

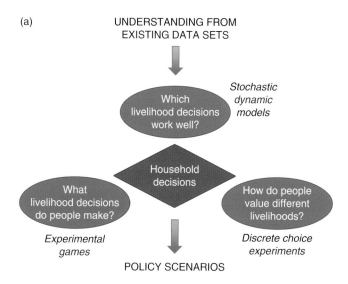

Figure 9.3 (a) Questions and tools used to explore the decision-making process in pastoralist households. (b) An economic game where the blue counters represent wealth held either as cash or livestock. The blue square on the board represents the bank: this player chooses to hold only a small part of his wealth there. The green square represents his own land, with more of his wealth held as livestock on this private grazing. The largest share of his wealth is held as livestock grazing illegally in the Maasai Mara National Reserve, represented by the red square on the board. These games and choice experiments revealed the extent to which pastoralists' strategies (and theoretically optimal choices) shift according to household wealth and land tenure context, as well as individual characteristics (age, gender, education ...), and the unanticipated outcomes of poorly-conceived conservation policies. These context-sensitive strategies represent an important dimension of response diversity crucial to the internal resilience of the pastoralist SES. (A black and white version of this figure will appear in some formats.)

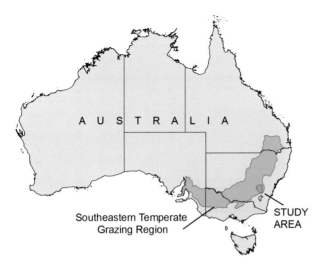

Figure 11.1 Case study area in the southeastern 'sheep–wheat' belt, the temperate grazing region of Australia. (A black and white version of this figure will appear in some formats.)

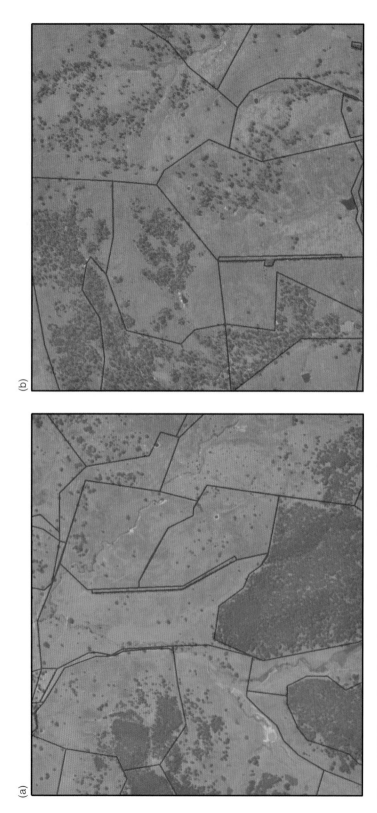

Figure 11.2 Examples of woody vegetation cover for subsets of an archetypal (a) conventional (spared) farm and (b) holistic (shared) farm from the study area. Red lines represent fencelines (fencelines do not necessarily represent spared land). (A black and white version of this figure will appear in some formats.)

Figure 11.3 Examples of (a) a spared patch of trees and (b) a 'shared', holistically grazed landscape to the left of the fence, compared with conventional grazing to the right. Both photos were captured by Lachlan catchment farmers when asked to photograph significant elements of their farm landscape (method explained in, for example, Sherren et al., 2012b). (A black and white version of this figure will appear in some formats.)

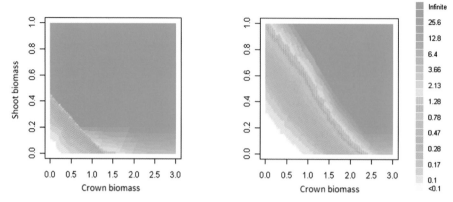

Figure 13.6 Two sections ($\gamma_g = 0.65$ (left) and $\gamma_g = 0.9$ (right)) of resilience values of the rangeland described by dynamics (13.3) for the property (13.4) toward drought events causing a sudden reduction in shoot biomass. Parameter values are $r_s = 3$, $a_c = 0.1$, $r_c = 1$, $\gamma_g = 0.65$, $\underline{s} = 0.1$, $\bar{u} = -\underline{u} = 0.05$ and $\bar{a} = 0.5$. (A black and white version of this figure will appear in some formats.)

PART II

Integrating biodiversity and building resilience into agricultural systems

CHAPTER TEN

Delivering sustainability in agricultural systems: some implications for institutional analysis

IAN HODGE
University of Cambridge

Introduction

Conventional agricultural systems will have to change in order to meet future demands and constraints. The extent and nature of these changes remain, and will remain, unclear – it is a shifting target within an evolving context – but a central tenet of the changes required is that agriculture will need to be more 'sustainable'. Over time, we can identify a series of shifts in perspective from seeing agriculture as an activity that impacts on the environment to multi-functionality that characterises the joint production of commodities and environment, and then to an ecosystems perspective that sees the agricultural system as being part of the environment. The most recent characterisation is in terms of a search for sustainable intensification. The ecosystems approach in particular leads to a focus on resilience that challenges assumptions about optimisation and raises potential differences between private and social resilience. In this context, this chapter considers possible interpretations of sustainability and their implications for the ways in which analysis and institutions may be developed in order to meet future challenges.

Environmental impacts of farming systems

The development of modern agricultural systems has tended to emphasise control and simplification, particularly through the importation of fossil fuel energy in the use of machinery, fertilisers and pesticides. Darnhofer et al. (2010a) characterise this as the 'engineering' approach to farm management with an objective of providing the best possible conditions for crops and stock and excluding competition from pests and disease. This has been highly successful in increasing levels of production per unit area and in reducing variability in levels of production over time, but it has done so at the expense of externalising costs through water and air pollution and reductions in the quality of the landscape in terms of habitats and amenity. The agricultural system has also become increasingly reliant on externally sourced inputs, particularly those derived from fossil fuels that are themselves sources of increasing external costs in their production. Agriculture in many developed countries has also been subsidised through agricultural policy. This has had apparently divergent environmental effects depending on the circumstances. Support for output prices and amelioration in the risks faced by farmers have encouraged investment involving technical change and greater production intensity along the intensive margin, generally increasing external costs. At the other extreme, support for marginal farming systems in areas of low productivity has prevented land abandonment and maintained High Nature Value landscapes (Plieninger & Bieling, 2013). Such systems are generally characterised as providing public goods. Nevertheless, even in these areas subsidies can stimulate excessive intensity causing environmental damage, such as by overgrazing or soil erosion.

Concerns for the environmental impacts of farming, often in the past in the face of agricultural surpluses and rising costs of government support, led to the development of agri-environment policy in the European Union (EU) from the mid-1980s (e.g. Hodge, 2014) and the concept of multi-functionality (OECD, 2001). These approaches were developed in a context of agricultural protection supporting product prices above world market levels. At least in the early stages of agri-environment policy within the EU, the conjunction of supported prices and environmental impacts made schemes introduced to restrain the intensity of agricultural production particularly attractive in that they could simultaneously reduce both environmental harm and government expenditure. The adjustments motivated by agri-environment schemes have mainly taken place around the margins of farming systems, such as by reduced grazing intensity, provision of uncropped buffer strips, expansions of wetland or woodland areas and the better management of boundary features such as hedgerows and stone walls. While overall production levels will have been diminished to some extent, the farming systems themselves have not changed fundamentally. There have been some gains for the environment, although their extent remains controversial. Kleijn et al. (2001) suggested that some schemes simply did not achieve their objectives or possibly even had adverse ecological effects, although this was countered by Stoate and Parish (2001) and Carey (2001). Kleijn and Sutherland (2003) reviewed the available evidence and concluded that it was often inadequate to provide reliable results. Recent analysis of studies of the effectiveness of agri-environment schemes (AES) finds that most syntheses demonstrate general increases in farmland biodiversity in response to AES, with the size of the effect depending on the structure and management of the surrounding landscape (Batáry et al., 2015).

The ecosystem context

The development of an ecosystems approach shifts the emphasis from agriculture as an activity that has an impact on the environment towards

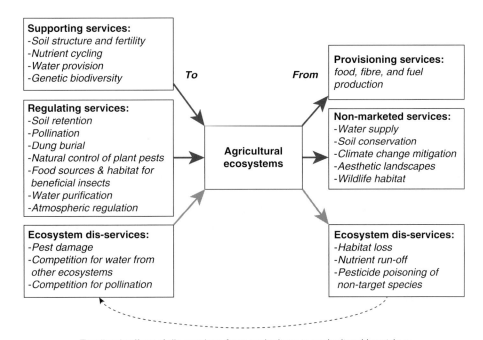

Figure 10.1 Agricultural systems within ecosystems. *Source*: Zhang et al. (2007).

a perspective of the agricultural system as operating within an ecosystem. Agricultural systems are set within a wider context of ecosystems, both being impacted by and impacting on ecosystem services. Farm management then may be seen as manipulating the ecosystem in order to enhance the output of provisioning services. This is illustrated by Figure 10.1 from Zhang et al. (2007).

In this context, it is important to be explicit about the counterfactual. It is commonly asserted, as suggested in the figure, that agriculture makes both negative and positive contributions to ecosystem services. Power (2010), for instance, characterises the mitigation of greenhouse gas emissions, such as by conservation tillage, as an ecosystem service from agriculture. The counterfactual being adopted here is that of conventional agricultural practices. But presumably, if agriculture was not undertaken at all, the level of greenhouse gas mitigation would be likely to be significantly greater than that arising under conservation tillage. This question is important because it effectively defines a reference level against which judgements are being made about property rights and hence where responsibilities lie. The implication of terming conservation tillage as generating an ecosystems service here is that land managers have a right to allow the level of greenhouse gas emissions associated with 'conventional' agricultural systems and that any reduction from that level represents the delivery of a benefit. Alternatively, we might take the unfarmed environment as the counterfactual and conclude that conservation tillage rather reduces the level of dis-service, with the policy implication that the farmer might face a lower penalty as compared with conventional farmers rather than receiving a payment.

This reflects a social judgement that needs to be made explicit in advance of the introduction of any policy towards ecosystem services, based on a definition of 'good agricultural practice' setting out a judgement as to what constitutes 'good' or 'acceptable' landownership. However, it is a judgement that will change over time, as has been the case with other impacts, such as with nitrogen leaching. In Europe, farmers were initially paid to reduce levels of nitrate leached from farmland, but leaching is now treated as pollution under Nitrate Vulnerable Zone regulations. Assuming that there is a generally increasing expectation of a responsibility to mitigate greenhouse emissions, we can anticipate that the reference level will shift, perhaps towards a baseline representing the position that would apply in the absence of agricultural production. This principle applies to all ecosystem services and is subject to debate. Rodgers (2009), for instance, has argued for the introduction of a general duty of stewardship as an attribute of property ownership in the law of England and Wales. This would shift the reference level towards a duty to achieve a higher standard of environmental conservation.

Towards sustainable intensification

With more recent predictions of a looming world food crisis (e.g. Godfray et al., 2010; Wheeler & von Braun, 2013), the emphasis on the way in which agricultural systems should develop has shifted again. Projections of increasing global populations and income levels indicate a demand for increases in agricultural output but within a largely fixed total area of agricultural land and subject to increasingly severe constraints on resource availability and utilisation. This will have significant implications for commodity prices, and in fact Piesse and Thirtle (2009) suggested that the trend towards higher commodity prices may have begun even before the price hikes experienced in 2007–08. The most recent easing of commodity prices after 2011 might, of course, reflect a return to long-run decline, but this combination of global factors does seem likely to have implications across the food system (Foresight, 2011), including for patterns of demand under the influence of changing relative prices. Specifically with regard to agriculture, there have been calls for 'sustainable intensification' (Royal Society, 2009; Garnett et al., 2013) in what is likely to be a more unstable environment, both physically and economically. Precisely what this involves or how it is to be achieved remains uncertain, but there is clearly a requirement for technical research,

Table 10.1. *Rebalancing systems for sustainable intensification.*

Reduced inputs (reduced ecosystem dis-services)	Fossil fuels
	Inorganic fertilisers
	Chemical pesticides
Substitute inputs	Technology (e.g. genetic material, machine efficiency)
	Information (e.g. precision farming)
	Recycling materials
	Local ecosystem services (e.g. pollination, pest management)
	On-farm plant nutrients (e.g. waste management, crop rotations)
	Renewable energy (e.g. anaerobic digestion)
External benefits (enhanced ecosystem services)	Space and management for biodiversity
	Carbon sequestration
	Landscape
	Public access

development and implementation and to make the best use of the technology already available in reducing the yield gap (Van Ittersum et al., 2013).

The degree to which farm systems will need to change is unclear but some of the sorts of farm level changes that may be expected in the context of sustainable intensification are suggested in Table 10.1. We can assume that there will be pressures to reduce the use of inputs that derive from fossil fuels and are imported onto the farm. Logically, then, resources have to be used more efficiently through technical change and/or some other material or change in the system has to substitute for this, presumably by recycling within the farm system itself, or within the wider local area. It also requires substitutes for the fertility and pest management services provided by chemical inputs. This seems likely to arise from changes to farming systems, perhaps in crop rotations or a return to mixed farming and increased use of ecosystems services. Kremen and Miles (2012) argue that diversified farming systems, those that incorporate agrobiodiversity across multiple spatial and/or temporal scales, offer a range of benefits over conventional systems, such as those associated with soil fertility and pest control. Bommarco et al. (2013) characterise the replacement of anthropogenic inputs and/or including regulating and supporting ecosystems services management as 'ecological intensification'. At the same time, there are demands for agricultural land to generate higher levels of ecosystems services that are valued by other sectors and this is likely to introduce a further limitation on the room for manoeuvre in designing agricultural systems for sustainable intensification. In case studies of farms achieving sustainable intensification (Firbank et al., 2013), farmers were found to have introduced changes in order to reduce waste of energy and nutrients and to manage risk through increasing on-farm capacity to grow animal feed, store water or to generate renewable energy. The motivations for the changes were primarily financial, but sustainable intensification also raises wider social issues. Fish et al. (2014) argue for the need to establish a new interdisciplinary framing of the ecological basis of sustainable intensification, beyond the valuation of ecosystem services in terms of their contribution to agricultural production systems.

Sustainability of agricultural systems

In this context, it is appropriate to review approaches towards sustainability. Pretty (2008) suggests that sustainability incorporates criteria of both persistence and resilience. To these, the Royal Society (2009) adds 'Autarchy', the capacity to deliver outputs from inputs and resources within key system boundaries, and 'Benevolence', the capacity to produce outputs while sustaining the functioning of ecosystem services and not causing depletion of natural capital (e.g. minerals, biodiversity, soil and clean water). With regard to the former, autarchy seems less critical in principle, provided that implications of drawing inputs from a wider system are fully taken into account in terms of the other criteria. Benevolence may be interpreted as already being encompassed by persistence and resilience, provided that landholders are regarded as having a duty to protect wider ecosystems and natural capital. If there is demand for agricultural systems to go further and enhance ecosystem functioning and/or natural capital for the benefit of other stakeholders, then we can expect positive incentives for this to be provided.

The emphasis in discussions of sustainability until recently has tended to be on persistence, maintaining some level of capital indefinitely into the future. Five types of capital are conventionally identified: natural, social, human, manufactured and financial. Within this, it is important to prioritise the maintenance of capitals for which there are no close substitutes, whose loss would be irreversible and around which there are significant levels of uncertainty or ignorance. In this context, there are likely to be significant option values. This offers some guidance as to which aspects of an agricultural system might warrant greatest protection: perhaps land from irreversible development, soils from erosion, or habitat from major and especially irreversible losses of biodiversity. The same sorts of argument can apply to human knowledge or social networks among farmers in terms of human and social capitals, but this does not provide guidance as to the scale or specific level at which they should be maintained. For instance, it might be acceptable to deplete habitat on one farm provided that it is enhanced to some equivalent extent nearby. Would this apply if the offsetting enhancement was in another region, in another country or for a different type of habitat or species? Such decisions need to be made within a specific context and institutional arrangements are required within which constraints can be set.

What is important is the capacity of the system in some sense rather than the need to set minimum levels of individual units or categories of assets. The criterion of persistence is less obviously applied to the management of ecosystems where the requirement is to maintain multiple functions. So, for instance, the maintenance of biodiversity within an agricultural ecosystem generally depends not only on the maintenance of certain elements of natural capital but also on the implementation of specific management practices. In this context, the emphasis shifts to resilience. Resilience is often considered in terms of the capacity to maintain a certain structure against external shocks, but this implies a rather static view. Folke (2006, p. 259) defines it as 'the capacity of a system to absorb disturbance and re-organize while undergoing change so as to still retain essentially the same function, structure, identity and feedbacks'. This points not just to a capacity to absorb change, but also to the potential to reorganise and adapt. Resilience thus refers to the capacity of a system to absorb disturbance and reorganise while undergoing change (Folke et al., 2005). It is not simply about preventing change; disturbance opens up opportunities for renewal and new trajectories. It can be possible to make changes to agricultural production activities without compromising the maintenance of ecosystem functions. Resilience thus includes ideas of adaptation, learning and self-organisation. A key approach is through adaptive management that recognises the high degree of uncertainty and ignorance and undertakes initiatives as experiments, incorporating the results into future management strategies. Institutions are required that build knowledge of the systems and apply this knowledge to management practices. They need to be flexible, dealing with perturbations and surprise, and operate across multiple levels. They also need to bring social values into decision-making processes. The Stockholm Resilience Centre (2014) suggests seven principles as guidance for building resilience:

- maintain diversity and redundancy,
- manage connectivity,
- manage slow variables and feedbacks,
- foster complex adaptive systems thinking,
- encourage learning,
- broaden participation, and
- promote polycentric governance.

Note that this approach implies that we cannot expect to identify an 'optimal' solution, even as an objective. We do not have sufficient information to be able to determine a single 'best' outcome, especially in the long term. A resilient system will be likely to include greater diversity than a system that seeks to maximise expected net present value on the basis of current knowledge and projections. Holling et al. (1995) argue that policies and management that apply fixed rules for achieving constant yields lead to systems that increasingly lack resilience, i.e. ones that may suddenly break down in the face of disturbances that previously could be absorbed. The 'engineering' approach to agriculture, noted above, reflects this type of management. A resilient system is also likely to include apparently redundant elements that may only become of value in the face of shocks and surprises, although it is, inevitably, difficult to know exactly what is to be preserved (Hanley, 1998). The maintenance of future options implies that not all resources can be used efficiently at any one time; 'excess' capacity will be retained 'in case' of unexpected changes (Darnhofer et al., 2010a). In the absence of knowledge as to what shocks may be experienced, we cannot be certain as to which elements are or are not redundant. The position is explained by Darnhofer (2014, p. 476): 'Understanding the ubiquity of change and accepting that many shocks are unpredictable implies that farmers spend fewer resources on identifying slacks and improvements in efficiency under the current situation or quantifying the impact of expected developments. Rather, resources are allocated to strategies that allow reducing the impact of a wide variety of potential (currently unknown) events and on identifying emergent opportunities'. This perspective then challenges the application of standard optimisation models in farm management and cost–benefit analysis in the evaluation of alternative projects. It also challenges the development of indicators of sustainability where the status of an organisation is measured against a standard based on efficiency, such as using Data Envelopment Analysis.

Sustainability at a farm level

While the focus of resilience analysis is generally at the ecosystem scale, it is also possible to characterise a farm as a complex system with similar attributes (Darnhofer et al., 2010b) and hence to apply this approach to the sustainability of individual farms. A farm may then be conceptualised as a set of agro-ecological, economic and social subsystems that interact and are subject to their own dynamics. Within a varying and uncertain environment, these systems cannot be held static; sustainability requires both the persistence of capital elements and adaptability and change. Resilience then covers buffer capability, adaptive capability and transformative capability (Darnhofer, 2014).

Darnhofer (2010) has examined the strategies of a sample of Austrian family farms with regard to four clusters of factors that have been identified as building resilience in social–ecological systems. These are: learning to live with change and uncertainty, nurturing diversity in its various forms, combining different types of knowledge and learning and creating opportunity for self-organisation and cross-scale linkages. She concludes that there are considerable parallels between the approaches taken by the small farmers who participated in the research and the types of strategy that are advocated in order to build resilience. The strategies adopted by this type of farmer have enabled farm businesses and households to survive through several generations. Farmers maintain flexibility and adaptability and build extensive networks to diversify information and income sources. They take a long-term perspective and avoid excessive debt or dependence on factors outside of their own control. This does not mean that they are static. Rather, they respond to opportunities both on and off the holding. They collaborate with other farmers and engage with local community institutions. There is also an ethical element to this perspective. All the farmers who were involved in the study had inherited land from the previous generation and all were aiming to pass it on to their children. This reinforced a commitment to use natural resources sustainably, balanced against financial considerations. There is thus a trade-off between short-term efficiency and long-term resilience.

Does this amount to more than the risk aversion that is often found among farmers or small businesses? It does hold many similarities, and the limits of simple profit-maximising models for the guidance of practical farm management decisions are well known (see also Ramsden & Gibbons, Chapter 5). However, a resilience approach seems to engage more proactively with issues of sustainability and uncertainty, reflecting perhaps a strategy that might emerge from a combination of risk aversion, a high significance attributed to future costs and benefits and an ethical approach towards the environment. Darnhofer et al. (2010b, p. 187) comment that 'focusing on farming system resilience thus builds on the recognition that it is necessary to move away from the analytical assumptions of equilibrium thinking, centred on linearity, predictability, optimisation, homogeneity and simplification'.

Various factors have been associated with resilience in other studies. Carlisle (2014) assessed the characteristics of a group of farmers in north-central USA who demonstrated greater resilience to drought through low-input management of diverse agroecosystems, diversification of markets and cooperation across farms. They also benefited from a network of civil society groups giving advocacy and technical support and public safety nets and supportive policy. Peerlings et al. (2014), on the basis of self-reported responses, identified large, more specialised farms with young and highly educated farm heads as being most resilient.

The example of apparent stability and long-term planning in Austria contrasts with the historical experience of the agricultural sector in the UK which has been through periods of significant change and adaptation. Would the pursuit of resilience at the individual farm level prevent periods of adjustment and structural change that can open up new opportunities and approaches

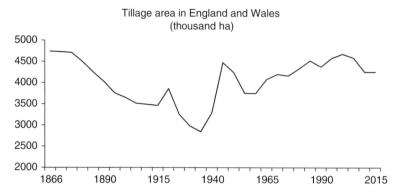

Figure 10.2 Area of tillage in England and Wales. *Source*: data from Ministry of Agriculture, Fisheries and Food (1968); Department for Environment, Food and Rural Affairs and StatsWales.

that could be advantageous at a social level? What institutional arrangements may affect the capacity of farms to adjust to changing physical and market conditions? Agriculture has been through various periods of recession or depression when farm businesses have experienced severe pressures, often promoting innovation and radical change. In the UK, such periods include the late nineteenth century and the late 1920s and 1930s, separated by a relatively brief respite through the First World War. The changing area of tillage in England and Wales is illustrated in Figure 10.2. Over these periods substantial numbers of farmers went out of business and areas of land were taken out of cereal production and fell into extensive grazing or were abandoned (Brown, 1987). The farm businesses that survived, those that were most resilient, tended to be those with greater reserves of capital or that managed to reduce costs or shift into alternative forms of production. Some farmers were reluctant or unable to change, some were committed to 'high' or intensive farming systems, in the hope or expectation that circumstances would soon return to what they regarded as the 'normal' state of affairs. They were less likely to survive.

In the agricultural depression in the nineteenth century, a high proportion of land was tenanted on large landed estates and landlords were often willing to reduce rents either from a commercial perspective in order to keep the land under production or from social concerns. In this way, the institution of the landlord/tenant system increased the resilience of the individual farm businesses. Some farmers embraced change and this may have been associated with the introduction of new individuals into farming. New entrants bring different approaches and expectations. In parts of East Anglia around the turn of the twentieth century, substantial numbers of migrants from Scotland and the West Country introduced different methods and a willingness to accept lower incomes. Fletcher (1961) reports comments made by a Lancashire farmer touring Essex in 1896 that 'new-comers are going in for milk, cheese, butter, fruit, and sheep but with the average Essex farmer it is corn, corn, corn'. In fact, Hunt and Pam (2001) argue that in Essex major changes were underway prior to the in-migration of farmers, but it is apparent that the new farmers did introduce different agricultural systems and managed to operate at lower cost. Fletcher (1961, p. 431) concludes that the agricultural depression 'turned English farmers in a new direction, that is to the production of quality livestock products. It inflicted suffering on a particular section of farmers, the large corn growers. Output figures shew [sic] that agriculture was not ruined; on the contrary an important internal revolution was effected during a period of falling prices'.

The period of depression in the 1920–1930s saw an even steeper decline in the area of land under tillage. Brown (1987) comments that the turnover of farms in the 1920s and early 1930s far exceeded that of the previous depression in the 1880s and 1890s. While there are many factors that may explain this difference, one relates to the way in which land was owned. The nineteenth century depression saw agricultural rents fall from a peak in the mid-1870s through the remainder of the century (Orwin & Whetham, 1964). These were helpful to tenant farmers, but low rents were combined with a series of other challenges to the ownership of landed estates. Death duties were introduced on agricultural land in 1894 (Cannadine, 1996) and while there were exemptions, this represented a problem that was exacerbated by the deaths of the owners and heirs of many landed estates in the First World War. While there were relatively few sales during the War, the years between 1918 and 1925 saw a very extensive turnover of land primarily in the sales of land to tenants. Thompson (1963, p. 332) estimated that between 1918 and 1921 between 2.6 million and 3.2 million hectares changed hands in England, approaching a quarter of the total area. While the reliability of the data has been questioned (Beckett & Turner, 2007; Thompson, 2007), the scale of the transfers from landlords to tenants is clear. In 1914, 11 percent of the agricultural land in England and Wales was occupied by its owners; this rose to 36 percent in 1927 (Sturmey, 1955). Thompson (1963, p. 333) comments that these transfers 'marked a social revolution in the countryside, nothing less than the dissolution of a large part of the great estate system and the formation of a new race of yeomen'. In consequence, by the time of the later depression in the late 1920s, a much larger proportion of farmers were owner-occupiers, often with significant levels of debt taken up in order to acquire property from their landlords. They were thus less resilient and more vulnerable to the pressures of the depression and large areas of land were substantially abandoned, not to be reclaimed until the demands for greater domestic food production in the Second World War. Brassley (2007) has reviewed British farming between the wars. He finds a period of both decline and regeneration. Some farmers successfully introduced new techniques, lowering costs and changing products, and overall, output rose, but other farmers went bankrupt and on average net farm incomes were relatively low. Similarly, Tranter (2012) has noted the stimulus of the 1921–1938 agricultural recession in the Berkshire Downs in promoting innovations in producing milk and rearing poultry and prompting reductions in casual labour and in fixed costs. This is not to say that all of the innovations made were successful; greater publicity is given to successes than to failures. More significant is that the periods of depression promoted a higher level of experimentation, some of which was successful in identifying new approaches that were adopted more generally, representing an adaptive management approach.

Major shocks also arise from outbreaks of animal and plant diseases. There have been several outbreaks of foot and mouth disease in the UK in the past that have had major impacts on the agricultural and rural sectors. The outbreak in 2001 had a major effect on farms and the agricultural sector, with some 4.2 million livestock being culled on over 13,000 holdings. For many farms their business was substantially wiped out and it must be doubtful as to whether the adoption of a resilience approach, other than possibly the implementation of strict biosecurity measures at the farm level, could have offered much defence. It is likely though that farm businesses that were more diversified or had higher levels of assets would have been more resilient and better placed to reestablish themselves subsequent to the outbreak. The specialisation typical in farming through the twentieth century will be likely to have left farms more vulnerable and less resilient. It is not apparent that this issue has been the subject of research. However, the outbreak drew particular attention to the extent to which the impacts were felt in rural areas beyond farms on non-agricultural businesses. The net costs to agriculture, the food chain and supporting services were estimated to be £0.6 billion, while tourism and supporting services were estimated to have lost around £5 billion in revenues. The cost to the public sector was estimated at £3 billion (Comptroller and Auditor General, 2002). A study of the responses of rural micro-businesses in the north-east of England to the outbreak of foot and mouth disease in 2001 assessed the ways in which they responded to the outbreak (Phillipson et al., 2004). The analysis looked at how businesses coped with the shock and the factors that provided resilience. Most common were working extra hours, cutting back on staff employment and running down financial and other capital. Two other aspects were important, drawing on external help such as for advice, credit or personal support, and making adjustments in the household, including household members working longer hours, taking a smaller wage, cutting back household spending, spending personal savings, or household members looking for a job. Businesses that had greater reserves, that had stronger external networks and that were linked to households that had capacity to adjust had a greater degree of resilience. This suggests that there are multiple strategies for building the resilience of individual businesses and that they are supported through links across businesses and households and through stronger external networks, both formal and informal. Similar principles may be assumed to apply to farm businesses too.

This discussion of shocks to farms and rural businesses has a number of possible implications. Agricultural depressions and other major shocks to an economic sector can be associated with periods of substantial change. Firms with substantial assets and capital resources, more flexibility and strong external networks are more likely to survive. However, there is also adoption of new approaches. These new approaches may be more likely to be adopted by businesses that are well capitalised and potentially by new entrants into the sector who may bring new ideas and innovations. The depth of the depression may depend on the extent to which risks can be passed on to others, such as through the way in which land is owned. This contrasts with the apparent long-term intergenerational stability of family farms in Austria. Does this imply that there are potential innovations and options that have not been adopted in Austrian agriculture?

Optimal private and social resilience

A resilience approach seems likely to become increasingly relevant to farm management in the future. The decoupling of agricultural support under the Common Agricultural Policy, where payments linked to production have been converted to fixed payments, means that farmers are likely to face more variable commodity prices. Brexit introduces a further layer of policy uncertainty. At the same time, climate change is forecast to increase the variability of weather conditions so that farmers are more likely to have to contend with droughts and flooding and generally unstable weather. This implies that the 'optimal' buffer capability in farm management may be increasing, but some sort of balance needs to be maintained. We do not know whether the industry as a whole will face major shocks of the type discussed here. While the adoption of a resilience approach by individual farms may be in a farmer's private interest and may also seem to be in the wider social interest, it is likely that the 'best' private and social approaches towards management for resilience may differ. Private resilience will tend to maintain a given set of businesses in place and hence the set of skills, abilities and objectives of the present population of farmers. Does this mean that society will forego some options that could potentially deliver a greater social value?

From a social perspective it may not be optimal to freeze the set of individual farms and their attributes as they happen to be in the agricultural sector at any particular time, or the sector's overall structure. When there are major shocks or stresses, as illustrated in the historic examples discussed above, it may require fundamental changes in agricultural systems that the existing population of farmers or the current structure of farm businesses is unwilling or unable to deliver. From the perspective of the individual farmers, they may prefer or only be capable of continuing with the present arrangements and systems, limiting the range of regeneration options considered. As pointed out by Darnhofer (2014), the overall goal pursued by family farms is to ensure farm continuity and intergenerational succession. This may rely, where possible, on use of the buffer capability. However, a social perspective may prefer a greater degree of transformation. Further, the 'core functions, structure, identity and feedbacks' that are protected through resilience at the individual farm level may well be different from the 'core functions, structure, identity and feedbacks' that would be protected by the adoption of resilience at a larger social scale. Individual farms need to be able to generate a level of income that can sustain the farm household. From a social perspective, it may not be critical to maintain the presence of particular individuals or structure of farms provided that there is persistence of the critical levels and types of capital and of ecosystem functions. The restriction of adjustment options might have a substantial opportunity cost to society in preventing a shift towards more socially valuable alternatives. Social resilience is not simply the sum of privately resilient farms. It may, of course, be the case that the existing structure of businesses is indeed necessary in order to protect sufficient levels of critical natural or social capital. For instance, rapid farm structural change might promote changes in land management that irreversibly damage habitats or cause an irreversible depletion of social or human capital (see e.g. Sinclair et al., 2014). The argument here is not that major structural change is necessarily socially beneficial, but that it could be.

The question for the present situation, then, is whether a focus on resilience at the farm level might constrain a drive towards sustainable intensification or whether it could prevent the adoption of more radical approaches towards rural land management. We have noted that, historically, the periods of agricultural depression have seen major changes in agricultural systems and in farm structure. Does management for resilience at an individual farm level reinforce the potential constraints of path dependency and limit the options available to society more generally? Of course, such options may not be prevented permanently, it might be that resilient firms slow down rather than prevent a shift towards socially preferred systems, but this will entail some sort of social cost. Thus, for instance, if it were to be the case that the 'best' approach towards sustainable intensification was through the establishment of very large-scale operations, might the maintenance of larger numbers of much smaller but resilient businesses prevent the achievement of this option? This depends on whether there are significant economies of size associated with the technology to advance sustainable intensification. This is an empirical question amenable to research. Or, if the highest social value use of certain upland areas were to be in the form of large-scale rewilding, as argued for example by Monbiot (2013), would this option be effectively foreclosed by the enhanced resilience of individual farm businesses committed to conventional livestock production? Rewilding almost certainly requires a number of contiguous farms to fundamentally change their approach to land management.

This raises obvious parallel questions about the consequences of an agricultural policy that protects businesses from the rigours of a freer market. We should recall that the Austrian farmers in the sample were operating behind the protection of the Common Agricultural Policy. By protecting a particular population of farm businesses and farm structure, are we preventing the forces of creative destruction (e.g. Levin et al., 1998) from spawning new assumptions and approaches towards the management of rural land? It has been argued, for example, that the liberalisation of agricultural policy in New Zealand stimulated considerable innovation and adjustment in the agricultural sector (Rae et al., 2004; Vitalis, 2009).

We consider two aspects to this question: whether institutional arrangements can help to promote higher levels of flexibility and to hold down costs of adjustment in the face of shocks and stress, and the implications for the scale over which governance arrangements are implemented.

Capacity for sustainability at the farm scale

The possible difference between private (from the perspective of individual farmers) and social (from the perspective of society as a whole) approaches to resilience will be reduced where the individual farms have greater capacity to respond to shocks and stress and face lower costs of adjustment while continuing to operate within the constraints of ensuring the persistence of critical capital assets and ecosystem functions. Resilience here involves both survival and regeneration. The previous discussion of farms demonstrating resilience suggests characteristics that can support resilience:

- good information on the present position, longer-term prospects and possible options for development;
- access to capital to enable survival and as a base for investment in new ventures;
- the capacity to share risks;
- strong networks, both informal among peers and market partners and formal such as with government agencies;
- willingness to experiment and adapt to changing conditions.

Property relations also had significant implications for the performance of farms through the periods of stress in the past. It seems likely that one of the constraints on adjustment that led to greater areas of land being underused in the 1920s and 1930s was the inflexibility of farm structure due to the greater prevalence of owner occupation and the lack of opportunity for farmers to have their rental costs reduced. Under inflexible land tenure arrangements changing the occupation of an area of land generally involves either a change of land ownership or a new formal tenancy agreement. When one farmer fails to generate sufficient income, the transaction costs of change are relatively high. However, over the past few decades the range of potential arrangements between a landowner and an occupier has expanded with the wider use of contract farming agreements. These can allow farm businesses to change their scale of operation without formal operations in the land market. Thus, more profitable farmers can expand their area and higher-cost farmers can effectively withdraw from farming, while retaining occupation of the farm, without buying and selling land. Alternative institutional arrangements can lower the transaction costs of adjustment, effectively increasing the size of the resilience domain. Such an arrangement represents an example of institutional blending (Hodge & Adams, 2012) whereby property rights are exchanged between agents without the full transfer of ownership. In this context, the current land system may be said to be more resilient in that land is more likely to remain in use in the face of the fluctuating fortunes of the agricultural sector.

There may also be risks associated with this approach in that it might undermine the sense of responsibility that the person using the land and/or the person contracting out the farming operations feels towards the conservation of the environment. It is of course also possible that the contractor could feel a stronger commitment to environmental stewardship than the original farmer. Nevertheless, the more regular changes in the occupation of land emphasise the importance of regulations governing land use practices, such as reflected in the cross-compliance requirements under the Common Agricultural Policy.

Scale of governance arrangements for agricultural sustainability

Locating agricultural systems within the wider context of ecosystems points to the need to manage ecosystems and agricultural systems, including the human and social elements, as a single social–ecological system at a landscape scale (Benton, 2012). Further, a shift towards sustainable intensification would seem to indicate a greater reliance on collective actions among local farmers and a higher degree of local specificity. The priorities in a lowland, more intensively farmed area will differ from those in an upland High Nature Value area. The approaches taken towards sustainable

Table 10.2. *Arguments for coordinated scale in land management.*

- Linkages and scale in biodiversity conservation and agroecological services (e.g. pest control and pollination)
- Water quantity (water holding and storage, flood protection)
- Water quality (catchment, river protection)
- Economies of size in farm management (e.g. energy generation, machinery sharing)
- Marketing and value-added initiatives
- Collective agri-environment contracts
- Discussion groups, social learning

intensification are likely to be more locally specific than the approaches adopted in conventional farming systems. Some of the arguments for coordinated actions among farmers are illustrated in Table 10.2. This suggests that promoting sustainability at an individual farm level cannot guarantee resilience of the ecosystem as a whole. Ecosystem sustainability and resilience depends on the overall impact of all farms in combination, also taking account of the demands of other stakeholders. The interactions between the elements of this broader system and the uncertainty or ignorance as to the relationships among them need to be recognised in an approach towards sustainability. Such a social–ecological system will need to be managed adaptively.

Within this wider setting, management needs to take account of the interests and preferences of the variety of stakeholders (Sagoff, 2004). This implies a need for an institutional arrangement within which the preferences and trade-offs can be reconciled in some form of co-management (Carlsson & Berkes, 2005). The co-management organisation can be expected to engage in a range of planning, implementation, coordination and monitoring activities. Table 10.3 suggests an illustrative list (Hodge, 2007). Lindberg and Fahlbeck (2011) found evidence of collective organisations becoming involved in these sorts of co-management activities in Sweden (although the potential scope for a co-management organisation would be more all-encompassing). Some of these actions may be undertaken by groups of landholders and others will need to encompass stakeholders from a variety of different sectors and interests.

This ecosystem or landscape-scale governance is itself set within a hierarchy of governance levels that bridge both the ecological and economic perspectives (Veldkamp et al., 2011).

Conclusions

Sustainability comprises elements of both ensuring the persistence of critical capital assets in the physical and social environments and promoting resilience

Table 10.3. Co-management activities within a social–ecological system.

- Establishing a land-use planning framework reflecting local values and priorities
- Setting local guidance and support for voluntary environmental initiatives, such as agri-environment schemes
- Sharing information about potential initiatives and collaborations
- Coordinating land management decisions across holdings
- Building social capital and promoting trust in decision-making processes
- Identifying critical sites for public or non-profit landownership and supporting land purchase
- Building dedicated funds to support local environmental conservation
- Coordinating actions to secure external funding for local initiatives
- Levering voluntary contributions and commitment for stewardship
- Monitoring environmental variables and supporting enforcement

in social–ecological systems. It will thus depend on the characteristics of individual businesses and on the institutions and social capital associated with the governance of land, land uses and ecosystems. Rules are required to constrain the impacts on ecosystem functions and to coordinate activities among farms to achieve targets at a landscape or ecosystem scale. Optimisation as conventionally understood at the individual farm level cannot be guaranteed to lead to sustainability; for instance, resilience may require elements of redundancy and diversity that diverge from what might be sought in terms of conventional efficiency. Resilience may be a preferred objective for farmers in that it offers greater stability and security. Technological progress will be important in facilitating the future sustainable intensification of farm businesses, but social aspects are also critical. Technology can expand the range of production opportunities available to a farm business and household, but the technology needs to be integrated into a farm business in a way that meets the objectives of the individuals involved and the resources available. It is a matter of adaptation rather than simply adoption. However, more major changes of direction and renewal may demand more fundamental change to the system beyond changes within the individual farms. Reorganisation of the system will depend on availability of knowledge, entrepreneurship and institutions that enable adjustment alongside the stimulus of new technology. This may only be stimulated under conditions of stress and may require changes in the composition of the farm population and the structure of farm businesses. There may thus be a degree of conflict between the pursuit of resilience by individual farmers and social objectives for the development of the agricultural sector as a whole.

There is a role for government in supporting the maintenance of natural and social capitals and effective and flexible institutions, especially in the land market, that permit the structure of the sector to respond effectively to changing circumstances. Experience through historical periods of depression in farming illustrates the institutional arrangements that can facilitate appropriate adjustments. However, the increasing focus on ecosystem services and landscapes indicates a requirement for governance at a larger landscape and regional scale that is itself nested within governance arrangements at national and international scales.

References

Batáry, P., Dicks, L.V., Kleijn, D. & Sutherland, W.J. (2015). The role of agri-environment schemes in conservation and environmental management. *Conservation Biology*, **29**(4), 1006–1016.

Beckett, J. & Turner, M. (2007). End of the old order? F.M.L. Thompson, the land question, and the burden of ownership in England, c1880–c1925. *Agricultural History Review*, **55**(2), 269–288.

Benton, T. (2012). Managing agricultural landscapes for production of multiple services: the policy challenge. *International Agricultural Policy*, **1**, 7–17.

Bommarco, R., Kleijn, D. & Potts, S.G. (2013). Ecological intensification: harnessing ecosystem services for food security. *Trends in Ecology and Evolution*, **28**(4), 230–238.

Brassley, P. (2007). British farming between the wars. In: *The English Countryside between the Wars. Regeneration or decline?*, edited by P. Brassley, J. Burchardt & L. Thompson. Woodbridge: The Boydell Press, pp. 187–199.

Brown, J. (1987). *Agriculture in England. A survey of farming, 1870–1947*. Manchester: Manchester University Press.

Cannadine, D. (1996) *The Decline and Fall of the British Aristocracy*. Revised edition. London: Papermac, Macmillan Publishers Ltd.

Carey, P. (2001). Schemes are monitored and effective in the UK. *Nature*, **414**, 687.

Carlisle, L. (2014). Diversity, flexibility, and the resilience effect: lessons from a social-ecological case study of diversified farming in the northern Great Plains, USA. *Ecology and Society*, **19**(3), 45.

Carlsson, L. & Berkes, F. (2005). Co-management: concepts and methodological implications. *Journal of Environmental Management*, **75**, 65–76.

Comptroller and Auditor General. (2002). The 2001 Outbreak of Foot and Mouth Disease. HC 939 Session 2001-2002. London: National Audit Office.

Darnhofer, I. (2010). Strategies of family farms to strengthen their resilience. *Environmental Policy and Governance*, **20**, 212–222.

Darnhofer, I. (2014). Resilience and why it matters for farm management. *European Review of Agricultural Economics*, **41**(3), 461–484.

Darnhofer, I., Bellon, S., Dedieu, B. & Milestad, R. (2010a). Adaptiveness to enhance the sustainability of farming systems. A review. *Agronomy for Sustainable Development*, **30**, 545–555.

Darnhofer, I., Fairweather, J. & Moller, H. (2010b). Assessing a farm's sustainability: insights from resilience thinking. *International Journal of Agricultural Sustainability*, **8**(3), 186–198.

Firbank, L., Elliott, J., Drake, B., Cao, Y. & Gooday, R. (2013). Evidence of sustainable intensification among British farms. *Agriculture, Ecosystems and Environment*, **173**, 58–65.

Fish, R., Winter, M. & Lobley, M. (2014). Sustainable intensification and ecosystem services: new directions in agricultural governance. *Policy Sciences*, **47**(1), 51–67.

Fletcher, T.W. (1961). The Great Depression of British agriculture, 1875-1900. *Economic History Review, New series*, **13**(3), 417-432.

Folke, C. (2006). Resilience: the emergence of a perspective for social–ecological systems analyses. *Global Environmental Change*, **16**, 253-267.

Folke, C., Hahn, T., Olsson, P. & Norberg, J. (2005). Adaptive governance of socio-ecological systems. *Annual Review of Environment and Resources*, **30**, 441-473.

Foresight. (2011). *The Future of Food and Farming. Executive Summary*. London: The Government Office for Science.

Garnett, T., Appleby, M.C., Balmford, A., et al. (2013). Sustainable intensification: premises and policies. *Science*, **341**, 33-34.

Godfray, H.C.J., Beddington, J.R., Crute, I.R., et al. (2010). Food security: the challenge of feeding 9 billion people. *Science*, **327**, 812-818.

Hanley, N. (1998). Resilience in social and ecological systems: a concept that fails the cost-benefit test? *Environment and Development Economics*, **3**, 244-249.

Hodge, I. (2007). The governance of rural land in a liberalised world. *Journal of Agricultural Economics*, **58**(3), 409-432.

Hodge, I. (2014). European agri-environmental policy: the conservation and re-creation of cultural landscapes. In: *The Handbook of Land Economics*, edited by J.M. Duke & J. Wu. New York, NY: Oxford University Press, pp. 583-611.

Hodge, I.D. & Adams, W.M. (2012). Neoliberalisation, rural land trusts and institutional blending. *Geoforum*, **43**(3), 472-482.

Holling, C.S., Schindler, D.W., Walker, B.W. & Roughgarden, J. (1995). Biodiversity in the functioning of ecosystems: an ecological synthesis. In: *Biodiversity Loss: economics and ecological issues*, edited by C. Perrings, K.-G. Maler, C. Folke, C.S. Holling & B-O. Jansson. Cambridge: Cambridge University Press, pp. 44-83.

Hunt, E.H. & Pam, S.J. (2001). Managerial failure in late Victorian Britain: land use and English agriculture. *Economic History Review*, **54**(2), 240-266.

Kleijn, D. & Sutherland, W. (2003). How effective are European agri-environment schemes on conserving and promoting biodiversity? *Journal of Applied Ecology*, **40**, 947-969.

Kleijn, D., Berendse, F., Smit, R. & Gilissen, N. (2001). Agri-environment schemes do not effectively protect biodiversity in Dutch agricultural landscape. *Nature*, **413**, 723-725.

Kremen, C. & Miles, A. (2012). Ecosystem services in biologically diversified versus conventional farming systems: benefits, externalities and trade-offs. *Ecology and Society*, **17**(4), 40.

Levin, S., Barrett, S., Aniyar, S., et al. (1998). Resilience in natural and socioeconomic systems. *Environment and Development Economics*, **3**, 221-262.

Lindberg, G. & Fahlbeck, E. (2011). New forms of local collective governance linked to the agricultural landscape: identifying the scope and possibilities of hybrid institutions. *International Journal of Agricultural Resources, Governance and Ecology*, **9**(1/2), 31-47.

Ministry of Agriculture, Fisheries and Food. (1968). *A Century of Agricultural Statistics*. London: HMSO.

Monbiot, G. (2013). *Feral. Searching for enchantment on the borders of rewilding*. London: Allen Lane.

OECD. (2001). *Multifunctionality: towards an analytical framework*. Paris: Organisation for Economic Co-operation and Development.

Orwin, C.S. & Whetham, E.H. (1964). *History of British Agriculture 1846-1914*. London: Longmans.

Peerlings, J., Polman, N. & Dries, L. (2014). Self-reported resilience of European farms with and without the CAP. *Journal of Agricultural Economics*, **65**(1), 722-738.

Phillipson, J., Bennett, K., Lowe, P. & Raley, M. (2004). Adaptive responses and asset strategies: the experience of rural micro-firms and Foot and Mouth Disease. *Journal of Rural Studies*, **20**, 227–243.

Piesse, J. & Thirtle, C. (2009). Three bubbles and a panic: an explanatory review of recent food commodity price events. *Food Policy*, **34**, 119–129.

Plieninger, T. & Bieling, C. (2013). Resilience-based perspectives to guiding high-nature-value farmland through socioeconomic change. *Ecology and Society*, **18**(4), 20.

Power, A.G. (2010). Ecosystem services and agriculture: tradeoffs and synergies. *Philosophical Transactions of the Royal Society B*, **365**, 2959–2971.

Pretty, J. (2008). Agricultural sustainability: concepts, principles and evidence. *Philosophical Transactions of the Royal Society B*, **363**(1491), 447–465.

Rae, A., Nixon, C. & Lattimore, R. (2004). Adjustment to agricultural policy reform – issues and lessons from the New Zealand experience. NZ Trade Consortium Working Paper no. 35. Wellington: New Zealand Institute of Economic Research.

Rodgers, C. (2009). Property rights, land use and the rural environment: a case for reform. *Land Use Policy*, **26S**, S134–S141.

Royal Society. (2009). *Reaping the Benefits. Science and the sustainable intensification of global agriculture*. London: The Royal Society.

Sagoff, M. (2004). *Price, Principle and the Environment*. Cambridge: Cambridge University Press.

Sinclair, K., Curtis, A., Mendham, E. & Mitchell, M. (2014). Can resilience thinking provide useful insights for those examining efforts to transform contemporary agriculture? *Agriculture and Human Values*, **31**, 371–384.

Stoate, C. & Parish, D. (2001). Monitoring is underway and results so far are promising. *Nature*, **414**, 687.

Stockholm Resilience Centre. (2014). *Applying Resilience Thinking: Seven principles for building resilience in social–ecological systems*. Stockholm: Stockholm Resilience Centre.

Sturmey, S.G. (1955). Owner-farming in England and Wales, 1900-50. *Manchester School of Economic and Social Studies*, **23**(3), 245–268.

Thompson, F.M.L. (1963). *English Landed Society in the Nineteenth Century*. London: Routledge and Kegan Paul.

Thompson, F.M.L. (2007). The land market, 1880–1925: a reappraisal reappraised. *Agricultural History Review*, **55**(2), 289–300.

Tranter, R.B. (2012). Agricultural adjustment on the Berkshire Downs during the recession of 1921-38. *Agricultural History Review*, **60**(2), 214–240.

Van Ittersum, M., Cassman, K.G., Grassini, P., et al. (2013). Yield gap analysis with local to global relevance – a review. *Field Crops Research*, **143**, 4–17.

Veldkamp, T., Polman, N., Reinhard, S. & Slingerland, M. (2011). From scaling to governance of the land system: bridging ecological and economic perspectives. *Ecology and Society*, **16**(1), 1.

Vitalis, V. (2009). *Domestic reform, trade, innovation and growth in New Zealand's agricultural sector – Case Study No. 2*. OECD Trade Policy Papers, no. 74. Paris: OECD.

Wheeler, T. & von Braun, J. (2013). Climate change impacts on global food security. *Science*, **342**, 508–513.

Zhang, W., Ricketts, T.H., Kremen, C., Carney, K. & Swinton, S.M. (2007). Ecosystem services and dis-services to agriculture. *Ecological Economics*, **64**, 253–260.

CHAPTER ELEVEN

The resilience of Australian agricultural landscapes characterised by land-sparing versus land-sharing

DAVID J. ABSON
Leuphana University
KATE SHERREN
Dalhousie University
and
JOERN FISCHER
Leuphana University

Introduction

The notion of resilience has extended beyond its original application – describing ecological systems (Holling, 1973) – to one useful for understanding social–ecological interactions and interdependencies (Holling, 2001). A social–ecological system consists of a biophysical unit and its associated (tied) social actors and institutions. Social–ecological systems are complex and adaptive and delimited by spatial or functional boundaries surrounding a particular ecosystem (Redman et al., 2004); in practical terms, they often span landscapes or regions. We define the resilience of a social–ecological system as the ability to maintain its 'identity' – key functions, structures and roles within society that define the system – in the face of exogenous perturbations (Walker et al., 2004). A resilient social–ecological system should have the adaptive capacity to maintain its identity, if not all its original processes (Folke et al., 2004), where adaptive capacity describes the ability of social–ecological systems to change in response to changing circumstances.

This chapter combines conceptual reasoning with evidence gathered during several years of extensive empirical research in the livestock grazing landscapes of the upper Lachlan River catchment of New South Wales (NSW), Australia. Specifically, we consider how two agricultural landscape management approaches – holistic management (land-sharing) and conventional livestock management with protected patches (land-sparing) – create different social–ecological system properties that influence the resilience of each farming system.

Social–ecological resilience

Following Walker and Salt (2006), we differentiate between two types of resilience – general resilience and specified resilience. General resilience describes 'the general capacity of a social–ecological system … to absorb unforeseen disturbances' (Walker & Salt, 2006, p. 121). Three system properties have been suggested as playing important roles in ensuring general resilience, namely diversity, modularity and the tightness of feedbacks (Levin, 1998; Walker and Salt, 2006; see Box 11.1 for details).

> **Box 11.1 System properties associated with general resilience in social–ecological systems**
>
> Three general system properties typically associated with greater resilience are diversity, modularity and tightness of feedbacks. Diversity refers to the range of different structures, functions, people and institutions in the social–ecological system. Such diversity is assumed to aid general resilience by increasing flexibility and response options in the face of disturbances (Walker & Salt, 2006). Modularity relates to 'the manner in which the components that make up the system are linked' (Walker & Salt, 2006, p. 121). Shocks tend to travel rapidly throughout highly connected systems. In contrast, systems that have clearly identifiable subgroups with strong internal links, which are only loosely connected to each other, are more likely to withstand shocks. Such modular systems can keep functioning when a particular module fails, particularly if there is a diversity of modules providing similar functions within the system (i.e. functional redundancies). Finally, the tightness of feedbacks relates to how quickly and strongly the changes in one part of the system are felt and responded to in other parts of the system. Tight feedbacks enable rapid adaptive changes to system perturbations.

In addition to these three general system properties – diversity, modularity and feedbacks – we also consider access to capital asset bundles as important determinants of general resilience (e.g. Carney & Britain, 2003). Capital assets are stocks of tangible and intangible assets that can be accessed to provide and improve human livelihoods (Scoones, 1998). In our case study, we specifically focus on three key capital asset types – natural capital (the elements of nature that produce value to people), financial capital and social capital (the institutions, relationships, and norms that enable societies to function effectively) – that can help buffer shocks in different farming systems. We acknowledge that other capital types, such as human capital (the collective skills, knowledge and intangible assets of individuals that can be used to create economic value), may also be important determinants of social–ecological resilience. However, we suggest that natural, financial and social capital assets are more clearly linked to the choice of land-sparing or land-sharing land management strategies than other capital asset types. Therefore, these three capital asset types are the most important determinants of differentiated general resilience in our case study.

In contrast to general resilience, specified resilience is premised on understanding specific threats to a system and identifying the key variables related to those threats that may change the system's identity. Specified resilience is the resilience 'of what, to what' (Carpenter et al., 2001, p. 765). Regarding resilience 'to what', in this chapter we focus on two major exogenous perturbations relevant to our case study area, namely fluctuations in (1) climate (e.g. drought) and (2) agricultural input prices (we set aside the issue of output price fluctuations as we assume that both holistic and conventional farmers are equally impacted by such perturbations). Regarding the resilience 'of what', we consider both maintenance of system identity (i.e. economically viable, family-owned livestock farms within functioning ecosystems) and also the resilience of valued native species within the case study area.

Land-sparing and land-sharing

Green et al.'s (2005) paper, 'Farming and the fate of wild nature', introduced the notions of land-sparing versus land-sharing (originally termed 'wildlife-friendly farming') as two alternative land-use options for managing the trade-offs between biodiversity and food production within agricultural landscapes. Land-sparing involves intensive (high yield), specialised agricultural production on existing lands, thus (in theory) sparing land for non-agricultural activities such as biodiversity conservation (Lambin & Meyfroidt, 2011). We assume that spared land must be within the same ecosystem as the intensively farmed land. This is necessary both in order to make meaningful comparisons between land-sparing and land-sharing, and because the assumption that intensive land use in one place necessarily leads to protection of land elsewhere is largely untenable (Lambin & Meyfroidt, 2011). In contrast, land-sharing involves lower intensity (low yield), but more extensive, agricultural production that promotes 'wildlife-friendly' agricultural landscapes (Fischer et al., 2008). Land-sharing approaches favour minimising agrochemical inputs and support the notion of multi-functional landscapes, in which economic and non-economic goods are co-produced (Altieri, 2000; Fry, 2001).

While the notions of sparing land for nature (e.g. Waggoner, 1996) and wildlife-friendly agriculture (e.g. Krebs et al., 1999) predate the work of Green et al., the idea of 'land-sparing versus land-sharing' as contrasting options within a conceptual framework for assessing the co-production of biodiversity and agricultural goods has sparked considerable interest and debate (Grau et al., 2013). Support for sparing (e.g. Grau et al., 2008; Phalan et al., 2011; Egan & Mortensen, 2012) or sharing (e.g. Dorrough & Crothwaite, 2007; Perfecto & Vandermeer, 2010; Mastrangelo & Gavin, 2012) land-use strategies is driven to some extent by the underlying scientific research approaches, paradigms and worldviews of individual researchers (Fischer et al., 2008). Similarly, the type of system studied may favour either a sparing or sharing approach (von Wehrden et al., 2014). For example, intact primary habitat may lend itself to sparing strategies, while land-sharing may be more suitable in secondary habitats (Ramankutty & Rhemtulla, 2013). In addition to the context-dependent nature of the findings, the sparing/sharing framework has been criticised for implying that static optimisation is possible or desirable in dynamic, multi-functional systems and for assuming commensurability between the different types of 'goods' being traded-off (Fischer et al., 2014).

Despite these limitations, the sparing/sharing framework has provided a useful way of thinking about contrasting types of land-use patterns and management approaches in agricultural landscapes. Here, rather than engaging in the contentious debate regarding which strategy (sparing or sharing) provides better outcomes for biodiversity and food provision, we apply the useful conceptualisation of landscapes as gradients from sparing to sharing in order to investigate how the system properties associated with starkly different land-use patterns and management approaches in agricultural landscapes influence social–ecological resilience.

Table 11.1 provides a list of typical characteristics of social–ecological systems under land-sparing and land-sharing. We recognise that for some species or purposes, a different characterisation of landscapes would be more suitable. Our 'archetypes' serve a conceptual purpose, to delineate the ends of the land-sparing/land-sharing gradient, and must, therefore, be considered as stereotypes that over-extend the differences observable in the real world; many hybrid systems sit on the continuum between these two extremes.

Methods overview

The methods used in the various pieces of previously published research are explained in detail in the cited references – for convenience, we provide a brief summary here. Our research combined ecology, social science and policy research components via 33 shared case study farms and collaborative workshops in the study area (Sherren et al., 2008, 2010a, 2011a), and two large-scale landholder surveys to test our observations at a broader scale (Schirmer et al., 2012a, 2012b; Sherren et al., 2012a).

We mapped case study farms and their woody vegetation, and established grazing practices such as stocking rates and total annual grazing days for each paddock through discussion with each farmer. Ecological surveys targeted 2-ha plots in grazed woodlands, open paddocks and scattered tree sites, as well as ungrazed woodlands, inventory trees, seedlings, birds and bats. These data allowed grazing practices to be linked to tree cover and seedling recruitment,

Table 11.1. *Typical characteristics of 'archetypal' land-sparing versus -sharing systems as assumed in this chapter. Note that many real-world farming systems exist between these two archetypal extremes.*

Land-sparing	Land-sharing
Clear delineation between protected and productive components within the landscape (beyond the mere presence of field margins)	No clear delineation, but a diversity of 'wildlife-friendly' elements throughout the landscape (e.g. scattered trees, shrubs, field margins, streamside vegetation)
Intensive agricultural land use (typically including the use of agrochemicals)	Extensive agricultural land use (typically low use of agrochemicals)
Yield/profit maximisation, often in conjunction with agricultural specialisation	Minimisation of agrochemicals, often in conjunction with on-farm business diversification
Landscape homogenisation within intensively used areas is tolerated to capture economies of scale in agricultural production	Landscape heterogeneity is encouraged to maintain natural capital for agricultural production
Separate production of biodiversity and agricultural goods (binary view of landscapes with an emphasis on efficient allocation of resources between biodiversity and agricultural commodity production)	Co-production of biodiversity and agricultural goods (holistic view of landscapes with an emphasis on interdependencies between commodity production and biodiversity)
Natural capital conceptualised as a resource conserved for reasons other than its benefits for agricultural production	Natural capital conceptualised as a resource used for agricultural production

and the resulting woody vegetation to be associated with biodiversity (Fischer et al., 2009; Fischer et al., 2010a).

Twenty-five farmers participated in interviews that used their own photographs of 'significant' landscapes as prompts for discussion. Those photo-elicitation interviews allowed us to understand how farmers valued various aspects of their landscapes, including how their perceptions drove management practices and outcomes (Sherren et al., 2010b, 2011b, 2012b). More detailed interviews followed with a subset of farmers to explore their financial management, such as farm gate receipts and costs associated with specific management regimes. Three annual workshops held with local graziers, farm consultants, catchment managers and government representatives helped us interpret our observations and their implications (Sherren et al., 2008, 2010a, 2011a).

Throughout the remainder of this chapter, we draw on the findings of this previous work, as well on the conceptual understanding of the study system we developed throughout the research process.

The case study area

The study area (approximately 1 million ha) is within the grassy-box woodland ecosystems of the upper Lachlan river catchment of New South Wales (NSW), Australia, in which livestock grazing is the dominant agricultural activity. The eastern side of the study area is hilly and rocky in parts and livestock grazing is the only viable agricultural activity. Further west, slopes are gentler and the amount of cropping increases. Large patches of trees are largely confined to the hilltops, but scattered paddock trees are common throughout the region, and account for approximately one-third of the remnant tree cover on farms (Fischer et al., 2010b). Grazing covers approximately three-quarters of the study area, while 10–15 percent is under crop production; much of the remainder is covered by patches of woodlands and dry forests and is thus broadly reflective of the wider temperate grazing zone or 'sheep–wheat belt' (Figure 11.1). The study area is characterised by old, fragile and sometimes degraded soils containing relatively low levels of organic matter. Average annual precipitation is typically between 600 and 850 mm and is distributed relatively evenly during the year (van der Beek & Bishop, 2003). Interannual variability in precipitation is relatively high with both droughts (e.g. the 'Big Dry' that prevailed for most of the previous decade; Cai et al., 2009) and floods (e.g. 2010–2012; McClusky et al., 2012) over the past two decades.

Conventional management (land-sparing approach)

Conventional livestock management in Australia's temperate grazing zone involves keeping livestock in paddocks for extended periods, or even

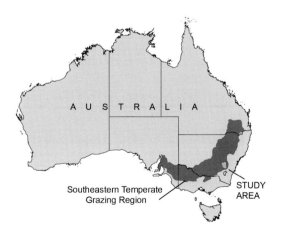

Figure 11.1 Case study area in the southeastern 'sheep–wheat' belt, the temperate grazing region of Australia. (A black and white version of this figure will appear in some formats. For the colour version, please refer to the plate section.)

year-round. The average annual stocking rates used by holistic versus conventional farmers do not necessarily differ (Fischer et al., 2009); the primary difference thus lies in the application of rotational grazing with long rest periods used by holistic managers versus more continuous grazing by conventional managers. Moreover, conventional practices typically involve the use of exotic or annual pastures, and regular applications of chemical fertilizers to maintain pasture productivity (Sherren et al., 2012b). External inputs such as fodder and fertilisers are used to bolster farms' carrying capacities when the natural variability in the climate, particularly lack of precipitation, reduces natural biomass production.

Conventional farmers seek to conserve biodiversity typically by creating clearly demarked 'fenced off' protected patches (typically measuring at least 5 ha, and often more) based on the assumption that the best way to protect native vegetation (especially trees and shrubs) is to completely exclude livestock (land-sparing, Figure 11.2a). For biodiversity conservation, 'sparing' farmers typically prioritise large patches because these are known to have the greatest value to biodiversity (e.g. exhibiting higher bird species richness than small patches; Watson et al., 2001). Many of these protected patches tend to be on relatively steep slopes and hilltops, which are not productive for livestock or cropping (Fischer et al., 2010b). Alternatively, they include riparian strips and fencelines. In both cases, the costs of protection have often been shared with local governments or catchment management authorities, with farmers providing labour and the materials needed for

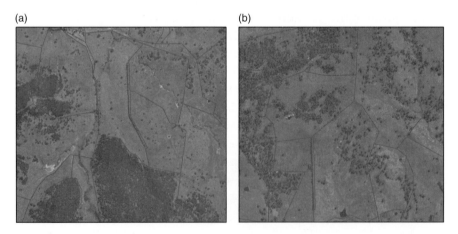

Figure 11.2 Examples of woody vegetation cover for subsets of an archetypal (a) conventional (spared) farm and (b) holistic (shared) farm from the study area. Red lines represent fencelines (fencelines do not necessarily represent spared land). (A black and white version of this figure will appear in some formats. For the colour version, please refer to the plate section.)

fencing often being subsidised by the state. Many conventional farmers are proud of their stewardship work of 'fencing off'; and protected woodland patches were twice as frequently captured by non-holistic than holistic farmers when they were asked to photograph significant elements of their farms (Figure 11.3a; Sherren et al., 2011b). Figure 11.2 shows the distribution of woody vegetation in subsets of archetypal conventional (Figure 11.2a) and holistically managed (Figure 11.2b) properties in the case study area.

Holistic management (land-sharing approach)

In contrast to conventional management approaches, livestock-based holistic management usually involves intense bursts of grazing pressure within a given grazing location, followed by extended recovery time (Stinner et al., 1997; Savory & Butterfield, 1999). Grazing occurs at very high stocking densities in a single paddock, with the livestock moved after a few days and the 'hard hit' paddocks given a long period of rest – often weeks or even months, informed by careful monitoring of feed levels – between grazing events. Rather than rely on protected patches of native vegetation, holistic management is intended to provide conservation benefits throughout the farm as the intensive grazing events are believed to be too infrequent to permanently damage native vegetation.

The holistic approach to grazing requires practitioners to experiment, manage adaptively and develop and monitor holistic goals related to the forms of production needed to support quality of life. Holistic management also includes landscape planning that protects and enhances biodiversity and supports ecosystem processes, such as succession, energy flows and hydrological and nutrient cycling (Stinner et al., 1997; Savory & Butterfield, 1999). In our study area, the implementation of holistic practices varied between individual farmers, but generally involved high-intensity, short-duration grazing, the reduction or elimination of artificial fertilisers, and an emphasis on pastures with native plant species (Sherren et al., 2012b). Within the holistic management philosophy, the health of the land and the natural resource base of the farmed landscape are considered fundamentally important for maintaining a profitable farm enterprise. The emphasis on supporting and building biodiversity and wildlife-friendly structures throughout the farmed landscape, and the relatively lower priority given to clearly delineated, protected patches, clearly mark holistic farming as an example of the land-sharing paradigm (Figure 11.2b and Figure 11.3b).

We note that while holistic management aspires to create landscape-wide biodiversity benefits, there is controversy as to the actual effectiveness of this approach. Nevertheless, there is some evidence of landscape-level benefits (e.g. for tree regeneration) in our case study area (Fischer et al., 2009).

Figure 11.3 Examples of (a) a spared patch of trees and (b) a 'shared', holistically grazed landscape to the left of the fence, compared with conventional grazing to the right. Both photos were captured by Lachlan catchment farmers when asked to photograph significant elements of their farm landscape (method explained in, for example, Sherren et al., 2012b). (A black and white version of this figure will appear in some formats. For the colour version, please refer to the plate section.)

Historical, environmental, socioeconomic and institutional setting

Prior to European settlement, our study area was dominated by grassy box woodland in the valleys and, to a lesser extent, dry sclerophyll forest on the hills. Both of these vegetation associations have an overstorey dominated by eucalyptus trees. In grassy box woodlands, trees are spaced widely and the crowns of individual trees often do not touch. The understorey of box woodland is dominated by grasses and forbs and to a lesser extent by shrubs. Dry sclerophyll forest is denser and typically contains more shrubs in the understorey. In both cases, trees, and to a lesser extent other types of woody vegetation, are the structural elements with which other native biota have coevolved. While a comprehensive assessment of biodiversity would need to consider a wide range of variables, the structure and spatial arrangement of woody vegetation have been shown to be particularly important (Law et al., 2000), including for birds (Watson et al., 2001; Hanspach et al., 2011), reptiles (Fischer et al., 2005; Brown et al., 2008) and bats (Law & Chidel, 2006; Hanspach et al., 2012). Over the last 150 years of European settlement and land clearing, approximately 80–95 percent of original tree cover has been lost, and the remaining cover occurs as small patches and as scattered trees (Fischer et al., 2009).

In 2006, farming of sheep, beef cattle and grain in the Cowra and Boorowa census districts (core to the study area) employed 11.7 percent and 36.6 percent, respectively, of the workforce, compared with 1.5 percent Australia-wide (Sherren et al., 2010a). In 2006, 55 percent of the region's AUD$130.6 million worth of production came from livestock. Farming within the Lachlan catchment persists in the face of volatile agricultural input and commodity prices, both within Australia (O'Donnell, 2010) and on global markets (e.g. Headey & Fan, 2008).

While the case study area is large, its farming community is small and interconnected through family and marriage; Cowra is the largest urban centre, at around 8000 people in 2011 (Australian Bureau of Statistics, 2013). Practices that are judged to be acceptable by trusted and well-connected members of the farming community are more likely to be taken up by the majority (Pannell et al., 2006). Thus, by far the most dominant strategy is that of conventional grazing, which is seen across many of Australia's farming regions (Sherren et al., 2012b). Courses teaching relatively novel practices such as holistic management can be expensive, and do not necessarily involve ongoing technical or social support, making change difficult in the first instance, and difficult to maintain (Box 11.2; Sherren et al., 2012b). Such systemic practices are also difficult to trial on small areas of a farm, which presents a risk to farmers (Pannell et al., 2006). Holistic management is thus still a marginal activity in much of Australia, estimated to be practiced by less than 10 percent of farmers (Sherren et al., 2012b). Finally, extended drought relief and investment in technological 'fixes' such as silage facilities for climate-related challenges have historically buffered conventional farmers from the need to reconsider their management practices.

In the case study area, as elsewhere in Australia, relatively large patches of native vegetation have typically been the priority for public-supported protection initiatives. Fencing costs can be reimbursed to protect compositionally and structurally intact woodlands. Many farmers have taken up such funding opportunities, thereby providing important core habitat for a wide range of species (Sherren et al., 2010b). Riparian fencing is also often supported, as well as the fences and seed to plant and protect vegetated strips along existing fencelines (Sherren et al., 2011a). More recently, however, some stewardship programmes have begun to include land-sharing principles: defining 'high-quality' habitat to

---

Box 11.2. Examples of the social difficulties facing graziers adopting holistic management practices in the case study area (unpublished data from photo-elicitation interviews in 2008, discussed, for example, in Sherren et al., 2012b).

'There's some BBQs that we go to that I'll actually go there thinking "Right, what are some topics that I can open up with?" You know, like at the moment it would be protein, screenings, the number of small grains yield per hectare, how late they've been working every night, that'd be the conversation. In January it'll be the amount of chemical, what rate per hectare and then what chemicals they're using to kill all the weeds in January because it's rained. Then all the way through autumn that would more or less be the same. Oh, and then, the cost of fertiliser prices and diesel. Then in May it's sowing, what are you sowing, how dry is it or how wet is it, what depth are you putting it all in. I don't do any of this anymore. I used to do all this stuff so I know what [conventional graziers are] talking about, I just think it's futile' (Husband, Farm 5).

'I came back [from the holistic management course] all enthused and everyone thought I'd lost the plot I think. I didn't say too much to a few because they were just – I mean most of them aren't open to the idea at all. Generally most graziers in Australia aren't open to the idea I don't think. It's too far out there for a lot of them. So I don't talk about it much. My brother, he went and did the course on my advice but he'd be one of the few I'd mention it to. ... I don't think the majority [that take the course] go on and do it. Even I found it difficult because when you come back no one else thinks like you're thinking. Unless you've got some support from somewhere you sort of just go ... There's so much push from everyone else around you, you gradually just sort of tend to go like a sheep and follow the rest and end up going back to your old ways' (Husband, Farm 17).

include scattered tree cover and native grasses, instead of just woodlands; and including payments for a range of habitat enhancements, including retaining coarse woody debris, ceasing chemical fertilisation, weed control, understorey plantings, as well as allowing occasional grazing or requiring only short-term stock exclusion (Sherren et al., 2012a). Several holistic managers tended to already do many of the things in this list (Sherren et al., 2012b) and some participated in the pilots of more recent 'land-sharing' stewardship schemes.

Assessing resilience in our study area

Having provided an overview of the study area and archetypes of farms managed holistically or conventionally, here we discuss which properties of these archetypal systems convey general or specific resilience to the system.

Comparing general resilience between farming systems

In our case study, the holistic management approach builds general resilience via dynamic adaptation to changing environmental conditions based on conserving natural capital. In contrast, the conventional management approach builds resilience by accumulating financial capital to mitigate the negative impacts of environmental perturbations (Table 11.2). These two general strategies might loosely be characterised as *internalised* (natural capital; holistic management) versus *externalised* (financial capital; conventional management). One potential advantage of the conventional (land-sparing) approach is that financial capital stocks can be converted more easily into other resources to deal with unexpected shocks than the natural capital stocks on which holistic (land-sharing) depends.

The land management options in our case study derive general resilience from different sources. The holistic management option relies on diversity, redundancy and adaptation as means of coping with shocks, building natural capital and increasing the adaptive capacity of the social–ecological system. By contrast, the conventional approach builds financial capital (perhaps at the expense of natural capital) to compensate for shocks to the system, effectively using financial capital accrued during good years to draw external resources into the social–ecological system to bolster resilience during perturbations.

One important difference in general resilience between the holistic management and conventional approaches in our case study area relates to the challenge of maintaining biodiversity (an important component of system identity). The conventional (land-sparing) approach tends to lead to relatively

Table 11.2. *Conceptual matrix to assess general resilience of the grazing regimes.*

Social–ecological system property	Holistic management (land-sharing)	Conventional management with protected patches (land-sparing)
Diversity	Biological diversity at farm level is intermediate, but spread across the landscape providing more adaptability to shocks. Income diversity can be on-farm (e.g. tourism; native plant sales) Experimentation as a key component of the management strategy Income diversification through on-farm income less resilient to some natural shocks (i.e. flood, fire, etc.)	Biological diversity at farm level potentially higher than for the holistic management approach. The isolation of biodiversity from production may be both a source of vulnerability (lack of redundancy) and resilience (lower connectivity to shocks). Income diversity often off-farm through secondary employment, or spousal incomes. Uniformity and control of the environment as key management strategies Off-farm income diversification increases resilience to on-farm natural shocks (i.e. flood, fire, etc.)
Modularity (redundancies)	Potentially high levels of redundancy in resource use (i.e. rested pastures and the use of diverse native grasses)	Relatively low level of redundancies in resource use (continued occupation of pastures)
Modularity (system connectivity)	Less reliance on imported fodder supplements and artificial fertilisers – reducing connectivity to markets	More reliance on imported fodder supplements and artificial fertilisers – increasing connectivity to markets. Dependency on inputs subject to price volatility reduces general resilience
Social–ecological feedbacks	Adaptive management that is sensitive to changes in ecological condition. Management approach premised on anticipation of/adaptation to shocks, and constant monitoring Limiting factor: available resources within the system, labour to monitor and move stock	Compensatory/buffering management that seeks to 'dampen' shocks by importing resources from outside the physical boundaries of the social–ecological system Limiting factor: the ability to command resources from outside the system
Financial capital assets	Financial capital not the primary means to buffer shocks	Financial capital the major source of buffering capacity within the system
Natural capital assets	The maintenance and enhancement of natural capital is a key concern, economic activity is kept within ecological carrying capacity, and natural capital is used as a substitute for financial capital as a buffer against shocks	Natural capital within the farmed areas is at risk of being degraded, with substitutions from financial capital used to maintain productivity. Protected areas are largely protected from economic shocks because they are separated from the economically productive components of the system

Table 11.2. (cont.)

Social capital assets	Holistic management farms were more frequently partnerships of married couples, collaborating on monitoring and stock rotation, and prioritising family time during goal-setting. The relationship within the couple might supplement the lack of support in the community for practices, especially with the dearth of formal ongoing supports for holistic practices/transitions	Conventional practices are strongly supported within social and community networks, giving a sense of legitimacy and validity. Such practices are likely to be perpetuated/replicated

high species diversity and intact vegetation condition within the protected patches (e.g. Fischer et al., 2010a). This relatively intact state, combined with the removal of economic pressure on the protected patches, may infer greater resilience of the ecological 'identity' of the landscapes (Fischer et al., 2010a). However, other (i.e. grazed) locations within conventional farms may be more susceptible to degradation through continuous livestock grazing, resulting, for example, in the loss of scattered trees (Fischer et al., 2009). The loss of scattered trees, in turn, would create a binary landscape with relatively intact patches becoming isolated within an increasingly hostile landscape context. If overgrazing is severe in a land-sparing system, there may be negative flow-on effects that could permeate throughout the landscape, also affecting the protected patches (e.g. reduced water infiltration capacity or soil erosion).

In contrast, biodiversity within a holistic management (land-sharing) landscape should be expected to be more evenly distributed, but may be lower in total. Such lower landscape-level species diversity may reduce ecological resilience due to a lower number of functional redundancies among species – for example, if a given pollinator goes extinct, in the absence of functional redundancy pollination services would be lost entirely (Walker, 1992, 1995). However, the more even distribution of woody vegetation throughout the landscape may afford greater landscape connectivity for some species, thereby facilitating better population viability in the face of disturbances. Both types of landscape therefore have advantages and disadvantages for ecological resilience. Highly sensitive species such as specialised woodland birds probably benefit particularly from structurally complex, large patches being fenced off

(Watson et al., 2001). Many generalist species, however, can persist throughout landscapes of scattered trees (Fischer et al., 2010a).

Comparing specified resilience between farming systems

Qualitative research with local livestock farmers demonstrated that the internalised, dynamic adaptation approach to resilience (holistic management) and the externalised, dynamic control or buffering approach (conventional management with protected areas) in our study systems have different consequences for the specified resilience to climate variability and agricultural input price shocks (Sherren et al., 2012b).

In the face of climate variability – especially droughts – holistic (land-sharing) farmers adapt by matching their stocking rates with the climate (and therefore the productivity of their pastures). Holistic farmers rarely use supplementary feed to get through a drought (other than occasionally on small parcels of specifically 'sacrificed' land). Instead, they strongly rely on grazing charts and constant monitoring of pastures to anticipate livestock feed availability and match stocking levels to carrying capacity (Earl & Jones, 1996; Sherren et al., 2012b). Precautionary de-stocking – the sale of livestock when fodder requirements exceed the farm's current production capacity – is used to avoid the requirement to buy expensive supplementary animal feed, and to ensure that stocking levels do not exceed the land's current carrying capacity. Moreover, soil management and high levels of ground cover in the pastures also help to retain moisture in the soils and thus lessen the impact of potential droughts on biomass production within the pastures. Lower incomes during droughts are offset by keeping input costs low, although the need to de-stock may leave holistic farmers exposed to market price fluctuations when selling livestock (Sherren et al., 2012b). Similarly, occasional de-stocking may prevent holistic farmers from developing breeding stock, and thus participating in markets where breeding is critical, such as super-fine merino wool; some go so far as to specialise in agistment (i.e. they do not keep any of their own breeding stock) (Sherren et al., 2012b). There is some evidence that holistic management practices improve outcomes in dryland grazing systems in terms of higher mean income, lower variance and preserving pasture quality (Jakoby et al., 2014).

Conventional management (land-sparing) approaches to coping with drought are very different. The key strategy used here is to make high profits in good years, thereby building a buffering stock of financial capital. High profits are achieved in part via the application of fertilisers to maximise pasture productivity when there is sufficient rainfall. When there is a rainfall deficit, conventional farmers typically bring in feed for their livestock from external sources (paid for via the profits from better years), so that they do not have to de-stock (or not as much) (Sherren et al., 2012b). While constant stock maintenance has the advantage of reducing a given farmer's exposure to market price fluctuations (she or he can retain livestock when prices are low), if stocking rates are not adapted to climatic conditions natural capital stocks can be degraded and their renewable flow of benefits reduced. Long droughts especially may force conventional farmers to invest increasing amounts of financial capital to replace resource flows from the degraded landscape – causing expenses for the acquisition of supplementary feed as well as for its storage (e.g. silos) (Sherren et al., 2012b).

With regard to agricultural input price fluctuations, many holistic farmers in our study area limit their exposure by minimising the use of external inputs, instead seeking to draw as much of their resource use as possible from within their farms. Moreover, incomes on holistic farms are sometimes bolstered via on-farm activities such as farm tours, farm stays, or native plant sales (Sherren et al., 2012b). In contrast, conventional farmers relying on external (agrochemical) inputs are more exposed to input price fluctuations: in our study area, there appeared to be a greater prevalence of off-farm jobs as 'backup income' among the spouses of conventional farmers.

In this case study, conventional farmers practising the more locally accepted landscape management approach can draw on a wider range of institutional resources, and the knowledge and social capital of the wider farming community, to bolster resilience in the face of perturbations. Holistic management farmers, in contrast, rely more strongly on less extensive, but potentially more tightly connnected social capital assets (because they are 'outsiders', they often connect strongly to the relatively small number of peers who are also 'outsiders'; Box 11.2) (Sherren et al., 2012b). However, these differences in social capital are to some extent an artefact of the relative popularity of the two aproaches in our case study area, and in Australia more generally, rather than resulting from inherent system properties relating to land-sparing versus land-sharing.

Other potentially incidental issues, rather than inherent factors, may also affect the resilence of the two contrasting approaches in our case study. For example, older farmers tended to farm more conventionally, compared with holistic farmers who more often had young families (Sherren et al., 2012b). However, younger farmers were less likely to invest in measures to protect biodiversity than older farmers, even though young farmers were more likely to recognise this as a source of resilience, perhaps because they were busier and less financially secure than older farmers (Sherren et al., 2012a).

Conclusion

Although regional and farm-level idiosyncrasies are likely, our case study suggests that the two archetypal approaches to land management –

conventional grazing with protected patches (land-sparing) and holistic management (land-sharing) – do differ in their primary means of how farmers manage for social–ecological resilience. The holistic management approach appears to primarily rely on dynamic adaptive management in the face of perturbations, often via experimentation and maintaining tight feedback loops between ecological conditions and livestock stocking density. The primary source of resilience in the conventional grazing approach lies in the careful use and management of diverse natural capital assets (maintaining redundancy and modularity). The holistic land-sharing approach of graziers in the Lachlan catchment therefore provides a social–ecological approach to resilience that relies primarily on on-farm resources. In contrast, the land-sparing approach to resilience in the case study area is typifed by the use of stocks of financial capital (accumulated during stable periods) to buffer shocks, and maintain productivity and income via the application of external inputs to the system. Here we should note that additional income might be obtained from the spared areas themselves (e.g. via recreation, wood fuel, conservation grants) which would provide additional financial buffers within this system. In the sparing approach, biodiversity conservation is much more separated from economic activity and is not managed as the primary safeguard for maintaining the farming system in times of perturbation, such as the drought events experienced in the study regions over the last decade.

Our findings suggest that it is difficult to say whether archetypal land-sparing or land-sharing is the inherently more resilient social–ecological management approach at the farm level. Notably, the shared landscapes appear more resilient in a social–ecological sense, in that the famers in such landscapes place greater emphasis on building general resilience via redundancy, modularity and tight social–ecological feedbacks. Moreover, they recognise the fundamental dependence of human livelihoods on natural capital as a productive input for resilient agricultural practices. However, the relative lack of financial buffers may lead holistic management approaches to more vulnerability to intense system perturbations. In contrast, the buffering approach used in the conventional, spared landscapes may lead to greater short-term resilience due to access to more institutional support and a greater reliance on more 'convertible' financial capital stock. However, the relative lack of emphasis on natural capital within the agriculturally productive land in the spared landscapes potentially leaves the social–ecological system exposed to long-term declines in natural capital (erosion, soil health and biodiversity on productive lands) and exhibits a potentially dangerous reliance on external inputs that are based on non-renewable resources.

One important point to note is that there may be profound effects in the transition from one archetypal approach to the other. In particular, a move from land-sharing to land-sparing may be problematic in that the sharing approach is less likely to accrue the financial and infrastructure capital over the years necessary to convert to intensified farming, meaning that such transitioning systems are likely to be less resilient, at least in the short term. Similarly, holistic management aspires to be – as its name suggests – a system-wide management practice, and is difficult to trial on small areas of the farm. Moreover, the time required to build the natural capital on which this farming approach depends may leave farmers in transition to such practices vulnerable. In the context of building social–ecological resilience, institutional support to ease the transitions between these two archetypal approaches (as well as potential hybrid practices) may be preferable to attempting to institutionalise a 'winner' in the sparing versus sharing debate. At the regional scale, the differing strengths and weaknesses of the sparing and sharing approaches may be complementary in maintaining overall system identity in the face of multiple system shocks. Thus, it may be that, at the regional scale, a mixture of land-sparing and land-sharing might provide the highest level of social–ecological resilience. A heterogeneous social–ecological region would provide a diversity of commodity options, as well as ecological structures, mixing continuous well-connected and evenly distributed 'shared' biodiversity with high value, species-rich 'spared' patches. Similarly, having a mixture of conventional and holistic farms within the upper Lachlan River catchment provides a diversity of socioeconomic management strategies to cope with perturbations, drawing on differing capital asset bundles and adaptive mechanisms – buffering (sparing) and dynamic management (sharing) – that may confer greater resilience for the region than if only a single approach was favoured.

Acknowledgements

KS and JF are grateful to the numerous farmers they worked with in the Lachlan catchment. Previous publications by the authors on this study area, on which this chapter draws (see Methods overview), were funded by the Australian Research Council and Australia's Commonwealth Environmental Research Facility, 2008–2010.

References

Altieri, M.A. (2000). Multifunctional dimensions of ecologically-based agriculture in Latin America. *International Journal of Sustainable Development and World Ecology*, **7**, 62–75.

Australian Bureau of Statistics. (2013). *2011 Census QuickStats*. Australian Bureau of Statistics.

Brown, G.W., Bennett, A.F. & Potts, J.M. (2008). Regional faunal decline – reptile occurrence in fragmented rural landscapes of south-eastern Australia. *Wildlife Research*, **35**, 8–18.

Cai, W., Cowan, T., Briggs, P. & Raupach, M. (2009). Rising temperature depletes soil moisture and exacerbates severe drought conditions across southeast Australia. *Geophysical Research Letters*, **36**, L21709.

Carney, D. & Britain, G. (2003). *Sustainable Livelihoods Approaches: progress and possibilities for change*. London: Department for International Development.

Carpenter, S., Walker, B., Anderies, J.M. & Abel, N. (2001). From metaphor to measurement: resilience of what to what? *Ecosystems*, **4**, 765-781.

Dorrough, J.M. & Crosthwaite, J. (2007). Can intensification of temperate Australian livestock production systems save land for native biodiversity? *Agriculture, Ecosystems and Environment*, **121**, 222-232.

Earl, J. & Jones, C. (1996). The need for a new approach to grazing management – is cell grazing the answer? *The Rangeland Journal*, **18**, 327-350.

Egan, J.F. & Mortensen, D.A. (2012). A comparison of land-sharing and land-sparing strategies for plant richness conservation in agricultural landscapes. *Ecological Applications*, **22**, 459-471.

Fischer, J., Fazey, I., Briese, R. & Lindenmayer, D.B. (2005). Making the matrix matter: challenges in Australian grazing landscapes. *Biodiversity and Conservation*, **14**, 561-578.

Fischer, J., Brosi, B., Daily, G.C., et al. (2008). Should agricultural policies encourage land sparing or wildlife-friendly farming? *Frontiers in Ecology and the Environment*, **6**, 380-385.

Fischer, J., Stott, J., Zerger, A., et al. (2009). Reversing a tree regeneration crisis in an endangered ecoregion. *Proceedings of the National Academy of Sciences*, **106**, 10386-10391.

Fischer, J., Zerger, A., Gibbons, P., Stott, J., & Law, B.S. (2010a). Tree decline and the future of Australian farmland biodiversity. *Proceedings of the National Academy of Sciences*, **107**, 19597-19602.

Fischer, J., Sherren, K., Stott, J., et al. (2010b). Toward landscape-wide conservation outcomes in Australia's temperate grazing region. *Frontiers in Ecology and the Environment*, **8**, 69-74.

Fischer, J., Abson, D.J., Butsic, V., et al. (2014). Land sparing versus land sharing: moving forward. *Conservation Letters*, **7**, 149-157.

Folke, C., Carpenter, S.R., Walker, B.H., et al. (2004). Regime shifts, resilience and biodiversity in ecosystem management. *Annual Review of Ecology, Evolution and Systematics*, **35**, 557-581.

Fry, G.L.A. (2001). Multifunctional landscapes – towards transdisciplinary research. *Landscape and Urban Planning*, **57**, 159-168.

Grau, H.R., Gasparri, N.I. & Aide, T.M. (2008). Balancing food production and nature conservation in the Neotropical dry forests of northern Argentina. *Global Change Biology*, **14**, 985-997.

Grau, R., Kuemmerle, T. & Macchi, L. (2013). Beyond 'land sparing versus land sharing': environmental heterogeneity, globalization and the balance between agricultural production and nature conservation. *Current Opinion in Environmental Sustainability*, **5**, 477-483.

Green, R.E., Cornell, S.J., Scharlemann, J.P.W. & Balmford, A. (2005). Farming and the fate of wild nature. *Science*, **307**, 550-555.

Hanspach, J., Fischer, J., Stott, J. & Stagoll, K. (2011). Conservation management of eastern Australian farmland birds in relation to landscape gradients. *Journal of Applied Ecology*, **48**(3), 523-531.

Hanspach, J., Fischer, J., Ikin, K., Stott, J. & Law, B.S. (2012). Using trait-based filtering as a predictive framework for conservation: a case study of bats on farms in southeastern Australia. *Journal of Applied Ecology*, **49**(4), 842-850.

Headey, D. & Fan, S. (2008). Anatomy of a crisis: the causes and consequences of surging food prices. *Agricultural Economics*, **39**, 375-391.

Holling, C.S. (1973). Resilience and stability of ecological systems. *Annual Review of Ecology and Systematics*, **4**, 1-23.

Holling, C.S. (2001). Understanding the complexity of economic, ecological, and social systems. *Ecosystems*, **4**, 390-405.

Jakoby, O., Quaas, M.F., Müller, B., Baumgärtner, S. & Frank, K. (2014). How do individual farmers' objectives influence the evaluation of rangeland management strategies under a variable climate? *Journal of Applied Ecology*, **51**(2), 483–493.

Krebs, J.R., Wilson, J.D., Bradbury, R.B. & Siriwardena, G.M. (1999). The second Silent Spring? *Nature*, **400**, 611–612.

Lambin, E.F. & Meyfroidt, P. (2011). Global land use change, economic globalization, and the looming land scarcity. *Proceedings of the National Academy of Sciences*, **108**, 3465–3472.

Law, B.S. & Chidel, M. (2006). Eucalypt plantings on farms: use by insectivorous bats in south-eastern Australia. *Biological Conservation*, **133**, 236–249.

Law, B.S., Chidel, M., & Turner, G. (2000). The use by wildlife of paddock trees in farmland. *Conservation Biology*, **6**, 130–143.

Levin, S.A. (1998). Ecosystems and the biosphere as complex adaptive systems. *Ecosystems*, **1**, 431–436.

Mastrangelo, M.E. & Gavin, M.C. (2012). Trade-offs between cattle production and bird conservation in an agricultural frontier of the Gran Chaco of Argentina. *Conservation Biology*, **26**, 1040–1051.

McClusky, S., Tregoning, P. & McQueen, H. (2012). *GRACE observations of 2010/2011 eastern Australian floods: producing precise GRACE gravity fields in the absence of satellite accelerometer observations*. In: EGU General Assembly Conference Abstracts, 8260.

O'Donnell, C.J. (2010). Measuring and decomposing agricultural productivity and profitability change. *Australian Journal of Agricultural and Resource Economics*, **54**, 527–560.

Pannell, D.J., Marshall, G.R., Barr, N., et al. (2006). Understanding and promoting adoption of conservation practices by rural landholders. *Animal Production Science*, **46**, 1407–1424.

Perfecto, I. & Vandermeer, J. (2010). The agroecological matrix as alternative to the land-sparing/agriculture intensification model. *Proceedings of the National Academy of Sciences*, **107**, 5786–5791.

Phalan, B., Balmford, A., Green, R.E. & Scharlemann, J.P.W. (2011). Minimising the harm to biodiversity of producing more food globally. *Food Policy*, **36**, S62–S71.

Ramankutty, N. & Rhemtulla, J. (2013). Land sparing or land sharing: context dependent. *Frontiers in Ecology and the Environment*, **11**, 178–178.

Redman, C., Grove, M.J. & Kuby, L. (2004). Integrating social science into the Long Term Ecological Research (LTER) Network: social dimensions of ecological change and ecological dimensions of social change. *Ecosystems*, **7**, 161–171.

Savory, A. & Butterfield, J. (1999). *Holistic Management: a new framework for decision making*. Washington, DC: Island Press.

Scoones, I. (1998). *Sustainable Rural Livelihoods: a framework for analysis*. Brighton: Institute of Development Studies.

Schirmer, J., Clayton, H. & Sherren, K. (2012a). Reversing scattered tree decline on farms: implications of landholder perceptions and practice in the Lachlan catchment, New South Wales. *Australasian Journal of Environmental Management*, **19**(2), 91–107.

Schirmer, J., Dovers, S. & Clayton, H. (2012b). Informing conservation policy design through an examination of landholder preferences: a case study of scattered tree conservation in Australia. *Biological Conservation*, **153**, 51–63.

Sherren, K., Fischer, J., Dovers, S. & Schirmer, J. (2008). Leverage points for reversing paddock tree loss in Upper Lachlan grazing landscapes – a workshop report. *Ecological Management and Restoration*, **9**, 237–240.

Sherren, K., Fischer, J., Clayton, H., Schirmer, J. & Dovers, S. (2010a). Integration by case, place and process: transdisciplinary research for sustainable grazing in the Lachlan River catchment, Australia. *Landscape Ecology*, **25**, 1219–1230.

Sherren, K., Fischer, J. & Price, R. (2010b). Using photography to elicit grazier values and management practices relating to tree survival and recruitment. *Land Use Policy*, **27**(4), 1056–1067.

Sherren, K., Fischer, J., Clayton, H., Hauldren, A. & Dovers, S. (2011a). Lessons from visualising the landscape and habitat implications of tree decline – and its remediation through tree planting – in Australia's grazing landscapes. *Landscape and Urban Planning*, **103**, 248–258.

Sherren, K., Fischer, J., Pink, J., et al. (2011b). Australian graziers value sparse trees in their pastures: a viewshed analysis of photo-elicitation. *Society and Natural Resources*, **24**, 412–422.

Sherren, K., Yoon, H.-J., Clayton, H. & Schirmer, J. (2012a). Do Australian graziers have an offset mindset about their farm trees? *Biodiversity and Conservation*, **21**, 363–383.

Sherren, K., Fischer, J. & Fazey, I. (2012b). Managing the grazing landscape: insights for agricultural adaptation from a mid-drought photo-elicitation study in the Australian sheep–wheat belt. *Agricultural Systems*, **106**, 72–83.

Stinner, D.H., Stinner, B.R. & Martsolf, E. (1997). Biodiversity as an organizing principle in agroecosystem management: case studies of holistic resource management practitioners in the USA. *Agriculture, Ecosystems and Environment*, **62**, 199–213.

Van der Beek, P. & Bishop, P. (2003). Cenozoic river profile development in the Upper Lachlan catchment (SE Australia) as a test of quantitative fluvial incision models. *Journal of Geophysical Research*, **108**(B6), 2309.

Von Wehrden, H., Abson, D.J., Beckmann, M., et al. (2014). Realigning the land-sharing/land-sparing debate to match conservation needs: considering diversity scales and land-use history. *Landscape Ecology*, **29**, 941–948.

Waggoner, P.E. (1996). How much land can ten billion people spare for nature? *Daedalus*, **125**, 73–93.

Walker, B.H. (1992). Biodiversity and ecological redundancy. *Conservation Biology*, **6**, 18–23.

Walker, B.H. (1995). Conserving biological diversity through ecosystem resilience. *Conservation Biology*, **9**, 747–752.

Walker, B. & Salt, D. (2006). *Resilience Thinking: sustaining ecosystems and people in a changing world*. Washington, DC: Island Press.

Walker, B., Holling, C.S., Carpenter, S.R. & Kinzig, A. (2004). Resilience, adaptability and transformability in social–ecological systems. *Ecology and Society*, **9**(2), 5.

Watson, J., Freudenberger, D. & Paull, D. (2001). An assessment of the focal-species approach for conserving birds in variegated landscapes in southeastern Australia. *Conservation Biology*, **15**, 1364–1373.

CHAPTER TWELVE

Ecological–economic modelling for designing cost-effective incentives to conserve farmland biodiversity

MARTIN DRECHSLER
Helmholz Centre for Environmental Research–UFZ
and
FRANK WÄTZOLD
Brandenburg University of Technology

Introduction

Since the middle of the 1950s agricultural intensification and the abandonment of marginal areas has led to a homogenisation of farmland in Europe, resulting in an ongoing loss of habitat and species diversity. For example, Voříšek et al. (2010) estimate that the populations of 36 selected farmland birds in 21 European countries declined by more than 50 percent during the period from 1980 to 2006. In order to halt this loss of biodiversity and, more generally, mitigate negative impacts from farming on the environment, so-called agri-environment schemes (AESs) have been developed in Europe and other parts of the world.

In Europe, AESs are designed to compensate farmers for income foregone for voluntarily carrying out land-use measures which are costly to them but lead to more biodiversity, the enhancement of ecosystem services, or other environmental improvements (Wätzold & Schwerdtner, 2005). Although AESs have become the key policy instrument to incentivise farmers to conserve endangered species and habitats, their success is mixed at best (Kleijn & Sutherland, 2003; Kleijn et al., 2011).

One major reason for this lack of success is certainly that designing AESs in a way that conservation goals are maximised for given financial resources requires an integration of ecological and economic knowledge. Although there is some integrated research in the field of AES design (see examples throughout this chapter), the vast majority of research is still disciplinary and dominated by ecology.

The need for integrated research increases when the resilience of agricultural production is also considered. In this context, understanding both the contribution of ecosystem services (e.g. pollination, pest control, water regulation) to the agricultural production cycle and the contribution of agricultural land management to the supply and conservation of ecosystem services is important, as both are central to the long-term viability and resilience of agriculture (Bommarco et al., 2013). The contribution and value of both elements should be reflected and integrated in the design of AESs (Ekroos et al., 2014).

Given the lack of much-needed integrated research and the aim of the book to foster cooperation among ecologists and economists, we focus in this chapter on the integration of ecological and economic knowledge in models (henceforth referred to as ecological–economic modelling) to improve the design of AESs for biodiversity conservation.[1]

We start by outlining some background information about the policy instrument of AESs in Europe and the concept of cost-effectiveness in the next section. The following section deals comprehensively with ecological–economic modelling which has been shown to be an appropriate method for integration (Wätzold et al., 2006; Drechsler et al., 2007a; Armsworth et al., 2012). The section starts with a brief discussion on benefits and challenges of integrated modelling; it then describes the three components which typically exist in integrated models for designing AESs, namely, ecological models, economic models and optimisation procedures, and finally provides a case study of ecological–economic modelling. In the next section, three promising future research areas are identified, with the first two being motivated by a resilience perspective: (i) the integration of social learning and adaptive governance in AES design to adjust AESs to global change, (ii) the joint consideration of ecosystem services and biodiversity in AES design and (iii) the development of policy-relevant software-based decision support tools. The following section concludes.

Policy background
Agri-environment schemes as a policy instrument in Europe

Since medieval times, agricultural landscapes in most parts of Europe consisted of many different types of land-use systems which generated substantial habitat and species diversity (Gerowitt et al., 2003). This changed with agricultural intensification – increased use of machinery, fertilisers and other inputs – which led to a homogenisation of the European rural landscape (Hampicke, 1991; Benton et al., 2003).[2] More recently, land abandonment in marginal areas with extensive farming systems is of growing importance.

---

[1] Other important areas of – mostly purely economic – research of how to improve the design of AESs are auctions and performance-based payments. We refer the interested reader to Ferraro (2008) for a discussion of auctions and to Burton and Schwartz (2013) for a discussion of performance-based payments.

[2] It has recently been argued that agricultural intensification is tolerable if sufficient other land is conserved as nature areas. We refer the interested reader for this so-called land-sparing versus land-sharing debate to Tscharntke et al. (2012) and Abson, Sherren & Fischer, Chapter 11.

Such farming systems have created semi-natural habitats; for example, mountain meadows, which are valuable habitats for many increasingly endangered species (Kleijn et al., 2011).

Mitigating the impacts of intensification and reversing the trend of land abandonment is costly (Wätzold & Schwerdtner, 2005). Although EU regulation requires farmers to comply with 'good farming practice',[3] it has been decided that, in general, farmers should not be forced to bear the costs of measures to conserve biodiversity which is threatened due to the impacts of land abandonment and intensification (see Buckwell, Chapter 15). Instead, farmers are compensated for measures which are costly to them but beneficial for biodiversity in the context of agri-environment schemes. Following Regulation 2078/92, AESs have been implemented in the 1990s all over the EU, and are now governed by Regulation 1698/2005 (Article 39). From 2007 until 2013, the EU allocated 22.2 billion euro to support AESs in the Member States (European Court of Auditors, 2011). Typically, AESs are of a voluntary nature and have the form of contracts in which farmers commit themselves for a certain period of time to carry out some predefined measures which benefit biodiversity or the environment in general. Although the EU regulations provide general guidelines on how to design AESs, details are left to the Member States.

Cost-effectiveness of AESs

A key challenge is to design AESs in a cost-effective way. Cost-effectiveness is understood here as maximising the achievement of conservation targets set by society for a given financial resource.[4] Alternatively, and depending on the conservation policy setting, cost-effectiveness may also be defined as designing AESs to minimise the financial resources used to achieve a given set of conservation targets (Wätzold & Schwerdtner, 2005).

When designing AESs in practice one needs to bear in mind that different types of costs arise. Following Wätzold and Schwerdtner (2005), one may differentiate between production, implementation and decision-making costs. Production costs refer to the costs associated with the actual conservation activities, such as profit losses of farmers due to an extensification of farming. The compliance of farmers with contract requirements cannot be taken for granted. In order to ensure compliance with AES requirements, monitoring and enforcement activities are necessary; these costs are referred to as implementation costs. Decision-making costs arise because

---

[3] This is laid down in Regulation (EC) No. 1750/1999, Article 28, which is very general and leaves it to the Member States to set out verifiable standards for good farming practice.
[4] Note that the term cost-effectiveness is not used uniformly in the literature. For a different use of the term and a differentiation between cost-effectiveness and a similar term – budget efficiency – see Wätzold and Drechsler (2014).

of the need to acquire information necessary to make appropriate decisions, including, for example, scientific knowledge and information on stakeholders' preferences in the case of conflicting goals. Decision-making costs also include costs that come from organising decision-making processes if different stakeholders are involved. Implementation and decision-making costs form transaction costs. In the following discussion, we only consider production cost, as integrated work between economists and ecologists on the cost-effectiveness of AESs focuses almost exclusively on production costs.

Identifying cost-effective AESs can be simple if there are only a few available conservation options and only one conservation goal. However, often there are many different conservation options (for example, grassland mowing regimes, which may differ depending on the number of cuts, the amount of fertiliser allowed, the inclusion of mowing strips, etc.), and goals (e.g. conservation of different meadow birds, grassland insects and grassland types). Moreover, the impact of the different measures on the different species may differ depending on when and where they are carried out (e.g. the effect of different mowing dates on species with low dispersal abilities in isolated or connected meadows). In addition, costs in terms of income foregone may also differ depending on when and where the conservation measures are carried out (mowing dates and spatially heterogeneous soil productivity may change the quantity and quality of hay and silage, cf. Mewes et al., 2015). Moreover, the response of farmers to an AES which may contain different measures needs to be understood.

Under these circumstances, identifying the set of agri-environment measures and the appropriate size of the payment for each measure to maximise the conservation of nominated species and habitats within a given financial resource is a non-trivial optimisation problem. To solve this problem requires (i) a quantitative assessment of spatially and temporally differentiated costs of measures, (ii) a reasonable prediction of farmers' behaviour in response to different payment sizes for different measures, (iii) a quantitative assessment of the spatially and temporally differentiated impact of different measures on different species and (iv) numerical optimisation to identify the cost-effective solution. Ecological–economic models have proven useful approaches to solve such optimisation problems (Wätzold et al., 2006).

Ecological–economic modelling for cost-effective AESs

The structure of most ecological–economic models designed to improve the cost-effectiveness of AESs is similar (e.g. Johst et al., 2002; Bamière et al., 2011; Armsworth et al., 2012). They consist of (i) an ecological model to quantify the impact of conservation measures on species, (ii) an agri-economic model to quantify the costs of these measures and (iii) an optimisation module to identify the cost-effective AES. However, integrating ecological and economic knowledge and models is challenging. We start this section by briefly analysing selected challenges of integrated scientific work, and go on to discuss in turn ecological modelling, agri-economic modelling and optimisation. Finally, we present a case study with an ecological–economic model to improve the design of AESs.

Bridging two disciplines – selected challenges and pitfalls

Wätzold et al. (2006) analyse challenges and pitfalls that arise if economists and ecologists work together and identify three requirements for a successful cooperation: an adequate identification and framing of the problem to be analysed, an in-depth knowledge of both disciplines involved and a joint understanding of modelling and scales.[5]

The adequate framing of the conservation problem to be analysed is challenging because ecologists and economists are trained to look at real-world phenomena from different perspectives. When looking at the same conservation problem they consider different aspects to matter and identify different research needs.[6] In such a setting, there is a certain danger that the view of one discipline dominates and knowledge from the other discipline is of an 'add-on' type. Only if the different perspectives are integrated on an equal basis and as early as possible in the research process can a comprehensive view on the conservation problem be developed and the appropriate research questions and design be identified.

If an in-depth knowledge of the concepts, ideas and methods of both disciplines is missing, the full potential of benefits arising from integrating economic and ecological knowledge and models cannot be achieved. Ecologists are increasingly aware of the importance of integrating conservation costs in models. However, other economic concepts such as transaction costs, asymmetric information between policymakers and land-users, property rights and risk aversion of economic agents are often also important for designing cost-effective conservation policies. If such concepts are not included in the analysis, the analysis may well be inferior to an analysis that includes these concepts. Equally important, and with the same negative consequences for the analysis, economists often restrict their analysis to spatially homogeneous, static, or scalar descriptions of ecological systems and

---

[5] This section is based on Wätzold et al. (2006) and Drechsler et al. (2007b).
[6] For example, in evaluating AESs, ecologists will most likely focus on their impact on species. Economists, by contrast, may just assume that the conservation goals are achieved, and will be more concerned to reduce 'extra-profits' (in economic terms: information rents) of farmers that arise because payments to farmers exceed their opportunity costs of conducting conservation measures (Ferraro, 2008).

processes, and are unaware of the knowledge and understanding of ecologists about the spatial, temporal and functional structure of ecosystems.

A common understanding of modelling and spatial and temporal scales is important as economists and ecologists have different approaches to modelling and scale. Drechsler et al. (2007b) compared differences in ecological and economic models related to biodiversity conservation. They found that economic models are often of a conceptual nature and of low complexity. Moreover, economic models tend to be formulated and solved analytically, are predominantly static and mostly ignore uncertainty. If space and time are included in economic models, the scales are mostly of an abstract nature. In contrast, many ecological models are specific to a particular species and region, and of high complexity. They are often rule-based models which are simulated step-by-step to model ecosystem dynamics and uncertainties. With respect to modelling space and time they frequently use concrete spatial and temporal scales such as square kilometres and days. However, there are also economic and ecological models which use similar approaches to modelling and scales (Drechsler et al., 2007b). With respect to cooperation between the two disciplines, it is important to remember that when speaking about modelling and scales, ecologists and economists do not necessarily mean the same thing. This may lead to significant misunderstandings that hamper the development of ecological–economic models.

To overcome these challenges, Wätzold et al. (2006) recommend learning about each other's discipline and engaging in comprehensive communication as early as possible and on an equal basis. They also suggest that development of ecological–economic models is facilitated if the joint research work starts from a concrete conservation policy problem, because the latter provides a focus for discussion of issues such as appropriate scales and modelling approaches.

Ecological modelling

As outlined above, ecological models are used to determine the ecological impacts of conservation measures. Ecological models may be distinguished into two different classes: mechanistic models and statistical models. Mechanistic models, as the term indicates, attempt to predict mechanistically the impacts of conservation measures. They do so by considering the effect of the measures on the essential processes that govern the dynamics of a species population. Statistical models analyse correlations between the observed presence of a species in a particular study area and certain landscape attributes observed in that area. The attributes may include natural as well as anthropogenic factors. Knowledge of these correlations enables predictions to be made of the effect of changes in these attributes on species.

Starting with the mechanistic models, the oldest and simplest model is the logistic growth equation that describes the rate of change of a single population, dn/dt, as given by

$$\frac{dn}{dt} = rn(1 - n/K)$$

where n is population size, r the intrinsic growth rate and K the carrying capacity. This model considers that populations cannot grow indefinitely, and that the sign of r determines the sign of population change. For positive r the logistic growth equation has one stable fix point to which the dynamics approach: $n^*=K$. The logistic growth model ignores a number of factors relevant in real populations. Here we want to emphasise three of them: stochasticity, stages and space. They are all relevant in the present context, because they can be influenced by AESs.

The dynamics of a population are often affected by stochastic factors such as weather and food abundance. Within the framework of the logistic growth equation that means that growth rate and carrying capacity may fluctuate stochastically. The population size n therefore does not approach a fixed value $n^*=K$, but fluctuates around that value. As consequence, the population size may reach very small values and go extinct, despite the existence of the stable fix point of the (deterministic) logistic growth equation at $n^*=K$. This fact opens a whole research field: population viability analysis, where the viability of stochastic populations is analysed as a function of natural and anthropogenic factors (see Martin, Chapter 13). Typical measures of viability are the risk of becoming extinct within a particular time frame and the expected time to extinction.

Most species have different life stages such as eggs, larvae and imagos in the case of insects, or seeds, seedlings, juveniles and adults in the case of plants. Each stage has a certain transition time or probability to develop into the next stage. Also, individuals in each stage have a particular survival probability which generally differs between stages. Moreover, the final stage(s) has a particular rate of producing offspring (eggs, seeds, etc.). Such a stage-based model may enable variation in the effects of conservation measures on different life stages to be included (Caswell, 2001).

The living conditions for species are generally spatially heterogeneous. Two main approaches are used in ecological research to consider spatial heterogeneity: patch-based metapopulation models and grid-based models. In patch-based metapopulation models the landscape is assumed to consist of patches of suitable habitat surrounded by a hostile matrix. Species can disperse between patches. Local populations on individual patches may go extinct, but empty patches may be recolonised by immigrants from other local populations. By this the ensemble of local populations, also termed the metapopulation, can survive for a long time (Hanski, 1999).

In grid-based models the landscape is modelled as a 'raster grid' where each raster cell is characterised by a set of parameters, such as vegetation cover, food abundance for species, density of predators, etc. (Kramer-Schadt et al., 2004). All these may be influenced by conservation measures. Grid-based models are usually used together with individual-based models where the fate of each individual such as birth, reproduction and death is modelled explicitly (Grimm & Railsback, 2005). Each individual is characterised, among others, by its spatial location. Depending on the local conditions the individual performs certain activities such as mating, hunting, feeding, moving to another location, etc. Its activity may again affect the local conditions, such as food abundance.

Individual-based mechanistic models have been proposed as alternatives to equation models, because they often allow the consideration of qualitative biological knowledge more easily. Also if the behaviour and interaction of individuals depends on their spatial locations or if the individuals differ in many attributes such as size, age, sex or nutritional state, the use of equation models for the description of the population dynamics is limited. The disadvantage of individual-based models, however, is that they are harder to handle and analyse than equation models, because they need to be analysed by numerical simulation, which is often more tedious than solving a set of mathematical equations.

Still a major challenge in ecological modelling is the consideration of multiple interacting species, because this requires a lot of knowledge about the nature of these interactions. Even on a qualitative level there are numerous possibilities of how species can interact, because one species can have a positive, neutral or negative influence on another, so for each pair of species there are nine possible different types of interaction. An example of a well-elaborated multi-species model has been developed by Finnoff and Tschirhart (2003) for eight marine species, one of which is harvested by humans.

In addition to mechanistic models, statistical models allow the prediction of the impacts of land-use measures, as well. Habitat suitability models correlate information of presence and absence of a species with parameters characterising the area the species inhabits. For this the study region is overlaid with a grid of points for each of which the presence or absence of the species is known, as well as the level of a number of variables to predict presence and absence. Presence/absence is linked through (logistic) regression to the predictor variables. The parameterised model is able to estimate the probability that the species is present or absent as a function of the local conditions: these local conditions may include land-use types incentivised by AESs, such as intensity of pesticide use or frequency of mowing.

On a larger geographic scale, species distribution models predict the range of species as a function of geographic variables such as altitude, temperature, etc. Particular land-use measures are usually not considered (but could be in principle), but the models can predict in which areas conservation measures for a particular species are meaningful.

In principle, all types of models can be used for ecological–economic modelling, although mechanistic equation-based models seem to be used more often than others. The reason may be that compared to statistical models they appear to be more flexible and compared to individual-based models they are easier to handle.

Economic modelling of land use

Agroeconomic models are used to estimate the costs associated with different conservation measures. In ecological–economic modelling two types of agroeconomic models are mostly used. The first type are simple gross margin calculations to assess the opportunity cost of conservation measures (e.g. Drechsler et al., 2007a) and the second type are somewhat more sophisticated linear programing (LP) models (e.g. Bamière et al., 2011). Less frequently, econometric and agent-based models are also used (see also Ramsden and Gibbons, Chapter 5).

In gross margin calculations, the opportunity costs of conservation measures are calculated by comparing the profit loss that arises if the farmer implements a conservation measure i on an area j of a specific size (usually one hectare) instead of the profit-maximising land-use alternative pm. The profits of the two alternatives depend on the price p of the produced agricultural good, the yield y of that good, and the cost c of producing the crop. The price is usually assumed to be exogenous. The yield depends on the conservation measure i and a vector $\mathbf{u}_j$ with the characteristics of the area j, which may include, for example, soil fertility, climate and humidity level. Production costs include costs of labour, machinery and other inputs and also depend on the conservation measure i as well as characteristics $\mathbf{v}_j$ of the area j like its size, shape and spatial location. Formally, the opportunity costs are then calculated as:

$$\Pi_{pm,j} - \Pi_{i,j} = py_{pm}(\mathbf{u}_j) - c_{pm}(\mathbf{v}_j) - py_i(\mathbf{u}_j) - c_i(\mathbf{v}_j).$$

The functions y_i and c_i are context-specific and can be determined, e.g. from data tables available from regional agricultural agencies, etc.

LP models can include more farm or farm-type specific details as well as resource, technical and policy constraints. In these models, farmers are assumed to maximise profits under some specific constraints. Consider as an example the model by Bamière et al. (2011) which takes into account

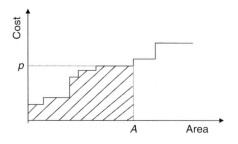

Figure 12.1 Cost per hectare as a function of area with biodiversity-friendly land use. If a payment of magnitude p is offered, conservation measures are carried out on an area A of fields.

field-level, farm-level and landscape-level constraints. Field-level constraints include, for example, the soil type which restricts the agricultural activities and cropping techniques which can be chosen; farm-level constraints consist of policy (e.g. milk quotas) and technical constraints (e.g. feed requirements); and landscape-level constraints refer to spatial constraints for field selection from the AES. With appropriate data, the model can determine which fields participate in an AES for a certain per ha payment, cp_r.

The knowledge about conservation costs can be aggregated into a conservation cost function. For this the fields are ordered by increasing cost and the cost is plotted as a function of the total field area in the region (Figure 12.1). This cost function permits, for example, calculating the cost that arises if on a number of fields with total area A certain conservation measures are applied. This cost is given by the area under the curve up to the level A (Figure 12.1). Combining this information with an ecological model that tells how much biodiversity is obtained for a given area A allows us to determine how much biodiversity is obtained for a given cost.

In the above, we assume that landowners decide in an informed and rational manner, i.e. they precisely know their cost of conservation and maximise their profit. This is an idealisation because in reality landowners have more than the simple objective of profit maximisation and they may not be able to estimate their costs and predict market prices precisely. Econometric studies allow a more realistic prediction of the response of landowners on a payment scheme. In these models, similar to the statistical models in ecology, observed land-use behaviour of landowners is correlated to a number of predictor variables, which include the payment and other parameters such as attributes of the farm and attitudes of the landowner (Nelson et al., 2008).

Generally, the costs and the characteristics of the farms and their owners will be distributed unevenly and non-randomly in the region, which implies that land-use measures will be distributed unevenly and non-randomly too. Economic land-use models therefore need to be able to predict not only the total area with conservation measures (as in Figure 12.1) but also the spatial distribution of the land-use measures. Such spatially explicit models are generally analysed through numerical simulation (Drechsler et al., 2007a; Nelson et al., 2008).

Furthermore, landowners may change their attitudes over time, acquire new information or interact with other landowners through communication or mutual learning. In these cases static models like those outlined above are not sufficient to predict the decisions of the landowners. Instead, agent-based models – equivalent to the individual-based models used in ecology – are more appropriate. These models may involve a complex and adaptive set of rules that describe the behaviour of each landowner and his or her response to an AES (Hare & Deadman, 2004). Due to their complexity, agent-based models are usually analysed through numerical simulation.

As might have become clear, there is some similarity between ecological and economic modelling. The simple ecological and economic models can be formulated with the help of mathematical equations and are often solved analytically. If space matters or the behaviour of entities (individuals in the case of individual-based models or agents in the case of agent-based models) becomes more complicated, simulation models are more adequate to deal with the complexity of the modelled phenomena. Also, in both disciplines mechanistic and statistical models play important roles when modelling the observed behaviour of individuals or agents. Historically, ecologists have a higher preference for complex and simulation-based models than economists (Drechsler et al., 2007b), but examples like Nelson et al. (2008) indicate that the modelling approaches of ecologists and economists may converge – at least in the field of land-use modelling.

Furthermore, these models vary in their ability to capture issues that are key to the concept of resilience such as non-linear dynamics, thresholds, uncertainty, surprise and adaptive governance (Folke, 2006). In ecology, many models consider thresholds and non-linear dynamics and individual-based models are also able to address the issue of emergence (Grimm et al., 2010). In economics, static models such as gross margin calculations, LP models and econometric models are only to a limited extent suitable to analyse change. The only possibility is to compare the outcomes of a static model with different parameters or constraints (for example, the outcome of an LP model before and after a policy change which is reflected by different constraints) which provides a certain understanding of how external changes affect individuals' behaviour and in this way land use and biodiversity conservation. In contrast, agent-based models are much better suited to analysing change, and they are often made explicitly for the purpose of investigating learning or adaptive and changing behaviour (Cong et al., 2014).

Optimisation methods

In order to derive a cost-effective design of AESs (i.e. one for which conservation goals for given costs are maximised or costs for given conservation goals are minimised), optimisation methods are needed which integrate the data from the economic and ecological models. Exact optimisation methods include linear and integer programming (Schrijver, 1998). These methods, however, cannot always be applied – e.g. when the effect of a land-use measure on one site is affected by the land-use measures carried out at other sites, or more generally, when there are spatial interactions. If such interactions are present, heuristic methods such as simulated annealing are frequent choices.

Simulated annealing (Kirkpatrick et al., 1983) may be regarded as a simple variant of a genetic algorithm (Goldberg, 1989). To understand its general principle, consider N sites, each of which may take one of two measures, A and B. To start the algorithm one may consider a random allocation of the two measures on the N sites. This land-use pattern is evaluated with the ecological and economic models. The land-use pattern is then slightly varied randomly, e.g. by randomly selecting one of the N sites and swapping the associated land-use measure. This new land-use pattern is again evaluated with the models. If it leads to a better outcome (e.g. a higher biodiversity level at lower or equal economic costs) it is taken with certainty. If it performs worse it is taken with a particular probability. This means that a temporary decline in performance may be accepted, which ensures that the algorithm does not get stuck in a local optimum (Figure 12.2). As the algorithm proceeds, the probability of accepting a lower performance is reduced gradually, which ensures that once the land-use pattern is close to the optimal solution it converges to it. This gradual reduction of the probability of accepting a lower performance is similar to the gradual cooling of a liquid to a perfect solid state, which gives the algorithm its name.

Case study on ecological–economic modelling

As outlined above, ecological–economic modelling involves the integration of ecological and economic modelling within an optimisation framework which may consider the spatial and temporal allocation of land-use measures as well as multiple objectives. In the following we will briefly present an application.

Many butterfly species require grassland to be managed in a way that is suboptimal with regard to agricultural profit. The ecological–economic model by Drechsler et al. (2007a) allows cost-effective mowing regimes to be determined that maximise butterfly abundance in a region in Germany for a given conservation budget. The model consists mainly of an ecological and an economic submodel.

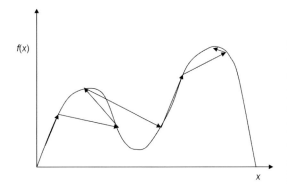

Figure 12.2 Maximisation of function *f(x)* through simulated annealing. Variable *x* is varied randomly around its current value. If *f(x)* for the new value of *x* exceeds *f(x)* of the old value, the new *x* value is chosen. In the beginning of the optimisation process slight reductions in *f(x)* are also accepted. This avoids getting stuck in the local maximum on the left.

The economic submodel assesses the profit losses a landowner has to bear when deviating from the profit-maximising mowing regime. These losses are estimated through gross margin calculations where the different costs for labour, machinery, fertiliser, etc. (that can be obtained, e.g. from agricultural agencies) are subtracted from the benefits (economic value of harvested grass). Altogether the profit-maximising mowing regime plus 111 alternative mowing regimes are considered. A mowing regime is defined by whether a meadow is mowed every single or every second year, by the time of the first cut, and by whether and when a second cut is applied.

The profit losses associated with each alternative mowing regime depend on abiotic parameters such as the meadow's size and soil quality. This information is taken from GIS data obtained from a satellite image. The satellite image has a resolution of 20×20 m and distinguishes between arable land, grassland, woodland, settlement and other major land-use types.

Once the economic cost assessment has been completed, the likely compensation demanded by a landowner for undertaking each alternative mowing regime for each meadow is known. A homogeneous payment is then offered – so that all landowners receive the same amount per hectare. To induce participation of landowners the payment has to exceed the profit loss plus transaction costs plus a random component that takes into account factors that were not explicitly modelled, such as landowners' preferences towards biodiversity conservation. The random component is drawn independently for each meadow from a uniform distribution. For a given budget and alternative mowing regime the magnitude of the required payment is determined as well as the meadows which apply the alternative mowing regime. The latter is determined by checking for each meadow whether the payment exceeds the sum of profit loss, transaction costs and the mentioned random component. Each meadow therefore is either managed conventionally or with the alternative mowing regime.

The ecological model simulates the life cycle of the considered butterfly species, the large blue (*Maculinea teleius*), with a stage-based model (cf. above, Johst et al., 2006) implemented in the programming language Delphi. The species reproduces between June and August. In that period the species deposits its eggs on a specific plant, the great burnet (*Sanguisorba officinalis*). The eggs then develop into larvae, and drop to the ground, where they are adopted by ants which raise them under the ground over the winter.

The aboveground processes are modelled on a weekly timescale. In each week, determination of the number of flying butterflies is based on a probability distribution and the number of pupae at the end of the winter. During its lifetime each butterfly lays about 80 eggs. Butterfly larvae survive on the host plant with a certain probability which is affected by the mowing regime because a cut that takes place before the adoption destroys the eggs and larvae.

Overwinter survival of larvae s considers their competition in the ant nest which depends on the density of larvae in the nest (the ratio of number of larvae l and the nest's carrying capacity K):

$$s = \frac{0.72}{1 + 3(l/K)^5}$$

(Bellows, 1981; Johst et al., 2006). At the end of the winter the life cycle starts again.

Depending on the time of the cuts, more or less of the local butterfly population is destroyed. If a local population becomes completely extinct the meadow can be recolonised by neighbouring meadows (metapopulation model: cf. above; Drechsler et al., 2007a). For this, butterflies emigrate from a high-quality meadow with probability 0.1 and from a low-quality meadow with probability 0.9. The chance of emigrants reaching a meadow depends on the distance and the type of land use in the space between the meadows. The larger the distance d_{ij} between two meadows i and j the lower the probability that butterflies can disperse between them. Also the higher the average landscape resistance $\bar{a}_{ij}$ between meadows i and j (e.g. settlements pose a higher resistance than open land) the lower the probability that butterflies can cross the distance. Altogether, the probability of reaching meadow j from meadow i is (Drechsler et al., 2007a)

$$\Phi_{ij} = \frac{exp(-\bar{a}_{ij}d_{ij})}{\bar{a}_{ij}d_{ij}^2} \sum_j \bar{a}_{ij}d_{ij}^2$$

Once a dispersing butterfly reaches a meadow it lays its eggs on the host plant and the above-described life cycle commences.

Based on these processes the ecological model determines the number of butterflies in the region for a given land-use pattern, where land-use pattern here means which meadow is managed in the profit-maximising manner and which meadow applies the alternative mowing regime. The analysis is done separately for all 111 alternative mowing regimes.

Altogether, the economic model shows, for each alternative mowing regime and budget level, which meadow applies the profit-maximising mowing regime and which meadow applies the alternative mowing regime. The ecological model uses this output to determine the number of butterflies in the region. Comparison of these outputs for the 111 mowing regimes for identical governmental (or 'regional') budgets allows us to identify the cost-effective alternative mowing regime.

Improving AESs – future research issues

We have shown with a case study that ecological–economic modelling is a powerful method to improve the design of AESs. Although some research has been carried out in this field, there are many more challenges that need to be addressed for a better design of AESs. We wish to highlight three research issues which seem particularly important to us, where research is currently lacking[7]: (i) social learning and adaptive governance to adjust AESs to global change, (ii) the joint consideration of ecosystem services and biodiversity when designing AESs and (iii) the need to develop policy-relevant software-based decision support tools for AES design. The first two research issues may be motivated by adopting a resilience perspective.

Adopting the resilience perspective
Social learning and adaptive governance
Within the context of a resilience perspective, Folke (2006) emphasises the importance of social and policy learning and the capacity of the governance structure to adapt to change. This perspective is particularly relevant to AESs because agricultural systems are strongly affected by global change. Consider as an example climate change which already has and increasingly will lead to a northward shift of habitats for species (Vos et al., 2008), and which will also change conditions for agricultural production. In the case of designing AESs for the conservation of the large blue butterfly (cf. the ecological–economic modelling case study), on the ecology side it has to be taken into account that the

---

[7] Another important research area is the question of how to design payments in a way that a certain spatial configuration like an agglomeration of habitat patches (e.g. Parkhurst et al., 2002) is achieved. However, there is already a substantial amount of research going on in this area (e.g. Drechsler et al., 2010; Bamière et al., 2011; Wätzold & Drechsler, 2014).

reproduction period might shift in time and that as a consequence the species requires different mowing regimes for its survival. Moreover, the species may not be any more able to persist in areas where it has been before, but may be able to colonise areas where it was not able to exist previously. On the economic side, the profit-maximising mowing regime and the profit losses of alternative mowing regimes may change due to climatic modifications and hence the payments given to incentivise farmers to implement such regimes may also need to change.

Such changes require an adaptation of AESs whose design in turn requires knowledge from different disciplines. This includes a prediction of regional climate change, sufficient ecological and economic expertise about the likely consequences of this change and knowledge about a governance structure which is able to make the necessary changes in the design of AESs. Currently, our understanding of how the cost-effective design of AESs should be changed in the face of climate change in particular and global change in general is very poor. There is also little understanding of how governance structures in the field of AESs need to be designed that allow for policy learning and for adapting to global change in a sufficient timeframe so that species can survive in agricultural landscapes. This, however, is urgently needed given the important challenges arising from global change to biodiversity conservation in agricultural landscapes.

Integrating protection of ecosystem services and biodiversity in AES design
Many conservation measures that are supported by AESs also have an impact on ecosystem services. These impacts may benefit farmers directly by lowering their production costs and/or increasing their yield, and in a more general sense, by enhancing the resilience of agricultural production.[8] Examples include semi-natural pastures and flower strips, which are beneficial for certain endangered species and enhance pollination and pest control in neighbouring fields (Lundin et al., 2013); and the reduction of fertilisers, pesticides and other agro-chemicals which improves the quality of ground and surface water and can have a positive impact on species, although some endangered birds actually profit from a certain amount of fertiliser on grassland (Johst et al., 2015). More generally, for many conservation measures dependencies, synergies and trade-offs exist between biodiversity conservation and ecosystem service provision.

These interrelationships, however, need to be taken into account in the design of cost-effective AESs to enhance the resilience of agricultural production and conserve biodiversity. An AES optimisation that aims to

---

[8] We refer the interested reader to Chapters 3, 4 and 6 of this book for more information about how ecosystem services may benefit farmers.

maximise the (weighted) achievement of all ecosystem service and biodiversity conservation goals for a given budget is at least as good as an optimisation that divides the budget in a way that each goal receives a certain part and then maximises the different goals separately under the respective constraints of individual budget parts. There is, however, a lack of integrated models that not only integrate ecological and economic models but also quantify the impact of conservation measures on ecosystem services in the context of AES design.

Developing decision support tools for policymakers
Existing ecological–economic models have contributed significantly to improving our understanding of how to integrate economic and ecological knowledge and models for cost-effective AES design. However, these models are of limited use as decision support. They are typically static, which means that they are quickly outdated if important economic or ecological parameters change (for example, a change in agricultural input and output prices or improved knowledge about the impact of conservation measures on species). Their relevance is further restricted by a focus on one or only few species and conservation measures and a specific (often rather small) region. However, conservation agencies involved in the design of AESs would prefer to have decision support tools which can be adapted to changing ecological and economic circumstances, which can be easily adopted to a region of their interest and which include a comprehensive set of conservation measures as well as species and habitats of conservation interest.

Such a decision support tool (named *DSS-Ecopay*) has recently been developed for AESs directed at grassland conservation (Mewes et al., 2012). *DSS-Ecopay* is software-based and has been developed to design AESs in the German federal states of Saxony and Schleswig-Holstein, but has also been applied to regions in Flanders, Belgium, and the German federal state of Brandenburg. It includes about 30 butterfly and bird species and seven habitat types. The ecological model of *DSS-Ecopay* is described in Johst et al. (2015), the economic model in Mewes et al. (2015) and an application to the analysis of an AES in Saxony in Wätzold et al. (2016). *DSS-Ecopay* is available for free on the internet (www.inf.fu-berlin.de/DSS-Ecopay/ecopay_main_eng.html) and can be adapted by users to their own region and to changing economic and ecological circumstances.

Conclusions

Cost-effectiveness, i.e. the maximisation of conservation outputs for given budgets, is a key criterion for the design of AESs. We hope we have shown that the integration of ecological and economic knowledge and models in the design of cost-effective AESs generates policy recommendations that are superior to recommendations based on purely disciplinary research. There is little research which quantifies efficiency gains of integrated research, but a seminal paper by Ando et al. (1998) on the selection of reserve sites in the USA suggests that they might be substantial. Ando et al. (1998) found that by integrating ecological and economic knowledge in a site-selection algorithm up to 50 percent more species could be conserved for given costs than if only ecological knowledge were considered in the algorithm.

Integrated research is even more important when adopting the broader perspective of the resilience of agricultural production. In such a perspective, social learning and adaptive governance to cope with global change have to be considered, as well as the resilience of agricultural production which requires the joint consideration of ecosystem services and biodiversity conservation in the design of AESs.

We do not mean to say that disciplinary research is not important. Excellent disciplinary research is a precondition for successful ecological–economic modelling as integrated research requires the integration of disciplinary state-of-the-art knowledge. However, given the substantial benefits of integrated research and that the lion's share of research on AESs is still disciplinary, we find that much could be gained if more research were done which integrates ecological and economic expertise. Next to the willingness and ability of researchers to do this type of research, appropriate funding opportunities and career rewards are essential.

References

Ando, A., Camm, J., Polasky, S. & Solow, A. (1998). Species distributions, land values and efficient conservation. *Science*, **279**, 2126–2128.

Armsworth, P.R., Acs, S., Dallimer, M., *et al.* (2012). The cost of policy simplification in conservation incentive programs. *Ecology Letters*, **15**(5), 406–414.

Bamière, L., Havlíka, P., Jacqueta, F., et al. (2011). Farming system modelling for agri-environmental policy design: the case of a spatially non-aggregated allocation of conservation measures. *Ecological Economics*, **70**(5), 891–899.

Bellows, T.S. (1981). The descriptive properties of some models for density dependence. *Journal of Animal Ecology*, **50**, 139–156.

Benton, T.G., Vickery, J.A. & Wilson, J.D. (2003). Farmland biodiversity: is habitat heterogeneity the key? *Trends in Ecology and Evolution*, **18**(4), 182–188.

Bommarco, R., Kleijn, D. & Potts, S.G. (2013). Ecological intensification: harnessing ecosystem services for food security. *Trends in Ecology & Evolution*, **28**(4), 230–238.

Burton, R.J.F. & Schwarz, G. (2013). Result-oriented agri-environmental schemes in Europe and their potential for promoting behavioural change. *Land Use Policy*, **30**, 628–641.

Caswell, H. (2001). *Matrix Population Models: construction, analysis, and interpretation*, 2nd edition. Sunderland, MA: Sinauer Associates.

Cong, R.-G., Smith, H.G., Olsson, O. & Brady, M. (2014). Managing ecosystem services for agriculture: will landscape-scale management pay? *Ecological Economics*, **99**, 53–62.

Drechsler, M., Wätzold, F., Johst, K., Bergmann, H. & Settele, J. (2007a). A model-based approach for designing cost-effective compensation payments for conservation of endangered species in real landscapes. *Biological Conservation*, **140**, 174–186.

Drechsler, M., Grimm, V., Mysiak, J. & Wätzold, F. (2007b). Differences and similarities between economic and ecological models for biodiversity conservation. *Ecological Economics*, **62**(2), 232–241.

Drechsler, M., Johst, K., Wätzold, F. & Shogren, J. F. (2010). An agglomeration payment for cost-effective biodiversity conservation in spatially structured landscapes. *Resource and Energy Economics*, **32**, 261–275.

Ekroos, J., Olsson, O., Rundlöf, M., Wätzold, F., & Smith, H.G. (2014). Optimizing agri-environment schemes for biodiversity, ecosystem services or both? *Biological Conservation*, **172**, 65–71.

European Court of Auditors (2011). *Is agri-environment support well designed and managed?* Special Report No. 7/2011. Luxembourg: Publications Office of the European Union. http://eca.europa.eu/portal/pls/portal/docs/1/8760788.PDF.

Ferraro, P.J. (2008). Asymmetric information and contract design for payments for environmental services. *Ecological Economics*, **65**, 810–821.

Finnoff, D. & Tschirhart, J. (2003). Harvesting in an eight-species ecosystem. *Journal of Environmental Economics and Management*, **45**, 589–611.

Folke, C. (2006). Resilience: the emergence of a perspective for social–ecological systems analyses. *Global Environmental Change*, **16**, 253–267.

Gerowitt, B., Isselstein, I. & Marggraf, R. (2003). Rewards for ecological goods – requirements and perspectives for agricultural land use. *Agriculture, Ecosystems and Environment*, **98**, 541–547.

Goldberg, D. (1989). *Genetic Algorithms in Search, Optimization and Machine Learning*. Reading, MA: Addison-Wesley.

Grimm, V. & Railsback, S.F. (2005). *Individual-based Modelling and Ecology*. Princeton, NJ: Princeton University Press,.

Grimm V., Berger, U., DeAngelis, D.L., et al. (2010). The ODD protocol: a review and first update. *Ecological Modelling*, **221**, 2760–2768.

Hampicke, U. (1991). *Naturschutzökonomie*. Stuttgart: Ulmer.

Hanski, I. (1999). *Metapopulation Ecology*. Oxford: Oxford University Press.

Hare, M. & Deadman, P. (2004). Further towards a taxonomy of agent-based simulation models in environmental management. *Mathematics and Computers in Simulation*, **64**, 25–40.

Johst, K., Drechsler, M. & Wätzold, F. (2002). An ecological–economic modelling procedure to design compensation payments for the efficient spatio-temporal allocation of species protection measures. *Ecological Economics*, **41**, 37–49.

Johst, K., Drechsler, M., Thomas, J.A., & Settele, J. (2006). Influence of mowing on the persistence of two endangered Large Blue (*Maculinea*) butterfly species. *Journal of Applied Ecology*, **43**, 333–342.

Johst, K., Drechsler, M., Mewes, M., Sturm, A. & Wätzold, F. (2015). A novel modelling approach to evaluate the ecological effects of timing and location of conservation measures in grassland at the landscape scale. *Biological Conservation*, **182**, 44–52.

Kirkpatrick, S., Gelatt, C.D. & Vecchi, M.P. (1983). Optimization by simulated annealing. *Science*, **220**(4598), 671–680.

Kleijn, D. & Sutherland, W.J. (2003). How effective are European agri-environment schemes in conserving and promoting biodiversity? *Journal of Applied Ecology*, **40**, 947–969.

Kleijn, D., Rundlöf, M., Scheper, J., Smith, H.G. & Tscharntke, T. (2011). Does conservation on farmland contribute to halting the biodiversity decline? *Trends in Ecology and Evolution*, **26**(9), 474–81.

Kramer-Schadt, S., Revilla, E., Wiegand, T. & Breitenmoser, U. (2004). Fragmented landscapes, road mortality and patch connectivity: modelling dispersal for the Eurasian lynx in Germany. *Journal of Applied Ecology*, **41**, 711–723.

Lundin, O., Smith, H., Rundlöf, M. & Bommarco, R. (2013). When ecosystem services interact: crop pollination benefits depend on the level of pest control. *Proceedings of the Royal Society B*, **280**(1753), 2012.2243.

Mewes, M., Sturm, A., Johst, K., Drechsler, M. & Wätzold, F. (2012). Handbuch der Software Ecopay zur Bestimmung kosteneffizienter Ausgleichzahlungen für Maßnahmen zum Schutz gefährdeter Arten und Lebensraumtypen im Grünland. 1/2012, -152. Leipzig, Helmholtz-Zentrum für Umweltforschung GmbH-UFZ. UFZ-Bericht.

Mewes, M., Drechsler, M., Johst, K., Sturm, A. & Wätzold, F. (2015). A systematic approach for assessing spatially and temporally differentiated opportunity costs of biodiversity conservation measures in grasslands. *Agricultural Systems*, **137**, 76–88.

Nelson, E., Polasky, S., Lewis, D.J., et al. (2008). Efficiency of incentives to jointly increase carbon sequestration and species conservation on a landscape. *Proceedings of the National Academy of Sciences of the USA*, **105**, 9471–9476.

Parkhurst, G.M., Shogren, J.F., Bastian, P., et al. (2002). Agglomeration bonus: an incentive mechanism to reunite fragmented habitat for biodiversity conservation. *Ecological Economics*, **41**, 305–328.

Schrijver, A. (1998). *Theory of Linear and Integer Programming*. Chichester: John Wiley and Sons.

Tscharntke, T., Clough, Y., Wanger, T.C., et al. (2012). Global food security, biodiversity conservation and the future of agricultural intensification. *Biological Conservation*, **151**, 51–59.

Voříšek, P., Jiguet, F., van Strien, A., et al. (2010). Trends in abundance and biomass of widespread European farmland birds: how much have we lost? BOU *Proceedings – Lowland Farmland Birds III* (www.bou.org.uk/bouproc-net/lfb3/vorisek-etal.pdf).

Vos, C.C., Berry, P., Opdam, P., et al. (2008). Adapting landscapes to climate change: examples of climate-proof ecosystem networks and priority adaptation zones. *Journal of Applied Ecology*, **45**, 1722–1731.

Wätzold, F. & Drechsler, M. (2014). Agglomeration payment, agglomeration bonus or homogeneous payment? *Resource and Energy Economics*, **37**, 85–101.

Wätzold, F. & Schwerdtner, K. (2005). Why be wasteful when preserving a valuable resource? A review article on the cost-effectiveness of European conservation policy. *Biological Conservation*, **123**, 327–338.

Wätzold, F., Drechsler, M., Armstrong, C.W., et al. (2006). Ecological–economic modeling for biodiversity management: potential, pitfalls, prospects. *Conservation Biology*, **20**(4), 1034–1041.

Wätzold, F., Drechsler, M., Johst, K., Mewes, M. & Sturm, A. (2016). A novel, spatiotemporally explicit ecological–economic modeling procedure for the design of cost-effective agri-environment schemes to conserve biodiversity. *American Journal of Agricultural Economics*, **98**(2), 489–512.

CHAPTER THIRTEEN

Viability analysis as an approach for assessing the resilience of agroecosystems

SOPHIE MARTIN

Irstea

Introduction

The word resilience dates back to the end of the nineteenth century, where it was used in the physics of materials to describe the ability of a metal to return to its original shape after deformation produced by a shock. Its use spread over the twentieth century to fields such as psychology, where the resilience of an individual relates to their ability to find a normal life after a trauma; to computer science, where the resilience of a system is its ability to function despite the failure of some components; and to areas of ecology, economics and social sciences. Objects change, but those that exhibit the ability to recover certain properties despite changes due to perturbations outside of their control can be described as resilient. For a physicist of materials, the system studied is a metal bar, the property its form, the disturbance – a shock and the disruption – a deformation; for ecologists, the system might be a lake, the property may be oligotrophy (to be in a state of clear water), the disturbance – rainfall events of high intensity and the disruption – the passage of the lake from oligotrophy to eutrophy (i.e. to be in a state of turbid water); alternatively, the ecological system might be a forest, the property – a minimum tree density to limit erosion and the disturbance – a storm. Whatever the system, the study of resilience involves the definition of the triplet: system, properties and disturbances.

The current interest in the issue of resilience in life sciences is explained by the fact that observations of eutrophication of lakes, bleaching of corals, and forests becoming savannas even after the end of cultivation, have weakened the representation of living systems as being dominated by stabilising forces which return the system to equilibrium following a disturbance. Alongside this representation is the notion that resistance is related to the intensity of the force necessary to move the system state a given distance away from the equilibrium.

In the field of ecology, the definition of resilience which enjoys a broad consensus today is the definition of Holling (1973): the ability of a system facing disturbances to maintain or recover some key properties. For example, when phosphorus discharges into a lake reach a critical level, the water becomes turbid and life (especially fish populations) is asphyxiated. In some cases, if phosphorus discharges are reduced, lake water can become clear again. In other cases, the damage is irreversible, and the reduction of phosphorus inputs is not sufficient to restore clear water.

The aim of the study of resilience is to avoid situations in which natural or man-made disturbances can lead to irreversible situations. It also facilitates the restoration of the essential properties, when possible.

Holling (1996) calls this definition of resilience 'ecosystem resilience' or 'ecological resilience'. Apart from its name, Levin et al. (1998) emphasise its application to the study of systems including both ecological and economic components, that is socio-ecological systems. The underlying objective is to maintain the system within certain limits rather than at a stable point.

'Ecological resilience' may be desirable or not: for example, a polluted resource or a dictatorship can both be very resilient. However, from the perspective of sustainability, increasing the resilience of desired properties reduces the intensity of damage caused by any disturbance.

The ability of a socio-ecological system to maintain its functioning mode depends on variables that control the boundaries of these different modes, on the intensity and the frequency of the disturbances under consideration, on the time scale (Carpenter et al., 2001), on the distribution of species or rather on the functions they perform within the same scale in time and space and between different scales (Peterson et al., 1998).

Following the interpretation of Holling (1973), resilience depends on: (i) the state of the system, (ii) the property of interest, (iii) anticipated disturbances, (iv) dynamics and available controls, (v) the cost associated with the effort needed to restore the property of interest if it is lost following a disturbance (this cost may be the time required for recovery, for example) and (vi) the time horizon (Carpenter et al., 2001).

Several measures of resilience have been proposed to evaluate resilience in socio-ecological models.

When the model is made of differential equations, most indices are related to the eigenvalues of the linearised system near equilibrium. Pimm and Lawton (1977) measured resilience as being inversely proportional to the eigenvalue of the largest norm which represents an asymptotic property; the rate of decrease of the distance to the equilibrium when time goes to infinity. To complete the resilience index as a measure of response to small perturbations when the system is close to equilibrium, Neubert and Caswell (1997) proposed indices that assess the intensity and duration of the transient behavior near an asymptotically stable equilibrium.

According to Beddington et al. (1976), resilience can be measured as the intensity of disturbance that a system can absorb without undergoing qualitative change. From a dynamic system view, a qualitative change can be interpreted as a jump into another attraction basin. Assuming that before the disturbance, the system is in an asymptotically stable equilibrium, resilience is then defined as proportional to the distance between this equilibrium and the boundary of its attraction basin (see for example Collings & Wollkind, 1990; van Coller, 1997; Anderies et al., 2002).

Considering that resilience variations are due to slow variables, Ludwig et al. (1997) studied the parameter space and proposed the inverse of the distance to bifurcation points as a measure of resilience. Indeed, at bifurcation points, equilibria change in nature or disappear (Casagrandi & Rinaldi, 2002; Lacitignola et al., 2007).

In the case of individual-based models, resilience has been studied by simulations as the inverse of the time required for the system to recover after disturbance to a state close to that before disturbance (Matsinos & Troumbis, 2002; Ortiz & Wolff, 2002).

Janssen and Carpenter (1999) studying the management of a lake showed that a choice of management can be qualified as more resilient than another when maximum values reached by the phosphorus concentration in the lake during simulation time are lower. In other words, a system is more resilient when the simulated curves are further away from dangerous areas.

In this chapter, we are interested in the issue of rangeland management using a model to describe grass dynamics (shoot biomass and crown biomass) according to the grazing pressure because an important part of rangeland management occurs through adjusting the stocking rate. It is necessary to define a control function and to specify the evolution rule for the grazing pressure as a function of time or grass biomass when using the measures of resilience described above.

Then, when the model consists of a set of differential equations, these measures of resilience give information:

- on the evolution of the system when time goes to infinity: namely on biomass and grazing levels of the attractor, the asymptotic rate of decrease of the distance to this attractor. However, transient behaviours can lead to states that are very far from the final attractor;
- on the size of the attraction basin of this attractor, but the attraction basin does not necessarily coincide with the property of minimal grazing pressure and minimal quantity of shoot biomass according to which we aim at assessing the rangeland resilience.

In the case of individual-based models, simulations can be used to evaluate the time needed for shoot biomass and grazing pressure to return to levels close to those observed before disturbance, but with no guarantee that these levels can then be preserved over time.

Thus, for the problem we address, definitions of resilience derived from the theory of dynamical systems or based on simulations suffer from three major limitations. First, they cannot be used when several policy actions are considered together, whereas the nature of socio-ecological systems is such that sets of policy actions may need to be identified in order to promote resilience. Second, these definitions limit the evaluation of the resilience of the properties of interest to attractors or unions of attraction basins, with no guarantee that these properties might actually coincide with these attraction basins. Finally, when other sets are considered, there is no guarantee that the evolution of the system will remain within these sets. However, as far as socio-ecological systems are concerned, the aim is not only to reach a satisfying state of the system but also to be able to keep this system in satisfying states over time.

The formalism of controlled dynamical systems allows us to consider from a given starting point sets of evolutions whose dynamics depend on the state of the system and also on exogenous controls which vary with time. Given prescribed properties, mathematical viability theory (Aubin, 1991; Aubin et al., 2011) develops methods and tools to determine controlled dynamical systems governing evolutions which satisfy these properties for ever or until the moment when objectives are achieved. Using this framework, which permits the design of control functions according to prescribed properties that are independent from the dynamics, Martin (2004) has proposed that the inverse of the cost associated with the effort needed to restore and preserve certain properties of the system following disturbances can be used as a measure of resilience.

Thus in a given state, for a given property and an anticipated set of disturbances:

- a system is infinitely resilient if the result of any occurrence of anticipated disturbances is that this property can be preserved;
- a system has a finite but non-zero resilience value if following an occurrence of an anticipated disturbance, this property is lost but can be restored and then preserved. Moreover, for any occurrence of anticipated disturbances, this property may be lost but can be restored and the cost associated with the restoration is bounded by the inverse of the resilience value;
- a system is not resilient if at least one anticipated disturbance causes a permanent loss of the given property (no restoration is possible).

This quantitative methodology for measuring and evaluating resilience in socio-ecological systems can be used to explore the effect of proposed policies on system resilience before they are implemented, and generates material that can help policymakers choose an appropriate policy action. In the next section, we use this viability-based measure to evaluate the resilience of rangeland management approaches that aim to preserve both a minimal shoot biomass (ecological component) and a minimal grazing pressure (economic component). We present this case study as a tutorial illustrating how viability analysis can be applied to a socio-ecological problem. We then describe the general mathematical framework used to implement the viability-based measure of resilience. We finish with some conclusions and perspectives.

Case study as a tutorial illustrating how to apply viability analysis to evaluate resilience

The measure of resilience as the inverse of the cost associated with the effort to restore and preserve certain properties of the system following disturbances involves the description of:

(1) the evolution of the system state based on the state itself, but also in terms of management actions, called controls,
(2) the properties of the system under consideration,
(3) the anticipated disturbances,
(4) the cost function used to evaluate the effort of restoration and preservation.

The dynamics, the property under study, the anticipated disturbances and the cost function

The dynamics

To describe grass dynamics, we use the model of Anderies et al. (2002) in which the grass plant is modelled as two parts, the crowns and the shoots. Growth occurs through the interaction of these two parts. Incorporating the relationship between shoots and crowns yields the following model:

$$\begin{aligned} c'(t) &= r_s s(t) - c(t) \\ s'(t) &= (a_c c(t) + r_c c(t) s(t))(1 - s(t)) - \gamma_g(t) s(t) \end{aligned} \quad (13.1)$$

where c represents crown biomass, s shoot biomass, r_s, a_c and r_c are parameters that describe the rate at which crown or shoot biomass grows when crown and shoot are present. γ_g represents the grazing pressure.

In our study, grazing pressure is associated with rangeland management policies because an important part of rangeland management occurs through adjusting the stocking rate. Managers decide when to mate their ewes and rams and when to buy, sell and move their stock. However, pastoralists cannot adjust stocking rates instantaneously. Thus, we consider that the variations of the stocking rate are bounded:

$$\gamma_g'(t) = u(t) \in [\underline{u}; \overline{u}]. \quad (13.2)$$

While Anderies et al. (2002) considered two scenarios of constant utilisation rate and constant stocking rate, we consider all policies that satisfy bounded variations of the stocking rate.

Hence we use a controlled dynamical system with a three-dimensional state space (the crown biomass, $c(t)$, the shoot biomass, $s(t)$, and the grazing pressure, $\gamma_g(t)$) and a one-dimensional control space (the variation in grazing pressure, $u(t)$):

$$\begin{aligned} c'(t) &= r_s s(t) - c(t) \\ s'(t) &= (a_c c(t) + r_c c(t) s(t))(1 - s(t)) - \gamma_g(t) s(t) \\ \gamma'_g(t) &= u(t) \in [\underline{u}; \overline{u}]. \end{aligned} \qquad (13.3)$$

The property under study

Our aim is to design effective policies for delivering economically and environmentally resilient agricultural systems. The property for which we are assessing the resilience has two components: on the economic side, a minimal grazing pressure, and on the ecological side, a minimal quantity of shoot biomass. This property is described by the following constraint set which is a subset of the state space (Figure 13.1):

$$\begin{aligned} \gamma_g &\geq \underline{\gamma_g} \\ s &\geq \underline{s}. \end{aligned} \qquad (13.4)$$

The anticipated disturbances

Next we measure the resilience of rangeland to drought events.

A period of drought causes a sudden reduction in shoot biomass. We represent a drought event as a jump in the state space from (c,s,γ_g) to $(\tilde{c},\tilde{s},\tilde{\gamma}_g)$ where:

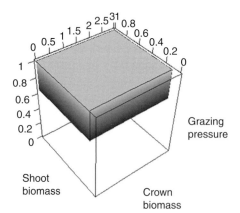

Figure 13.1 Constraint set described by (13.4) as a subset of the three-dimensional state space (c,s,γ_g) with $\underline{s} = 0.1$ and $\underline{\gamma_g} = 0.65$.

- $\tilde{c} = c$, we assume that the drought event does not affect crown biomass,
- $\tilde{s} = s - as$ where $a \in [0, \bar{a}]$ represents the severity of drought, the maximal anticipated severity is $\bar{a} \leq 1$,
- $\tilde{\gamma}_g = \gamma_g$, the drought event has no direct impact on the grazing pressure.

Thus, anticipated disturbances are jumps in the state space from (c, s, γ_g) to $(c, s - as, \gamma_g)$ where $a \in [0, \bar{a}]$ (Figure 13.2).

The cost function

The cost function measures the effort necessary to achieve a state from which the property under study is satisfied and can be preserved over time. In the rangeland management context, we choose to evaluate this effort in terms of time necessary to achieve a safe position, where both shoot biomass and grazing pressure are over the minimal acceptable values and for which there are rangeland management policies that ensure that the evolution of the state remains within these minimal levels over time.

Resilience evaluation

The implementation of the measurement of resilience proposed by Martin (2004) involves two steps described in the general case in the section below (the viability-based measure of resilience). The results presented in this section for our case study are derived from calculations performed using the software of Patrick Saint-Pierre that implements the algorithm of Saint-Pierre (1994).

The first step consists in studying the compatibility between grass and grazing pressure dynamics described by equations (13.3), and the property of interest of the rangeland described by equation (13.4).

Actually, even if the levels of crown biomass, shoot biomass and grazing pressure are such that the property of interest is satisfied, there is no reason why this property should be maintained over time because the system state

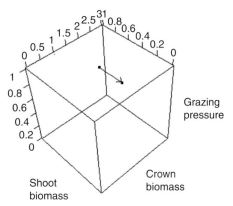

Figure 13.2 Representation of the consequence of a drought event in the three-dimensional state space. The system state jumps from the state $(c = 2, s = 0.7, \gamma_g = 0.8)$ to the state $(c = 2, s = 0.35, \gamma_g = 0.8)$. The shoot biomass has been divided by 2; the severity, a, of such a drought equals 0.5.

evolves with time and its state can exit the constraint set representing the property under study.

The viability kernel (a fundamental concept of mathematical viability theory, see below) is the subset of the constraint set gathering all the combinations of shoot biomass, crown biomass and grazing pressure such that from an initial state belonging to the viability kernel, the rangeland can support a minimal grazing pressure while preserving a minimal quantity of shoot biomass over time if accurate stocking rate adjustments are fulfilled.

The size of the viability kernel depends on the possibility of stocking rate adjustments: the quicker the stocking rate variations can be performed (the bigger $|\underline{u}|$ and $|\bar{u}|$ are), the more numerous are the situations from which the property can be preserved, and consequently, the larger the viability kernel is.

In the extreme case where variation in the stocking rate is not possible ($|\underline{u}| = |\bar{u}| = 0$), the grazing pressure $\gamma_g(t)$ remains constant over time. Nevertheless, crown and shoot biomass evolves with time governed by the dynamics given in equations (13.3). Even if the initial grazing pressure and shoot biomass are over the minimal thresholds, the shoot biomass may become smaller. Given an initial grazing pressure which will remain constant, the viability kernel gathers all initial values of crown and shoot biomass such that shoot biomass remains above the threshold despite this constant grazing pressure. Clearly, as grazing pressure increases, the viability kernel will decrease in size. Figure 13.3 displays in the two-dimensional plane (c,s) viability kernels for different values of grazing pressure. Over a given value of grazing pressure, the viability kernel is empty, indicating that whatever the initial values of crown and shoot biomass, at this constant level of grazing pressure, the minimal threshold of shoot biomass is doomed to be crossed in finite time.

If the rangeland manager is now given the possibility of adjusting his stocking rate, as the crown and shoot biomass change, the third state variable, the grazing pressure, may evolve with time. The viability kernel is a three-dimensional set, a subset of the constraint set displayed in Figure 13.1 and shown in Figure 13.4.

Obviously, as the maximal speed of grazing pressure variations increases, the viability kernel increases since there are more management opportunities. To illustrate this point, we show on the same graph in Figure 13.5 sections of viability kernels for $\gamma_g = 0.9$ and different values of the bounds $\underline{u}$ and $\bar{u}$. The smallest section corresponds to no grazing pressure variation and then the section surface increases as the absolute values of the bounds do.

The second step consists of evaluating the resilience of the rangeland from the impact of drought events on its ability to preserve minimal levels of shoot biomass and grazing pressure.

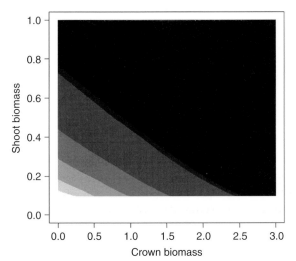

Figure 13.3 Viability kernels for dynamics (13.1) with constant grazing pressure and property (13.4) for different values of the grazing pressure. The viability kernel increases as the value of the grazing pressure decreases: for $\gamma_g = 0.905$, the viability kernel is coloured black, for $\gamma_g = 0.9$, the viability kernel extends to the darker grey area, and so on for $\gamma_g = 0.8, 0.7, 0.6$, the viability kernel extends to the lighter grey area for $\gamma_g = 0.5$. The viability kernel is empty for $\gamma_g \geq 0.91$. Parameter values are $r_s = 3$, $a_c = 0.1$, $r_c = 1$ and $\underline{s} = 0.1$.

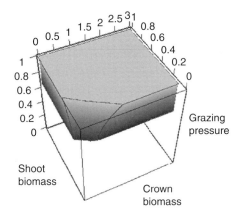

Figure 13.4 Viability kernel for dynamics (13.3) and property (13.4). Parameter values are $r_s = 3$, $a_c = 0.1$, $r_c = 1$, $\gamma_g = 0.65$, $\underline{s} = 0.1$ and $\bar{u} = -\underline{u} = 0.05$.

A period of drought causes a sudden reduction in shoot biomass. Thus, anticipated disturbances are jumps in the state space from (c,s,γ_g) to $(c,s - \alpha s,\gamma_g)$ where $\alpha \in [0,\bar{\alpha}]$ (Figure 13.2). Even if the state (c,s,γ_g) belongs to the viability kernel, the state $(c,s - \alpha s,\gamma_g)$ may not, implying that the property of shoot biomass and grazing pressure preservation cannot be satisfied over time whatever the stocking rate adjustments.

The damage associated with a drought event that causes a jump outside the viability kernel is measured by the time necessary to re-enter the viability kernel, if possible. This time depends on the stocking rate adjustment policy which is implemented. Thanks to viability theory tools, computing the minimal time necessary to re-enter the viability kernel enables

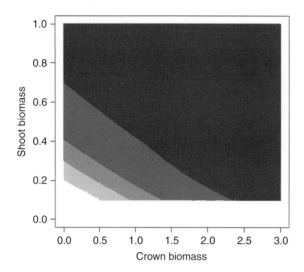

Figure 13.5 Sections of viability kernels for dynamics (13.3) and property (13.4) for $\gamma_g = 0.9$ and different values of the bounds on the grazing pressure variations. In black, the section of the viability kernel when the grazing pressure cannot be modified, $\bar{u} = -\underline{u} = 0$ (we get the same result as Figure 13.3). The viability kernel increases as the value of the bounds on the grazing pressure variations increase: for $\bar{u} = -\underline{u} = 0.03$, the section of the viability kernel extends to the darker grey area, and so on for $\bar{u} = -\underline{u} = 0.05$; the viability kernel extends to the lighter grey area for $\bar{u} = -\underline{u} = 0.1$. Other parameter values are $r_s = 3$, $a_c = 0.1$, $r_c = 1$, $\gamma_g = 0.65$ and $\underline{s} = 0.1$.

us to derive the associated stocking rate policy (see Mathematical Viability Theory below). Following such a policy, the time spent by the rangeland system with shoot biomass or grazing pressure lower than the minimal threshold will be the lowest possible one; and the resilience will be measured from this value.

The resilience measure of the rangeland is then evaluated at any point of the state space as the inverse of the minimal restoration cost following a sudden shoot biomass reduction due to a drought event of maximal anticipated severity $\bar{a}$.

Figure 13.6 displays resilience values for two different values of grazing pressure: a satisfying but relatively low grazing pressure ($\gamma_g = 0.65$) and a high one ($\gamma_g = 0.9$). From this figure we can compare the effect of level of grazing pressure on the resilience of rangeland subjected to drought events.

Figures such as Figure 13.6, which compare the resilience values associated with two values of grazing pressure, facilitate the appreciation of the impact of a decision to increase grazing pressure in terms of the loss of rangeland resilience against possible drought.

For both values of γ_g, there is a region of state space (c,s) in which the system's resilience to drought causing a halving of grass biomass is infinite. However, this surface is twice as large when the grazing pressure is $\gamma_g = 0.65$ as when it is $\gamma_g = 0.9$ (Figure 13.7: right). Similarly, in both cases, there is a space area in which the restoration time following such a drought would be more than 10 years. However, this surface is more than ten times bigger

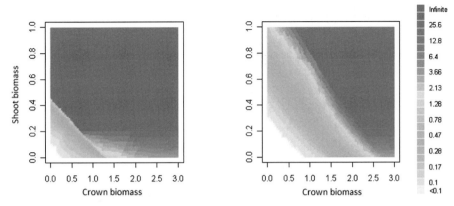

Figure 13.6 Two sections ($\gamma_g = 0.65$ (left) and $\gamma_g = 0.9$ (right)) of resilience values of the rangeland described by dynamics (13.3) for the property (13.4) toward drought events causing a sudden reduction in shoot biomass. Parameter values are $r_s = 3$, $a_c = 0.1$, $r_c = 1$, $\gamma_g = 0.65$, $\underline{s} = 0.1$, $\bar{u} = -\underline{u} = 0.05$ and $\bar{a} = 0.5$. (A black and white version of this figure will appear in some formats. For the colour version, please refer to the plate section.)

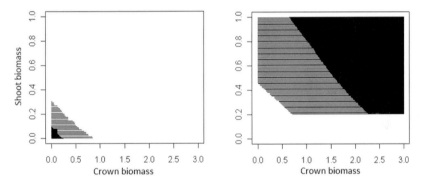

Figure 13.7 Left: the coloured black area corresponds to grass states for which resilience to drought events is smaller than 0.1 when the grazing pressure is relatively small ($\gamma_g = 0.65$). With a high grazing pressure ($\gamma_g = 0.9$), the area of resilience smaller than 0.1 increases and includes both the black and the hatched areas. Right: the coloured black area corresponds to grass states for which resilience to drought events is infinite when the grazing pressure is high ($\gamma_g = 0.9$). With a smaller grazing pressure ($\gamma_g = 0.65$), the area of infinite resilience increases and includes both the black and the hatched areas.

when grazing pressure is high ($\gamma_g = 0.9$) than when it is lower ($\gamma_g = 0.65$) (Figure 13.7: left).

Actually, even with grazing pressure $\gamma_g = 0.9$, the viability kernel is not empty, suggesting that from well-chosen grass states the minimal thresholds of shoot biomass and grazing pressure can be guaranteed over time; however, resilience against drought events is dramatically reduced compared to a lower grazing pressure of $\gamma_g = 0.65$.

Evolution, management and constraints are common features of socio-ecological systems. Evaluating their resilience using the viability-based measure may encourage the use of resilience thinking in environmental policy analysis. The next section focuses on the presentation of the general implementation framework.

What do we mean by viability analysis?
Mathematical viability theory
Aim and scope

Viability theory deals with the control of dynamical systems under constraints. Two reference books are Aubin (1991) and Aubin et al. (2001).

The theory has two aspects: as a mathematical theory and as a provider of mathematical metaphors of evolution of real systems.

In viability theory, the system under study is described by its state made up of a finite number of variables gathered in the finite-dimensional vector x. These variables evolve with time, such that at time t, the state of the system is described by $x(t)$.

We can distinguish direct models and inverse problems. In direct models the phenomenon under study is modelled by differential equations or rules, e.g. 'if then' in individual-based models. Once this description is completed, the study of existence, of uniqueness, of various types of stability or asymptotic stability of solutions provides answers to questions on the future behaviour of the system. In an inverse problem, once the list of the prescribed properties and objectives is established, the aim is to determine dynamical systems governing evolutions that satisfy these properties for ever or until the moment when the objectives are achieved. Such questions are crucial if we no longer assume that the dynamical model for the system is well known as in Physics or Mechanics, but has to be built, as in socio-ecological systems.

Among the prescribed properties systematically studied by viability theory are properties of constraint satisfaction. For example, in the energy field, these constraints may be pollution thresholds that are not to be transgressed, or minimal requirements of energy supply. They may be satisfied at any time or until a finite, or prescribed, or minimal time when the evolution reaches a given target.

Main concepts

Given some dynamics and a constraint set, there is no reason why the evolutions governed by these dynamics should be compatible with the constraints. There are two ways of insuring viability: reduce the state space and modify the dynamics.

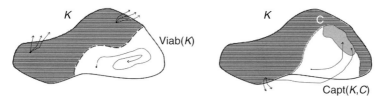

Figure 13.8 Diagrams of a set of constraints *K* delimited by a black plain line. Inside it and coloured white is a viability kernel (left) and a capture basin (right) of target *C* (light grey area).

We consider here the first way – reducing the state space. The dynamics are given, for instance, in the form of a controlled dynamical system, the constraint set is given as a subset of the state space, and the objective is to find points inside this constraint set at which the constraints can continue to be satisfied in the future. In the terminology of viability theory, the viability kernel is the subset (possibly empty) of states within the constraint set from which at least one viable evolution (remaining all the time in this constraint set) starts. From a point inside the viability kernel there exists an evolution that remains in the constraint set. From a point outside the viability kernel, all the possible evolutions leave the constraint set in finite time (Figure 13.8: left).

If an objective is added in the form of a target to be reached inside the constraint set, viability theory uses the concept of a capture basin. This basin is the subset of states in the constraint set from which at least one evolution starts, which is viable in this constraint set until it reaches the target in finite time (Figure 13.8: right).

It is worth noting that finding the viability kernel or capture basin allows us to design feedbacks that govern evolutions so as to maintain viability until a target, if present, is captured.

Intertemporal optimality
Starting from a viability kernel or a capture basin, several evolutions might be viable. We can use the classical optimal control theories to select viable evolutions that minimise a given intertemporal criterion. Using appropriate mathematical techniques, the search for optimal evolutions becomes the search for viable evolutions for an auxiliary controlled dynamical system composed of the original system with an additional dimension which corresponds to the cost. Viability techniques can provide the same results as dynamical programming (Hamilton–Jacobi–Bellman equations), but in the presence of state constraints and for a wider class of problems (free boundary problems with obstacles).

Crisis management
Even perfect management cannot avoid crisis when some constraints are violated; for example, following a disturbance that causes a jump in the

state space. Again introducing an auxiliary dynamical system, one can find control functions that allow viability to recover in the best way, for instance by minimising the crisis time (Doyen & Saint-Pierre, 1997), or, if that is not possible, by confirming the irreversibility of the constraints violation.

Exit time
Viability methods also allow us to determine for each initial position, the first time when constraints are violated. This is called the exit time function.

Available numerical tools
Few exact descriptions of viability kernels are available. Three exact descriptions of viability kernels in a two-dimensional state space are: the model called population growth in a limited space (Aubin & Saint-Pierre, 2006), the model called consumption (Aubin, 1991) and the model of Abrams Strogatz for language competition (Chapel et al., 2010). There is also the exact description of a viability domain for a three-dimensional state space (Bernard and Martin, 2012).

The possibility of exact determination of a viability kernel is studied on a case-by-case basis. The use of approximation algorithms is essential. Viability algorithms have existed since the 1990s.

The first of them by Saint-Pierre (1994) used discrete approximations in the Lipschitz case to build viability kernels. This algorithm involves two steps:

- the approximation of the viability kernel of the continuous system by kernels of discrete time systems,
- then the approximation of the viability kernels of discrete time systems by kernels of discrete systems in time and space.

Since then, other algorithms have been designed using classification procedures which are less memory-intensive than the regular grid points (Deffuant et al., 2007; Alvarez et al., 2013). Nevertheless, all these algorithms require large amounts of memory because the number of points of a regular grid increases exponentially with the dimension of the state space. Moreover, at each point of the grid, the discretised values of the set of admissible controls have to be tested; the computation time is then exponential with the dimension of the control space.

Consequently, with current calculation capacities the number of variables of the state space is in practice limited to seven.

Survey of applications
Viability theory is used in many areas: in genetics, Bonneuil and Saint-Pierre (2000) used the concept of the viability kernel to determine the initial

frequencies that lead to the maintenance of polymorphism; in demography, using the tools of viability theory, Bonneuil and Saint-Pierre (2008) answered questions about lifestyle choices especially with regard to children, which can be made to guarantee a certain standard of living; in finance, the management of portfolios of financial assets can also be assessed by the tools of viability theory (Aubin et al., 2005a); in aeronautics, Tomlin et al. (2003) determined the thrust and angle of attack to be applied to an aircraft to make it land under safe conditions; viability theory was also used to control food processes with the aim of identifying the set of all possible actions that make it reach a quality target with respect to manufacturing constraints (Sicard et al., 2012; Mesmoudi et al., 2014).

Considering the current forest area in the world and the current amount of CO_2 present in the atmosphere, Andrès-Domenech et al. (2011) studied the reforestation rate and CO_2 emissions required to satisfy the standards for CO_2 in the atmosphere both now and in the future. Aubin et al. (2005b) have evaluated the transition cost for maintaining the concentration of greenhouse gases within specified boundaries. The framework of viability theory has also been used by Béné et al. (2001) to analyse renewable resources management and by Bruckner et al. (2003) to describe the Tolerable Windows approach.

Martinet and Doyen (2007) used the tools of viability theory to define sustainability as the points at which ecological, economic and social constraints were met at the same time and for any time. Thus, each of the three pillars of sustainable development was described by a set of constraints in the state space and sustainable development was defined as the development that occurs at and remains in the intersection of these three sets of constraints.

All criteria were treated in the same way, thus avoiding the problem of having to choose weights whose values are difficult to justify as in the case of criteria based on optimisation of an aggregated utility function. Looking for sustainable developments then involves finding the conditions under which the viability kernel is not empty (as in Rapaport et al., 2006). For instance, imposing constraints that state that the levels of consumption and the stock of exhaustible resources should never fall below certain threshold limits in a society that is described by indexes belonging to the viability kernel ensures that a certain level of consumption and resource conservation can be guaranteed for all generations to come. Martinet and Doyen (2007) then define sustainability as the ability to transmit a set of 'minimum rights' to future generations. Since then, several works have developed this point (De Lara & Martinet, 2009; Bernard & Martin, 2013; Wei et al., 2013; Perrot et al., 2016).

The viability-based measure of resilience

The measure of resilience proposed by Martin (2004) is the inverse of the minimal cost associated with the effort to restore and preserve some properties of the system following disturbances. Using this measure, the evaluation of resilience involves the description of:

- the evolution of the system state based on the state itself and, also on the management actions, called controls,
- the properties of the system under consideration,
- the anticipated disturbances,
- the cost function used to evaluate the effort of restoration and preservation.

The implementation in the mathematical viability theory framework

The implementation of the measurement of resilience proposed by Martin (2004) involves two stages. Given a controlled dynamical system and a property of this system, the first step is to study the system's ability to preserve this property over time. Given anticipated perturbations, the second step is to assess the impact of these disturbances on the system ability to maintain the property under consideration, which is evaluated using cost functions defined on the set of possible evolutions.

First step: the calculation of the viability kernel associated with the dynamics of the system and the desired property allows us to distinguish the states of the system from which the studied property can be preserved.

The first assumption we make is that the evolution of the system state is governed by a controlled dynamical system with changes depending on the state of the system and also on controls by an external manager.

The state of the system is described by an n-dimensional vector $x \in X \subset \mathbb{R}^n$; the controls through which an external manager can act on the system belong to a p-dimensional vector space, $u \in \mathbb{R}^p$. The controlled dynamical system S is described by the pair (U, f) where

- U is a set-valued map $\mathbb{R}^n \rightsquigarrow \mathbb{R}^p$ which associates any state of the system with the set of admissible controls. We use a set-valued map because the control function is not *a priori* defined. We only know that at each state of the system, several control choices are available, and all these possible choices are gathered in the set $U(x)$
- f is a function $\mathbb{R}^n \times \mathbb{R}^p \rightsquigarrow \mathbb{R}^n$ which associates any pair of system state and control with the variation of the system state.

An evolution $t \in [0, +\infty[\rightsquigarrow x(t) \in X$ which describes the state of the system over time is governed by the controlled dynamical system S when:

$$S \begin{cases} x'(t) = f(x(t), u(t)) \\ u(t) \in U(x(t)) \text{ for almost all } t \geq 0 \end{cases}. \quad (13.5)$$

It is worth noting that the control function $u(t)$ is not *a priori* defined. The only condition on $u(t)$ is that for all t $u(t)$ belongs to the set of admissible controls $U(x(t))$. Consequently, given an initial state x, there may be several evolutions satisfying (13.5) and corresponding to different control functions.

We note $S(x)$ the set of all evolutions starting at x and governed by S.

We then assume that the studied properties can be described as a subset of the state space of the system. Let K be a subset of X, we assume that the system has the properties under consideration when the state of the system belongs to K.

Thus, the system will preserve the property over time if its state follows an evolution that remains in K.

Such an evolution for which $\forall t \geq 0$, $x(t) \in K$ is called viable in the mathematical viability theory framework. The study of the compatibility between dynamics and properties is solved by computing the viability kernel, one of the fundamental concepts of the viability theory.

The viability kernel, $\text{Viab}_S(K)$, associated with a controlled dynamical system S and subject to a constraint set $K \subset X$ gathers all states from which there exists at least one viable evolution governed by S (Aubin, 1991):

$$\text{Viab}_S(K) = \{x_0 \in K | \exists x(.) \in S(x_0) \text{ such that } \forall t \geq 0, x(t) \in K\} \qquad (13.6)$$

Second step: measuring resilience assesses the impact of disturbances on the system's ability to preserve some of its properties.

Often, when a disturbance is considered, we do not know exactly what the state of the system will be after its occurrence, we can only define a set of possible states. We then assume that the perturbations under consideration can be described by a set-valued map in the state space, D, which associates with each state of the system the set of reachable states that can occur after one of these disturbances:

$$D : \mathbb{R}^n \rightsquigarrow \mathbb{R}^n. \qquad (13.7)$$

(The symbol $\rightsquigarrow$ denotes set-valued maps.)

Several disturbances can be considered such as uncertainties about the precise state of the system after their occurrence, hence the use of a set-valued map rather than a function. The anticipated disturbances are shocks in the state space.

The jump in the state space following a disturbance can be done out of the viability kernel and even outside of the set of constraints. In such cases, either the properties described by this set of constraints can be restored and preserved, that is to say that the previously calculated viability kernel can be reached by using appropriate control functions; or the properties cannot be restored, whatever the control functions.

The impact of the occurrence of a disturbance on the system's ability to preserve the properties under consideration is assessed by calculating the capture basin of the viability kernel. The capture basin is the second key concept of viability theory, which is more recent than the viability kernel. The capture basin gathers all initial states from which a target can be achieved while respecting the constraints.

The capture basin $\text{Capt}_S(K,C)$ associated with the controlled dynamical system S, with the target $C \subset X$ and subject to constraints $K \subset X$ is the set of all initial points from which there exists at least one evolution viable in K until it reaches C in finite time:

$$\text{Capt}_S(K,C) = \{x_0 \in K | \exists x(.) \in S(x_0), \exists T \geq 0 | x(T) \in C \text{ and } \forall t \in [0;T], x(t) \in K\}. \quad (13.8)$$

Thus, if after the occurrence of a disturbance the system state remains in the capture basin of the viability kernel calculated in the first step, the property can be restored and maintained. The set K which contains the set K (which represents the studied property) then represents ultimate limits beyond which the system dynamics are unknown.

From one point of the capture basin of the viability kernel, restoration is possible, but it can have a cost:

- the cost will be zero if, after the jump due to the occurrence of a disturbance, the system state remains in the viability kernel. Indeed, the property of interest will continue to be preserved,
- the cost will be non-zero but finite if the state after disturbance belongs to the capture basin of the viability kernel, because the property will necessarily be lost but can be restored and preserved,
- the cost will be infinite, if after the jump the property can not be preserved, i.e. if the system state is outside of the capture basin.

Thus, the most obvious example of cost function is the function that associates an evolution with the time spent out of the viability kernel. Let $x(.) \in S(x)$

$$c(x(.)) := \int_0^\infty (1 - \mathbb{1}_{Viab_S(K)}(x(\tau))) d\tau \quad (13.9)$$

where $\mathbb{1}$ denotes the indicator function of the set.

From the state $x \in X$ of the system, the minimum cost among all evolutions governed by S is:

$$c(x) = \min_{x(.) \in S(x)} \int_0^\infty (1 - \mathbb{1}_{Viab_S(K)}(x(\tau))) d\tau. \quad (13.10)$$

Anticipated disturbances when the system state is $x \in \mathbb{R}^n$ are described by the set $D(x)$ of reachable states after the occurrence of one of these disturbances (13.7). For the evaluation of resilience, the worst case is taken on, i.e. the jump in state space that leads to the highest cost of restoration.

The resilience of the system at state x facing disturbances described by D is equal to the inverse of the maximum cost among all jumps from x to $x_1 \in D(x)$:

$$R(x) := \frac{1}{max_{x_1 \in D(x)} c(x_1)}. \tag{13.11}$$

Scope and applications
The viability-based definition of resilience extends known measures to controlled dynamical systems and desired properties disconnected from equilibria.

Martin et al. (2011) have shown in the case of a savanna model of the literature (Anderies et al., 2002) that the definition of resilience in the context of viability theory generalises the definitions based on attractors.

In the case of measurement based on the size of the attraction basin, the resilience value is linked to the maximal intensity of the disturbance, measured in terms of sudden loss of biomass that the system at an equilibrium point can withstand while still remaining in the attraction basin of this equilibrium. The time to achieve this equilibrium is infinite.

The measure of Martin (2004) which uses a cost function associated with the return within the viability kernel of the constraint set representing the studied property is more informative. Indeed, the resilience value for any state of the system (which is not necessarily at equilibrium), gives information about the maximum time required to return the system to a given vicinity of this equilibrium following a disturbance. Moreover, this measure of resilience, together with the tools of viability theory, helps us to determine the policy actions that preserve the property of the system under consideration or, if possible, restore it within a minimal time (or minimal cost).

Viability-based measures of resilience in models of socio-ecological systems
The following studies are examples where viability analysis has been used to assess the resilience of a socio-ecological system.

In the case of lake eutrophication, Martin (2004) assessed the resilience of the oligotrophic property of a lake in the watershed of which agricultural activities were present, in the face of extreme rainfall events.

For grazing management, Martin et al. (2011) evaluated the cost associated with the restoration and preservation of a certain amount of grass biomass. Such a cost can be used to assess the resilience of the grazing system following disruptions caused by abrupt changes in livestock or drought events.

In the case of competition between languages, the evaluation of the cost associated with the restoration and the preservation of a minimum number of speakers of each language can be used to assess the resilience of the property of linguistic diversity in the face of disturbances caused by a sudden change in the proportion of speakers of each language (Bernard & Martin, 2012; Alvarez et al., 2013).

In the case of reconciling tourism and environmental quality, Wei et al. (2013) evaluated the cost associated with the restoration and preservation of situations with high standards of both tourism activity and environmental quality.

Conclusion and perspectives

The measure of resilience based on viability analysis provides an evaluation of the maximal impact of anticipated disturbances on the ability of the system to preserve or restore a given property. Hence, different agricultural systems can be compared according to their resilience values. Moreover, viability analysis can provide the management policy associated with the minimal cost of restoration.

Land management decisions need to involve all stakeholders in building the definition of the problem and its solution. This participatory management style focuses on the knowledge of each actor, discussions and negotiations. The development of methods and tools to support this participatory management process is an important area of research (Lynam et al., 2007). The modelling approach, originally proposed to help stakeholders to create a collective understanding of conflicts and to negotiate strategies for coping with them, uses multi-agent simulations to represent the evolution of environmental and economic resources and role-playing games played by stakeholders (Barreteau, 2003). Serious games are a new and efficient approach to explore and test the possibilities of changes in a realistic setting without cost or risk (Susi et al., 2007; Homewood et al., Chapter 9).

Integrating viability and resilience analysis into decision support systems for participatory management would provide stakeholders with knowledge of the resilience of the system they are currently building (Wei et al., 2012) and would enable them to examine the feasibility of using this knowledge to assess the quality of the adopted management policy.

References

Alvarez, I., de Aldama, R., Martin, S. & Reuillon, R. (2013). Assessing the resilience of bilingual societies: coupling viability and active learning with kd-tree. application to bilingual societies. In: *IJCAI 2013 AI and Computational Sustainability Track*. Menlo Park, CA: AAAI Press/IJCAI, pp. 2776–2782.

Anderies, J.M., Janssen, M.A. & Walker, B.H. (2002). Grazing management, resilience, and the dynamics of a fire-driven rangeland system. *Ecosystems*, **5**, 23–44.

Andrès-Domenech, P., Saint-Pierre, P. & Zaccour, G. (2011). Forest conservation and CO_2 emissions: a viable approach. *Environmental Modeling and Assessment*, **16**(6), 519–539.

Aubin, J. (1991). *Viability Theory*. Basel: Birkhauser.

Aubin, J. & Saint-Pierre, P. (2006). An introduction to viability theory and management of renewable resources. *Advanced Methods for Decision Making and Risk Management*, **44**, 52–96.

Aubin, J.-P., Pujal, D. & Saint-Pierre, P. (2005a). Dynamic management of portfolios with transaction costs under tychastic uncertainty. In: *Numerical Methods in Finance*, edited by M. Breton & H. Ben-Ameur. New York, NY: Springer, pp. 59–89.

Aubin, J.-P., Bernado, T. & Saint-Pierre, P. (2005b). A viability approach to global climate change issues. In: *The Coupling of Climate and Economic Dynamics*, edited by A. Haurie & L. Viguier, Vol. 22 of *Advances in Global Change Research*. Dordrecht: Springer Netherlands, pp. 113–143.

Aubin, J., Bayen, A. & Saint-Pierre, P. (2011). *Viability Theory: new directions*. New York, NY: Springer.

Barreteau, O. (2003). The joint use of role-playing games and models regarding negotiation processes: characterization of associations. *Journal of Artificial Societies and Social Simulation*, **6**(2).

Beddington, J., Free, C. & Lawton, J. (1976). Concepts of stability and resilience in predator-prey models. *Journal of Animal Ecology*, **45**, 791–816.

Béné, C., Doyen, L. & Gabay, D. (2001). A viability analysis for a bio-economic model. *Ecological Economics*, **36**, 385–396.

Bernard, C. & Martin, S. (2012). Building strategies to ensure language coexistence in presence of bilingualism. *Applied Mathematics and Computation*, **218**(17), 8825–8841.

Bernard, C. & Martin, S. (2013). Comparing the sustainability of different action policy possibilities: application to the issue of both household survival and forest preservation in the corridor of Fianarantsoa. *Mathematical Biosciences*, **245**(2), 322–330.

Bonneuil, N. & Saint-Pierre, P. (2000). Protected polymorphism in the two-locus haploid model with unpredictable fitnesses, *Journal of Mathematical Biology*, **40**(3), 251–277.

Bonneuil, N. & Saint-Pierre, P. (2008). Beyond optimality: managing children, assets, and consumption over the life cycle. *Journal of Mathematical Economics*, **44**, 227–241.

Bruckner, T., Petschel-Held, G., Leimbach, M. & Toth, F.L. (2003). Methodological aspects of the tolerable windows approach. *Climatic Change*, **56**, 73–89.

Carpenter, S., Walker, B., Anderies, J. & Abel, N. (2001). From metaphor to measurement: resilience of what to what? *Ecosystems*, **4**, 765–781.

Casagrandi, R. & Rinaldi, S. (2002). A theoretical approach to tourism sustainability. *Conservation Ecology*, **6**(1), 13.

Chapel, L., Castello, X., Bernard, C., et al. (2010). Viability and resilience of languages in competition. *PLoS ONE*, **5**(1), e8681.

Collings, J. & Wollkind, D. (1990). A global analysis of a temperature-dependent model system for a mite predator–prey interaction. *SIAM Journal of Applied Mathematics*, **50**(5), 1348–1372.

De Lara, M. & Martinet, V. (2009). Multi-criteria dynamic decision under uncertainty: a stochastic viability analysis and an application to sustainable fishery management. *Mathematical Biosciences*, **217**(2), 118–124.

Doyen, L. & Saint-Pierre, P. (1997). Scale of viability and minimal time of crisis. *Journal of Set-Valued Analysis*, **5**, 227–246.

Deffuant, G., Chapel, L. & Martin, S. (2007). Approximating viability kernels with support vector machines. *IEEE Transactions on Automatic Control*, **52**(5), 933–937.

Holling, C. (1973). Resilience and stability of ecological systems. *Annual Review of Ecology and Systematics*, **4**, 1–24.

Holling, C. (1996). Engineering resilience vs. ecological. In: *Engineering within Ecological Constraints*, edited by P. Schulze. Washington, DC: National Academy Press, pp. 31–43.

Janssen, M.A. & Carpenter, S.R. (1999). Managing the resilience of lakes: a multi-agent modeling approach. *Conservation Ecology*, **3**(2), 15.

Lacitignola, D., Petrosillo, I., Cataldi, M. & Zurlini, G. (2007). Modelling socio-ecological tourism-based systems for sustainability. *Ecological Modelling*, **206**, 191–204.

Levin, S., Barrett, S., Aniyar, S., et al. (1998). Resilience in natural and socioeconomic systems. *Environment and Development Economics*, **3**(2), 222–235.

Ludwig, J., Walker, B. & Holling, C. (1997). Sustainability, stability and resilience. *Conservation Ecology*, **1**(1), 7.

Lynam, T., Jong, W.D., Sheil, D., Kusumanto, T. & Evans, K. (2007). A review of tools for incorporating community knowledge, preferences, and values into decision making in natural resources management. *Ecology and Society*, **12**(1), 5.

Martin, S. (2004). The cost of restoration as a way of defining resilience: a viability approach applied to a model of lake eutrophication. *Ecology and Society*, **9**(2), 19.

Martin, S., Deffuant, G. & Calabrese, J. (2011). Defining resilience mathematically: from attractors to viability. In: *Viability and Resilience of Complex Systems*, edited by G. Deffuant and N. Gilbert. Heidelberg: Springer, pp. 15–36.

Martinet, V. & Doyen, L. (2007). Sustainability of an economy with an exhaustible resource: a viable control approach. *Resources, Energy and Economics*, **29**(1), 17–39.

Matsinos, Y.G. & Troumbis, A.Y. (2002). Modeling competition, dispersal and effects of disturbance in the dynamics of a grassland community using a cellular automaton. *Ecological Modelling*, **149**, 71–83.

Mesmoudi, S., Alvarez, I., Martin, S., Reuillon, R., Sicard, M. & Perrot, N. (2014). Coupling geometric analysis and viability theory for system exploration: application to a living food system. *Journal of Process Control*, **24**(12), 18–28.

Neubert, M. & Caswell, H. (1997). Alternatives to resilience for measuring the responses of ecological systems to perturbations. *Ecology*, **78**(3), 653–665.

Ortiz, M. & Wolff, M. (2002). Dynamical simulation of mass-balance trophic models for benthic communities of north-central Chile: assessment of resilience time under alternative management scenarios. *Ecological Modelling*, **148**, 277–291.

Perrot, N., Vries, H.D., Lutton, E., et al. (2016). Some remarks on computational approaches towards sustainable complex agri-food systems. *Trends in Food Science & Technology*, **48**, 88–101.

Peterson, G., Allen, C. & Holling, C. (1998). Ecological resilience, biodiversity, and scale. *Ecosystems*, **1**, 6–18.

Pimm, S. & Lawton, J. (1977). Number of trophic levels in ecological communities. *Nature*, **268**, 329–331.

Rapaport, A., Terreaux, J. & Doyen, L. (2006). Viability analysis for the sustainable management of renewable resources. *Mathematical and Computer Modelling*, **43**, 466–484.

Saint-Pierre, P. (1994). Approximation of viability kernel. *Applied Mathematics and Optimization*, **29**, 187–209.

Sicard, N., Perrot, N., Reuillon, R., Mesmoudi, S., Alvarez, I. & Martin, S. (2012). A viability approach to control food processes: application to a camembert cheese ripening process. *Food Control*, **23**(2), 312–319.

Susi, T., Johannesson, M. & Backlund, P. (2007). *Serious Games: an overview*. Skövde: University of Skövde.

Tomlin, C., Mitchell, I., Bayen, A. & Oishi, M. (2003). Computational techniques for the verification of hybrid systems. *Proceedings of the IEEE*, **91**(7), 986–1001.

Van Coller, L. (1997). Automated techniques for the qualitative analysis of ecological models: continuous models. *Conservation Ecology*, **1**(1), 5.

Wei, W., Alvarez, I., Martin, S., Briot, J.-P., Irving, M. & Melo, G. (2012). Integration of viability models in a serious game for the management of protected areas. In: *IADIS Intelligent Systems and Agents Conference 2012, Lisbonne, Portugal*, edited by A. Palma dos Reis & P.S.P. Wang.

Wei, W., Alvarez, I. & Martin, S. (2013). Sustainability analysis: viability concepts to consider transient and asymptotical dynamics in socio-ecological tourism-based systems, *Ecological Modelling*, **251**, 103–113.

CHAPTER FOURTEEN

Integrating economics and Resilience Thinking: the context of natural resource management in Australia

MICHAEL HARRIS[1]
*Australian Bureau of Agricultural and Resource Economics and Sciences,
University of Sydney*
GRAHAM R. MARSHALL
University of New England
and
DAVID J. PANNELL
University of Western Australia

Introduction

The blending of resilience thinking and economics is not a straightforward matter; the two paradigms can seem to be at odds. To incorporate resilience thinking into economics requires more than opportunistic use of terminology and a handful of 'cherry-picked' ideas; it requires a careful assessment of how elements of two paradigms that operate quite differently can be integrated in mutually beneficial ways.

To emphasise the challenges facing successful integration, imagine two highly stylised and oversimplified representations. First, economists typically rely heavily on formalistic methods that appear mechanistic and deterministic, often with assumptions made about existence of and movement towards a unique equilibrium state. Comparative statics, or simplified comparative dynamics,[2] tend to be emphasised. Those used to thinking in complex adaptive systems (CAS) terms view these tendencies with some scepticism, arguing that the concept of a unique equilibrium is misleading when applied to a complex and dynamic world and hence that mechanistic approaches cannot properly represent reality.

By contrast, adherents of the CAS view have sometimes been criticised for creating a 'cult of complexity' in which reductionist analytical approaches are dismissed out of hand as oversimplifying a world in which everything is connected to everything else, with useful analytical results consequently being

---

[1] Michael Harris now works for the Environmental Protection Authority, New South Wales.
[2] Comparative static analysis compares equilibria before and after some system shock or policy change. An example of a comparative static result from a demand–supply model would be 'banana prices are predicted to rise after a storm damages the crop'. Comparative dynamics do similar comparisons for timepaths of variables in explicitly dynamic models.

hard to come by. Critics note that the more complexity is incorporated into the analysis of a system, the fewer concrete insights into the system can be found.

These two – deliberately unflattering – representations will be expanded and critiqued in what follows. Our intention in this chapter is to contribute towards resolving this paradigm clash, highlighting complementarities rather than emphasising incompatibilities. Our starting point is that economies are in fact CASs, but that this does not always and everywhere render conventional analytical approaches from economics irrelevant or misleading, a position supported by Ostrom (2010, p. 659) in the context of markets for private goods:

> [W]hat is called 'rational choice theory' is not a broad theory of human behavior but rather a useful model to predict behavior in a particular situation: a highly competitive market for private goods.[*]

The appropriate boundaries between where reductionist methods are and are not useful in complex systems will continue to be contested and reasonable analysts will disagree. (The three authors of this chapter would not necessarily define those boundaries in exactly the same places.) However, testing where conventional approaches have validity, versus where they should be supplemented or abandoned, will be an important part of ongoing work in this interdisciplinary enterprise.

Readers less concerned with philosophical discussions of technical economics and its difficult, but improving, relationship with complexity thinking may wish to skip ahead to the section on practical blending of the approaches in the context of natural resource management.

Complication and complexity

The study of CAS is inherently the study of evolving, self-organising dynamic systems that are subject to a mix of exogenous pressures and shocks and internal interacting processes involving feedback loops that amplify or dampen the effects of shocks. Such systems often have particular stages within their life cycles, where those stages are characterised by states that can be analysed separately but not independently of each other. In the Resilience-Thinking literature, these cycles are referred to as adaptive cycles, with four main stages occurring in sequence: rapid growth, conservation, release and reorganisation (e.g. Folke, 2006). This is not meant to be a deterministic or prescriptive model of how any given system works, but rather a commonly observed description of the phases through which many systems pass. After the reorganisation phase, the system will typically move into a new growth phase, which may be quite similar or markedly different to the previous growth phase.

---

[*] Mainstream economics emphasises devices made by rational agents, hence rational choice theory.

Complex systems will also feature emergent behaviour or emergent properties, which are features of a system that arise from the interaction of individual elements or decision-making agents and cannot be predicted *a priori* by analysing the behaviours of those elements or agents in isolation.

A common definition of resilience concerns the likelihood that the system will be shocked from its existing regime of states into a new regime. When expressed in terms of regime thresholds, this can be quantified as the 'distance to the threshold'. Different systems will have different dynamics, including the degree of hysteresis or lock-in associated with the crossing of a threshold; in other words, it may be more or less difficult because of path dependence to 'cross back' into the original regime. The analysis of CAS also requires an explicit recognition of the interaction across spatial levels; resilience at a higher level may require transformation (regime change) at a lower level, for example. Finally, none of this should be inferred as presuming that resilience is always good (or, more carefully worded, that more resilience is always better). For example, an economy with persistently high unemployment (as in the Great Depression) would be considered resilient, but undesirably so. If the initial regime is less desirable than the post-flip regime, greater resilience means a desirable transition is further away and hence less likely to occur in a given time frame. Examples of where a desirable transition has been made could include a number of post-Communist states, including Poland and Slovakia, which have experienced more favourable economic conditions since 1990 than they did under Communist rule.

While economics – even what is broadly considered 'mainstream' economics – is a wide-ranging discipline with various subdisciplines employing a variety of analytical methods, standard textbook economics does not appear particularly consistent with the study of CAS. Applied economic analysis is – not always, but often – static in nature, ahistorical and focused on equilibrium states and conditions. The analytical tools used tend to be formal and quantitative, identifying relationships between variables and aggregating these relationships as necessary (e.g. from individual to market and from partial to general equilibrium). The use of optimising methods, assumptions of rational behaviour and presumptions of existence and stability of unique equilibrium states in much of economics research seems inconsistent with analytical paradigms employed in the study of CAS.

At this stage we can usefully invoke the management seminar cliché that distinguishes complicated problems ('building a satellite') from complex problems ('raising a baby'). Our caricatures above implicitly present economics as a discipline in search of a satellite to build and complexity analysis as a paradigm searching for a baby to raise. In the following sections, we explore and critique these caricatures more deeply in order to show that, in effect, economics is also a discipline focused on the analysis of self-organising systems (a category that includes CAS) and that while it has approached this task quite differently from how it has been approached by complexity scientists, there are many complementarities.

A microeconomic example

A defence of conventional economics might run as follows. *Theoretically*, reductionist approaches have demonstrated themselves robust and *practically* useful in particular contexts. When Cyclone Yasi hit Queensland, Australia in 2011, the qualitative predictions of the basic demand and supply model were consistent with observed reality: the storm greatly reduced the supply of bananas, prices rose sharply (from one equilibrium level to another, in effect), returning eventually as production was restored. Making quantitative predictions would require more extensive 'complicated' modelling, using short- and long-run supply and demand elasticity estimates to predict the expected price rise and the length of time before supply and the initial equilibrium was restored.

However, economic theory can make no definite *a priori* predictions that the initial equilibrium will be restored. A supply shift may occur (damaged trees not replanted), or a demand shift (consumer preferences change permanently), leading to a new combination of price and quantity until the next shock. To assume restoration of the original supply–demand balance requires a *ceteris paribus* assumption[3] – whether made explicitly or implicitly – to be imposed on the analysis, which may or may not reflect reality accurately. As it happened, the simple model provided accurate qualitative predictions of the impact of a supply shock, but alternative outcomes based on path dependence cannot be ruled out on theoretical grounds.

Looking beyond the effects on the banana market would require further supplementation of the model, for example into a computable general equilibrium framework (a 'complicated' model) to investigate impacts on related markets, prices and so on, while looking at social and environmental impacts may require the use of other disciplinary perspectives, as well as extending the range of economic tools employed (such as non-market valuation). Applied economists may – with at least some justification – feel they have some professional expertise in having workable models for tackling specific microeconomic problems.

Controversies in macroeconomics

Modelling an entire economy presents more significant challenges than modelling an individual market behind the safety of *ceteris paribus* assumptions. Prior to 2008, leading macroeconomists were confident that significant controversies in macromodelling lay in the past (Blanchard et al., 2010) and that practical problems of business cycle management had been solved (Quiggin, 2010, chapter 1).

---

[3] The assumption that all factors not represented within the current analysis are held constant.

In particular, growth paths of aggregate variables such as GDP were relatively stable, and understanding long-run trends was deemed a far more important area for intellectual effort than the management of minor deviations from trend. It appeared that macroeconomies displayed considerable resilience, in which shocks at lower levels (regional or sectoral) typically had only minor/short-term impacts at higher levels. Such macro resilience was a combination of the economy's own absorptive capacity and policy settings that act as shock absorbers, particularly elements of the tax and transfer system referred to as 'automatic stabilisers'.

Notwithstanding that real shocks (such as the closure of a major regional employer) could still have significant impacts both economically and politically at lower levels of an economic system (including cascading effects along a supply chain) without leading to a recession or depression, the 2008 Global Financial Crisis (GFC) provided a stark reminder that the macroeconomy may be vulnerable to particular shocks such as those emanating from the financial sector. National economies may be prone to threshold effects because of, *inter alia*, the economy-wide interconnectedness of the financial system, noting that this system functions on a particular form of social capital, namely, trust in the system itself. This trust can be stable under 'business as usual' conditions, but is subject to fragility if a systemic failure becomes thought to be possible (Diamond & Dybvig, 1983; Brunnermeier, 2009).

Post 2008, disagreements between prominent economists about the state of macroeconomics and the credibility of its key models and its methodology became far more pronounced and critiques of standard approaches to the subdiscipline were published by prominent mainstream economists (e.g. Caballero, 2010; Quiggin, 2010, chapter 3; Leijonhufvud, 2011). The dominant model used in academic macroeconomic analysis, the Dynamic Stochastic General Equilibrium (DSGE) model, while being both dynamic rather than static and stochastic rather than deterministic and allowing for multiple variants including various frictions and market imperfections, is regarded as being seriously flawed by its critics on a number of grounds. These include its reliance on a 'representative agent', diminishing its usefulness in analysing how multiple heterogeneous agents interact; its reliance on rational expectations in how the agent(s) in the model handle the unknown future; and its focus on dynamic equilibria (in contrast to explicitly allowing for and attempting to model disequilibrium states that persist through time).

In testimony before the US Congress, Nobel Laureate Robert Solow (2010) proclaimed:

Here we are, still near the bottom of a deep and prolonged recession, with the immediate future uncertain, desperately short of jobs, and the approach to macroeconomics that dominates serious thinking, certainly in our elite universities and in many central banks and other influential policy circles, seems to have absolutely nothing to say about the problem. Not only does it offer no guidance or insight, it really seems to have nothing useful to say.

Solow was not being flippant. Kay (2011) noted, disparagingly, that Nobel Laureate and pioneer of the 'new classical macroeconomics' Robert Lucas remarked that economists failed to predict the 2008 financial crisis because such events are not only inherently unpredictable, but that in fact this is the key insight to be learned from the study of modern macroeconomics! Economic theory itself predicts that such events cannot be predicted. According to what became famously known as the 'Lucas Critique', structural relationships are not invariant to policy changes. Put simply, the optimal choices and strategies for agents will change as policy settings change, meaning that hitherto stable relationships between policy variables break down. This limits our ability to understand the system (economy) as a set of stable and predictable interactions, hence hindering our ability to manage and control it (Ljungqvist, 2008). Recalling the stylised view of economics as 'mechanistic' relative to complexity studies, it is worth noting here that a high priest of orthodox macroeconomics is stating that the macro economy is so fundamentally complex that significant events will *always* be unpredictable.

There have been several strategies suggested for reforming mainstream macroeconomics post-GFC and the reader should not infer that those we present below are complete or mutually exclusive. One suggestion is to move back towards simplicity when appropriate. The idea is to use *less* complicated, 'middlebrow' models (see Krugman, 2000) to answer key real-world questions instead of insisting on 'highbrow' rigour in modelling strategies and/or insisting that everything is fundamentally unpredictable.

A second suggestion is that long-abandoned research strategies in macroeconomics need to be re-evaluated and revived (Gordon, 2009). Moving explicitly in the direction of complexity is the strategy of using computational methods to explore coordination problems among heterogeneous and (somewhat) myopic agents. For example, Peter Howitt describes an agent-based modelling research strategy in macroeconomics (Howitt & Clower, 2000; Howitt, 2006).

> We have been building simple, stylized models of economies in which, instead of behaving according to the conventional rules of rational behaviour, people follow simple myopic but adaptive rules using little information, and in which they trade, in and out of equilibrium, in markets that are created and operated by a self-organizing network of profit-seeking business firms. (Howitt, 2011, pp. 7–8.)

Another response has been to pay more attention to previously marginalised – arguably heterodox – ideas, particularly those associated with the post-Keynesian economist Hyman Minsky (Minsky, 1986), whose basic model of a debt-driven business cycle has elements of a Resilience Thinking 'adaptive cycle' about it, with periods of complacency followed by periods of crisis

(Eggertsson & Krugman, 2012). The 'Minsky moment' when asset values collapse represents the crossing of a financial threshold, analogous to the crossing of an ecological threshold. How the crossing of a financial threshold affects the real economy is unknowable in advance, depending upon many things including the reaction of monetary authorities. In 1987 a major stock market crash did not cause a recession; in 2008 a fall in US house prices did.

Reductionism and aggregation

At this stage it would seem that from our preceding discussions we are confirming a well-known narrative in economics, namely, that microeconomics is relatively functional and settled while macroeconomics is comparatively contested and disordered. Before taking the opportunity to make some comments about reductionism and aggregation, it is worth being reminded that the usefulness of conventional microeconomic analysis is contingent upon the placement of *ceteris paribus* assumptions to effectively insulate the wider system from the impacts of shocks at lower levels. Macroeconomics is difficult precisely because it has to cope with system-wide shocks and impacts.

Reductionism is, at its most basic, an analytical approach that defines a subproblem to be analysed in isolation. Economists by training are likely to be comfortable with this approach. The more challenging aspect of reductionism – particularly for economics – is the idea that the whole can be understood by examining all of the parts separately. Hence a complicated, mechanistic system can be 'pulled apart', reverse-engineered and then 'reassembled' with an understanding of how everything fits together and the whole system functions.

Complexity issues aside, some of the challenges of aggregating from micro to macro are presented in elementary textbooks. The simplest macroeconomic models warn of the traps from careless aggregation resulting from fallacies of composition; the 'paradox of thrift' is one example whereby the lessons from an individual saving more to increase their wealth cannot be generalised to an entire interlocked economy (if everyone saves more simultaneously, spending and hence incomes decline).

Things get more difficult as the analysis gets more complicated. Formalistic economics is quite clear on the point that aggregation from micro to macro is difficult (see Chiappori & Ekeland, 2011 for an overview), and that strong assumptions are required for behaviour in the large to mimic behaviour in the small. With sufficient diversity in a population of consumers, 'aggregation in the large tends to destroy any structure that may exist at the individual level' (p. 639), revealing a kind of complicated, formalistic version of emergent behaviour. A famous aggregation result from general equilibrium theory, the Sonnenschein–Mantel–Debreu (SMD) theorem, is sometimes called the 'Anything Goes Theorem' due to its lack of specific predictions. The theorem 'means that assumptions guaranteeing good behaviour at the microeconomic level do not carry over to the aggregate level or to qualitative features of the equilibrium. ... Subfields of economics that relied on well-behaved aggregate excess demand for much of their theoretical development, such as international economics, were also left in the lurch' (Rizvi, 2006, p. 230). The SMD theorem allows for multiple equilibria, an issue which has been pursued further in the context of infinite horizon models by Kehoe and Levine (1985), who explore cases of not only multiple, but indeterminate, equilibria.

Consequently, macroeconomics has been highly contested historically in large part because of debates over appropriate 'microfoundations'. Keynesian macroeconomics started out as a 'top-down' approach, with microfoundations to be filled out 'below'. Key contributors to this tradition from the 1940s to the 1960s included Hicks, Patinkin, Clower and Leijonhufvud, who sought to complicate microeconomics to make it more consistent with a Keynesian macroeconomic vision dominated by the possibility of aggregate market failure (see Weintraub, 1979 for an overview). Modern macroeconomics has since moved towards a 'bottom-up' approach, a tradition beginning notably with Friedman, Phelps and Lucas and leading to the eventual dominance of DSGE models (Quiggin, 2010, chapter 3), where macroeconomic models are built around optimising representative agents with rational (i.e. model-consistent) expectations who react to a variety of shocks. Keynesian aspects (such as price rigidities and financial frictions) can be incorporated into these models, but critics argue that reliance on this approach restricts how questions in macroeconomics can be tackled, with tendencies towards dynamic equilibrium emphasised, and coordination problems and disequilibrium states de-emphasised or assumed away completely.

Risk, stochasticity and uncertainty

Certain classes of models in economics are deterministic; the model predicts outcomes as consequences of actions (or shocks) with certainty. Generalising such analyses towards risky or uncertain situations can occur in stages. The first and most straightforward step would be to specify all outcomes and associated probabilities, and to assume all agents know this distribution. Another approach would be to allow agents to hold subjective probabilities that they update as outcomes are observed. A further extension is to consider situations in which agents know what could possibly happen but have no basis on which to assign probabilities to these outcomes; as distinct from *risk*, this is known as Knightian *uncertainty* (after Knight, 1921).

Because the CAS tradition gives a prominent role to uncertainty of process and outcome, the concept of Knightian uncertainty will resonate more readily with analysts familiar with this tradition than will deterministic models or models with well-specified risks. Economists have in practice worked extensively with expected utility models that are analytically tractable despite having well-recognised empirical weaknesses (Rabin & Thaler, 2001). However, Randall (2011, chapter 3) notes that even Knightian uncertainty presupposes a comprehensible ('stationary') system, in which agents can understand outcomes but are ignorant of likelihoods. The assumption, according to Randall, is that if only these agents could study and learn the system sufficiently, its secrets would be revealed. However, a more complex non-stationary system will create things that are unknowable, because the processes underlying the system are themselves subject to change. The system itself will seem incomprehensible.

Debates around 'market efficiency' in financial economics are driven not only by views on the 'rationality' or otherwise of market participants, but also by how much the system is presumed to be knowable in the first place. *Ambiguity* aversion (a preference to avoid not knowing how the model/system works) is thus distinct from *risk* aversion (a preference to avoid not knowing which outcomes from a known distribution will occur; Campbell, 2014). Although the different views of the 2013 Nobel Laureates in economics, particularly Robert Shiller and Eugene Fama, have been presented as disagreements on behavioural grounds, they can be seen as disagreements on perceptions of how the world works. Shiller's view is that people can't know the true model.[4]

It is worth highlighting here that, for economists, the possibility of *knowing the world* well enough to be able to model it is not just a problem for the analyst; it is also a problem for the agents under study. In economics, the agents of study are humans and their own knowledge of the system and its interactions will condition their behaviour. At the aggregate level this can be crucial to interpreting system dynamics. In stock markets, for example, prices are driven not just by exogenous events but also by players' expectations of how others will react and respond to events; hence there are controversies over the relevance of the Efficient Markets Hypothesis (Quiggin, 2010, chapter 2), in which 'events' that market players react collectively to include releases of information about forthcoming *actual* events. Similarly the aforementioned Lucas Critique is driven by the notion that people react to changes in policy settings consciously and with foresight, not simply reactively. Hence, announcement effects and policy signalling will have real and immediate impacts in human systems in ways that would not apply in systems dominated by, say, plants or ants.

The dominance of risk and uncertainty in the financial sector, the networked and interdependent nature of financial flows and commitments, and the potential for significant impacts on the real economy resulting from a breakdown in the financial system mean that increasing attention is being

---

[4] https://hbr.org/2014/04/why-those-guys-won-the-economics-nobels/ (accessed 28 July 2016).

paid to what are effectively the 'complex system' properties of financial systems. Accordingly, an emerging literature displays a greater focus on systemic risk and deeper uncertainty in the financial sector and its linkages with the overall macro economy, such as Brunnermeier and Krishnamurthy (2014) and Battiston et al. (2012).

Deeper issues of uncertainty and 'unknown unknowns' apply beyond the financial sector, not least where environmental and ecological impacts are involved. Randall (2011) provides a detailed account of where and how it might be appropriate to move beyond the conventional risk framework ('ordinary risk management').

The 'cult of complexity'

In some cases, an overly fervent or narrow adherence to a resilience-based approach has reduced the potential for constructive engagement between economists and scientists interested in complex systems. Economists have tended to resist a view expressed in some quarters that complexity thinking renders all existing economic approaches obsolete, or at the very least, vastly restricts their applicability. As one example, a blog discussion on definitions and interpretations of 'equilibrium' in economics (a term which has particular context-specific meanings in economics, differing from its use in other disciplines focused on biophysical systems) elicited this response from an anonymous commentator:

> Equilibrium in complex systems is non-existent. Complex systems are characterised by disequilibrium. If you start from the assumption that a complex system with [sic] reach any kind of equilibrium then you are not in the real world.[5]

This confusion about what is signified by 'equilibrium' carries over in attempts in the literature to construct new – allegedly complexity-based – models that supplant existing modes of analysis, in this case of a joint agricultural/ecological system:

> While ecological communities progress through stages of development following disturbance toward a 'climax' stage, they are *never* in equilibrium ... Even if this is found to be appropriate, the fact that ecosystems are so strongly coupled to the environment (i.e., 'open') and so far from equilibrium is likely to place them in the realm of nonlinear nonequilibrium thermodynamics.
>
> (Richardson et al., 2011, p. 173.)

Terminological confusion leading to conceptual confusion, as frustrating as it is, is only the beginning of the potential difficulties to be encountered. As far back as 1995, a *Scientific American* article titled 'From complexity to perplexity'

---

[5] Comment on http://noahpinionblog.blogspot.com.au/2013/04/what-is-economic-equilibrium.html (accessed 24 July 2016).

(Horgan, 1995) offered sceptical commentary on the 'cult of complexity', and in particular on the 'true believers' idea of finding a unified theory of complex systems, seen as an early goal of complexity researchers, including many at the Santa Fe Institute where much interdisciplinary work has been carried out. Horgan saw complexity theory as the latest in a series of C-word fads (after cybernetics, catastrophe theory and chaos theory), and like them doomed to eventual irrelevance after its period in the spotlight.

This has not been borne out. Complexity, adaptive management and resilience thinking have been developed and refined over the intervening period, and economics is one of a number of disciplines that has been taking complexity ideas seriously (see next section). However, the evangelical stance of 'true believers' remains a potential barrier to effective integration of complexity thinking with economics as well as other disciplines. Biological anthropologist and blogger James Holland Jones describes how the 'cult of complexity' problem occurs in anthropology in his discussion of 'Complexity and nihilism':

I am continually frustrated by anthropologists who, when confronted with complexity, throw their hands up and say it's too complex to make predictions, [so] why bother to do science or understand the principles underlying the system?[6]

Jones highlights with concern ecologists who appear to argue that formal analysis of complex social–ecological systems becomes near-impossible because of their inherent complexity and uncertainty. As he notes, sometimes traditional 'linear' methods are useful and provide explanatory power for particular problems over particular domains. He argues that in many cases, complex systems may feature 'local linearity'; the strength of the response (to a shock) may be muted; the perturbation may be small; and that the effect of random noise may be to make otherwise 'exotic' dynamics reduce to more conventional ones. These are empirical matters for case-by-case analysis, and nihilistic declarations that the system is, *a priori*, simply too complex to understand are to be avoided in the absence of plausible reasoning and/or evidence to support such a contention.

Complexity and resilience in economics

Resilience-related ideas have entered the economics literature for some decades now, particularly in the environmental and ecological economics fields. Examples of early engagement with modelling of resilience include Perrings and Walker (1997) and Perrings (1998). Specified resilience and thresholds have been a particular focus of economic research; Naevdal (2003), for example, presents a framework for dealing with environmental threshold effects. Walker et al. (2010) treat a particular form of resilience relative to a known

---

[6] http://monkeysuncle.stanford.edu/?p=949 (accessed 24 July 2016).

natural threshold as a capital stock, indicating how it can be quantified and priced and incorporated into larger measures of wealth for purposes of sustainability assessment.

Complexity analysis has also moved into the economics sphere, with books such as Miller and Page (2007), and a new journal, *Complexity Economics*, dedicated to studies in the complexity vein. Some authors have expressed the view that this marks a critical turning point in economics.

> [T]he neoclassical era in economics has ended and is being replaced by a new era. What best characterizes the new era is its acceptance that the economy is complex, and thus that it might be called the complexity era. The complexity era has not arrived through a revolution. Instead, it has evolved out of the many strains of neoclassical work, along with work done by less orthodox mainstream and heterodox economists. (Holt et al., 2011, p. 357.)

The remainder of the chapter will address issues of the integration of economics and resilience in the context of Natural Resource Management (NRM). We choose this as a suitable context to discuss our thoughts on this integration because it is an area where both paradigms have been applied extensively, sometimes apparently in competition, and because we perceive that a degree of integration is possible and desirable.

What economics and Resilience Thinking bring to natural resource management

Economics and Resilience Thinking bring different strengths to the assessment of NRM issues and threats in geographical areas devoted to agricultural production. Resilience Thinking brings a focus on social (including economic) and biophysical phenomena as closely intertwined elements of what it refers to as social–ecological systems (a subcategory of CAS). This perspective also emphasises linkages across the multiple scales and levels characterising a social–ecological system, and the possibility of particular thresholds and regime changes within the system. Economics brings a focus on opportunity costs, on benefits (social and private), and on the trade-offs associated with scarce resources (including public funds and human effort). These elements are useful whether in a 'well-behaved' (predictable) system, or in a 'complex' (less-predictable, or unpredictable) system. Other useful concepts from economics include diminishing returns, time preference, externalities, incentive compatibility and time consistency.

Integrating the two approaches into NRM will involve identifying and resolving tensions that are evident in both ideas/concepts, and in the language used. A key example is *efficiency*, an important concept in economics with no immediate analogue in Resilience Thinking. If we take a limited and specific concept of efficiency – achieving a given goal at lower cost (cost-effectiveness, basically) – then seeking ever-greater efficiency may lead to a reduction in resilience. The problem here is not one of economics (efficiency) versus Resilience Thinking; if anything, the problem might be economists' tendency to approaching the problem of achieving a goal with insufficient attention paid to risk and uncertainty; that is, assuming the goal can be achieved deterministically and then minimising the cost of doing so. Alternatively, the problem might be focusing on the efficient achievement of a particular goal without consideration for side-effects or externalities in other parts of the system. The idea of unintended consequences resulting from partial perspectives is certainly not unfamiliar to economists.

Walker and Salt (2012) note that resilience itself may be a goal, with associated costs and benefits of achieving greater (or reduced, if relevant) resilience. In particular circumstances, the value of increased specified resilience (where it is desirable) can be determined numerically where probabilities and costs of threshold crossing are known. These can be compared to costs of actions to reduce the risk of crossing.

Of particular relevance to NRM, Marshall (2013) proposed a framework which can be applied to optimise, in terms of cost-effectiveness, the choice of policy options for a particular social–ecological system. This framework extends the conventional cost-effectiveness metric by accounting for the lock-in costs of the options which follow from the path dependencies that typically arise from disturbing a complex adaptive system. The lock-in costs associated with a policy option are the additional costs that path dependencies triggered by the option would create for subsequent efforts to adapt or transform the option (e.g. in reforming water policy). Given that path dependencies are emergent phenomena, however, lock-in costs are typically subject to deep uncertainty. Marshall argues nevertheless that limited predictability of such costs in a particular case may be possible on the basis of inductive analysis of experiences in related cases and that optimisation may therefore be feasible on a boundedly rational basis. At the very least, however, as argued by Challen (2000), an optimisation framework of this kind provides opportunities to increase the transparency of existing, typically *ad hoc*, approaches to accounting for path dependencies in policy analysis.

Regional natural resource management in Australia

Land degradation emerged as an issue of political importance in Australia in the early 1980s, with establishment of the National Soil Conservation Program in 1983. During the 1990s, NRM in rural areas was driven primarily by the National Landcare Program, which focused on encouraging farmers to change their land management practices on a voluntary basis, supported by paid facilitators but largely unfunded in terms of compensating farmers for their opportunity costs. Issues addressed (all of which continue to be relevant) included dryland salinity, habitat for native species, soil acidity, soil erosion and water pollution.

During the 1990s, arguments developed that the locally based Landcare model was incapable of addressing significant landscape-scale issues, especially salinity. Its reliance on unpaid voluntary action failed when it came to issues like salinity that required landholders to undertake actions that would be highly costly to them while generating very much smaller private benefits to them. Notwithstanding the generation of public benefits, the limits of voluntary contribution had been reached. The rising interest in integrated catchment management (ICM) provided a conceptual vehicle for the development of a regional, catchment-based delivery model. The regional scale was considered the most appropriate to support NRM that is holistic (covers all landscape elements); systematic (considers the interactions between elements); and comprehensive (embraces the range of values attached to landscapes) (Bammer, 2005).

Around 2000, 56 regional NRM bodies were created across the country, motivated by the opportunity to obtain substantial funding from two major national programs, the Natural Heritage Trust and the National Action Plan for Salinity and Water Quality. Each regional body was responsible for liaising with local community members and land managers to develop a local plan for improved NRM and for development and funding of projects. The Catchment Action Plans thus developed were to be devised in the context of other state and national frameworks, including water sharing, environment and biodiversity strategies, and basin management (for the Murray–Darling Basin in particular).

In the first phase of this arrangement, the regional bodies had reasonably high levels of autonomy to select which projects they would support and were able to fund those projects out of funds they received from the programmes. In 2008, following a change of government and a change of national programmes to Caring for our Country, regional bodies were given less funding and less autonomy. Their role still included community consultation and development of projects, but now they had to bid for funds for those projects on a competitive basis.

Integrating economics and Resilience Thinking in regional natural resource management

Two approaches for regional NRM are discussed in this section. First is the 'Investment Framework for Environmental Resources'. Second is a system for applying Resilience Thinking assessments regionally.

Following criticism of the quality of project planning and project selection in the regional NRM system (e.g. Pannell, 2001; Pannell & Roberts, 2010), an attempt was made to develop economics-based tools and frameworks to assist the regional bodies. One tool, INFFER (Investment Framework for Environmental Resources) (Pannell et al., 2012) has been used to some degree by around half of the regional NRM bodies.

Subsequently, a system for applying Resilience Thinking in regional NRM was developed and offered to regional NRM bodies. Initially this system was most influential in New South Wales, but over time, interest has grown in other states.

The developers of both approaches were motivated by concerns about limitations of past approaches to the planning of NRM. Their efforts occurred at a time when there was a willingness to try new approaches to address the acknowledged limitations. Neither approach has dominated, but up until now they have been applied independently, as if they are alternatives, while we consider that there is potential for cross-fertilisation between the approaches. Each of the approaches utilises a step-wise consultative process, which we compare now.

The sequence of steps in INFFER, which takes an asset-based approach, is to:

(1) Prepare an inventory of all the environmental assets the community might be interested in investing in. There may initially be several hundred in an NRM region.
(2) Conduct some fairly simple filtering of those assets to reduce them to a much smaller number for which more detailed analysis is feasible. A simple scoring approach is utilised, using criteria like asset significance, environmental threat, and feasibility.
(3) Develop specific projects for those assets that remain in play after the filtering process and evaluate them quantitatively, considering a range of factors including values, threats, management effectiveness (depending on ecological processes), behaviour change, risks and costs.
(4) Identify the best projects based on the evaluation in (3).

The steps in a Resilience-Thinking assessment, using a systems-based perspective, are to:

(1) Identify the relevant system. In doing so, identify the component parts; the focal scale; what is valued; issues/problems/trends in play, and drivers of change.
(2) Assess resilience: specified; general; transformative needs and capacity; then design state-transition models.
(3) Identify/design interventions: feasibility; type/scale; sequencing; adaptive management programme; any need for transformation?

It is apparent that the two approaches have elements in common. They both involve the participation of experts and community members. They start relatively broadly, become more specific and detailed and produce information intended to support decision-making. There is significant overlap in their information requirements. They are both intended to be practical approaches that can assist bodies responsible for NRM. Both recognise the value of monitoring projects for the purpose of adaptive management.

There are also important differences. Interventions in INFFER are defined around environmental assets, which can be defined at any scale, but tend to be parts of landscapes, rather than whole landscapes. The Resilience-Thinking approach focuses more on processes at the whole-landscape scale. INFFER does not explicitly focus on the potential for ecological thresholds, whereas Resilience Thinking does. INFFER's process involves both qualitative and quantitative elements and produces a benefit:cost ratio for each project, to assist with prioritisation. Resilience Thinking is less quantitative and does not formally prioritise.

The best way to combine and integrate the two approaches is open to question and is likely to be a process that requires experimentation and experience. A partially joint, but not particularly integrated, approach would be to incorporate elements of one process into the other. For example, the potential for crossing thresholds could be considered when filtering the environmental assets (INFFER step (2)) or when evaluating the benefits of specific projects (INFFER step (3)).

More ambitiously, a joint integrated framework could be developed. Because each approach adopts a consultative and participatory process, where the first step is aimed largely at gaining relevant information and perspectives (whether by identifying key environmental assets of interest, as per INFFER, or by describing the system holistically, as per Resilience Thinking), these activities would ideally be done in an integrated fashion. While it is not trivial to do this, it might be that a consultation from a holistic systems perspective could be expected to yield a list of key assets of interest, an output of relevance to the INFFER process. Whether this is to be done all in one step, or whether in a particular sequence (work through system definition first, then develop a list of assets, for example) is something that could be explored in trials with environmental managers.

We note that the distinction between stocks and flows in economics has elements in common with the distinction between slow and fast variables in Resilience Thinking. A stock is an accumulated amount of a resource measured at a point in time, like water in a dam, while a flow is an amount of something measured over a period of time, such as how much water in a river flows past a given point in a day. On the other hand, a slow variable evolves gradually, driving changes in the 'state of the system' as measured by a fast variable – a variable that may respond relatively quickly. In many cases, the slow variable from Resilience Thinking can be classified as a stock in an economic framework. Stocks can have important implications for resilience and thresholds, and they may also be relevant to an economic evaluation. For example, the height of a saline groundwater table in an agricultural area clearly influences the resilience of that area (Walker et al., 2010) and also influences the expected benefits from intervening with salinity management practices (Pannell & Ewing, 2006).

Another integration of the two approaches might involve extending the process beyond the point of project selection and intervention design. For example, it could account for lock-in costs – the cost of being stuck on a particular path as a result of current decisions (Marshall, 2013). Policies that involve financial assistance to particular groups in society, such as payments for land stewardship would be a case in point. Such policies are politically and socially difficult to withdraw, even if they are found to be failing to achieve the original objectives of the policy. Efforts to change such policies inevitably incur costs. Such lock-in costs could be accounted for in a post-INFFER phase where stakeholders deliberate about whether differences between the lock-in costs of the various options are likely to be significant enough to overturn the priority rankings resulting from the INFFER process.

In considering the potential for integrating the approaches, an important contribution from economics would be recognition of the importance of considering transaction costs when designing systems for use in organisations. The designers of any such system need to carefully consider the trade-offs between the comprehensiveness of the system and the transaction costs involved in its use (e.g. Pannell et al., 2013). It may (or may not) be that the ideally integrated system may not provide sufficient additional benefits over a simpler system to be worth bearing the additional costs involved.

Conclusions

Economies are complex, self-organising, systems. In much applied work, economists treat the part of the system under study 'as if' it were mechanistic enough to be modelled without specific attention being paid to complexity issues such as thresholds and emergent behaviour. Individual opinions will vary on whether and when this approach has validity; 'all models are wrong', just as all maps are simplified and scaled-down versions of the territory they represent, and the question of whether a given model is useful and fit-for-purpose is one for a particular analyst to decide.

Economic *theory* is another matter. Multiple (indeterminate) equilibria, difficulties in aggregation, hysteresis, thresholds and emergent behaviour are all allowed for in theory, if sometimes under other names; and they may be observed in reality.

The fact is that both the economics discipline and the Resilience-Thinking paradigm involve the analysis of self-organising systems. In economics the system is dominated by humans who have the capacity to be self-aware and forward-looking, so that their understanding of the system and how it is managed feeds directly back into the operation of the system itself, one of the reasons macroeconomics remains so fraught with tension and disagreement.

This commonality between paradigms presents opportunities to think about the integration of economics and Resilience Thinking when applied to NRM at regional scale in agricultural landscapes. We have presented two complementary or integrated approaches (INFFER and Resilience Thinking); both involve consultative, participatory processes and both are aimed at identifying key elements of the system of interest in order to understand what matters, to identify management responses and to inform decisions about interventions. Harmonising and integrating the processes would be challenging but, we believe, worthwhile. Economists, ecologists and other scientists engaged in this integrative exercise stand to learn considerably from it, and from each other.

References

Bammer, G. (2005). Guiding principles for integration in Natural Resource Management (NRM) as a contribution to sustainability. *Australasian Journal of Environmental Management*, **12**(Suppl), 5–7.

Battiston, S., Gatti, D., Gallegati, M., Greenwald, B. & Stiglitz, J.E. (2012). Liaisons dangereuses: increasing connectivity, risk sharing, and systemic risk. *Journal of Economic Dynamics and Control*, **36**, 11–21.

Blanchard, O., Dell'Ariccia, G. & Mauro, P. (2010). Rethinking macroeconomic policy. *Journal of Money, Credit and Banking*, **42**(Suppl 1), 199–215.

Brunnermeier, M.K. (2009). Deciphering the liquidity and credit crunch 2007–2008. *Journal of Economic Perspectives*, **23**, 77–100.

Brunnermeier, M. & Krishnamurthy, A. (2014). *Risk Topography: systemic risk and macro modeling*. Chicago, IL: University of Chicago Press.

Caballero, R.J. (2010). Macroeconomics after the crisis: time to deal with the pretence-of-knowledge syndrome. *Journal of Economic Perspectives*, **24**, 85–102.

Campbell, J.Y. (2014). Empirical asset pricing: Eugene Fama, Lars Peter Hansen, and Robert Shiller. *Scandinavian Journal of Economics*, **116**(3), 593–634.

Challen, R. (2000). *Institutions, Transaction Costs and Environmental Policy: institutional reform for water resources*. Cheltenham: Edward Elgar.

Chiappori, P.A. & Ekeland, I. (2011). New developments in aggregation economics. *Annual Reviews in Economics*, **3**, 631–668.

Diamond, D.W. & Dybvig, P.H. (1983). Bank runs, deposit insurance, and liquidity. *Journal of Political Economy*, **91**, 401–419.

Eggertsson, G.B. & Krugman, P. (2012). Debt, deleveraging, and the liquidity trap: a Fisher–Minsky–Koo approach. *Quarterly Journal of Economics*, **127**(3), 1469–1513.

Folke, C. (2006). Resilience: the emergence of a perspective for social–ecological systems analyses. *Global Environmental Change*, **16**(3), 253–267.

Gordon, R.J. (2009). Is modern macro or 1978-era macro more relevant to the understanding of the current economic crisis? 12 September 2009 revision of a paper first presented to International Colloquium on the History of Economic Thought, Sao Paulo, Brazil, 3 August 2009. http://economics.weinberg.northwestern.edu/robert-gor

don/files/RescPapers/ModernMacro1978.pdf (accessed 24 July 2016).

Holt, R.P.F., Rosser, J.B. Jr. & Colander, D. (2011). The complexity era in economics. *Review of Political Economy*, **23**, 357–369.

Horgan, J. (1995). From complexity to perplexity. *Scientific American*, **272**, 104–109.

Howitt, P. (2006). The microfoundations of the Keynesian multiplier process. *Journal of Economic Interaction and Coordination*, **1**, 33–44.

Howitt, P. (2011). Comment on 'PSST: Patterns of sustainable specialization and trade' (by Arnold Kling). *Capitalism and Society*, **6**, 1–11.

Howitt, P. & Clower, R. (2000). The emergence of economic organization. *Journal of Economic Behaviour and Organization*, **41**, 55–84.

Kay, J. (2011). The map is not the territory: an essay on the state of economics, Institute for New Economic Thinking. Published online at http://ineteconomics.org/blog/inet/john-kay-map-not-territory-essay-state-economics (accessed 17 April 2014).

Kehoe, T.J. & Levine, D.K. (1985). Comparative statics and perfect foresight in infinite horizon economies. *Econometrica*, **53**, 433–53.

Knight, F. (1921). *Risk, Uncertainty and Profit*. New York, NY: Harper Torchbooks.

Krugman, P. (2000). How complicated does the model have to be? *Oxford Review of Economic Policy*, **16**, 33–42.

Leijonhufvud, A. (2011). Comment on Jordi Gali, Frank Smets and Rafael Wouters, 'Unemployment in an estimated new Keynesian model' at Research Workshop on 'Analyzing the Macroeconomy: DSGE versus Agent-based Modelling', Central Bank of Austria, 15–16 June 2011.

Ljungqvist, L. (2008). Lucas Critique. *The New Palgrave Dictionary of Economics*. 2nd edition, edited by S.N. Durlauf & L.E. Blume. Basingstoke: Palgrave Macmillan.

Marshall, G.R. (2013). Transaction costs, collective action and adaptation in managing complex social–ecological systems. *Ecological Economics*, **88**, 185–194.

Miller, J.H. & Page, S.E. (2007). *Complex Adaptive Systems: an introduction to computational models of social life* (Princeton Studies in Complexity). Princeton, NJ: Princeton University Press.

Minsky, H. (1986). *Stabilizing an Unstable Economy*. New York, NY: McGraw-Hill Professional.

Naevdal, E. (2003). Optimal regulation of natural resources in the presence of irreversible threshold effects. *Natural Resource Modeling*, **16**, 305–333.

Ostrom, E. (2010). Beyond markets and states: polycentric governance of complex economic systems. *American Economic Review*, **100**(3), 641–672.

Pannell, D.J. (2001). Dryland salinity: economic, scientific, social and policy dimensions. *Australian Journal of Agricultural and Resource Economics*, **45**, 517–546.

Pannell, D.J. & Ewing, M.A. (2006). Managing secondary dryland salinity: options and challenges. *Agricultural Water Management*, **80**(1–2–3), 41–56.

Pannell, D.J. & Roberts, A.M. (2010). The National Action Plan for Salinity and Water Quality: a retrospective assessment. *Australian Journal of Agricultural and Resource Economics*, **54**, 437–456.

Pannell, D.J., Roberts, A.M., Park, G., Alexander, J., Curatolo, A. & Marsh, S. (2012). Integrated assessment of public investment in land-use change to protect environmental assets in Australia. *Land Use Policy*, **29**, 377–387.

Pannell, D.J., Roberts, A.M., Park, G. & Alexander, J. (2013). Improving environmental decisions: a transaction-costs story. *Ecological Economics*, **88**, 244–252.

Perrings, C. (1998). Resilience in the dynamics of economy–environment systems. *Environmental and Resource Economics*, **11**, 503–520.

Perrings, C. & Walker, B. (1997). Biodiversity, resilience and the control of ecological–economic systems: the case of fire-driven rangelands. *Ecological Economics*, **22**, 73–83.

Quiggin, J (2010). *Zombie Economics: how dead ideas still walk among us*. Princeton, NJ: Princeton University Press.

Rabin, M. & Thaler, R.H. (2001). Anomalies: risk aversion. *Journal of Economic Perspectives*, **15**, 219-232.

Randall, A. (2011). *Risk and Precaution*. Cambridge: Cambridge University Press.

Richardson, C., Courvisanos, J. & Crawford, J.W. (2011). Toward a synthetic economic systems modeling tool for sustainable exploitation of ecosystems. *Annals of the New York Academy of Science*, **1219**, 171-184.

Rizvi, S.A.T. (2006). The Sonnenschein–Mantel–Debreu results after thirty years. *History of Political Economy*, **38** (annual supplement), 228-245.

Solow, R. (2010). Building a science of economics for the real world. Prepared statement to the House Committee on Science and Technology, Subcommittee on Investigations and Oversight, United States Congress, 20 July 2010.

Walker, B. & Salt, D. (2012). *Resilience Practice: building capacity to absorb disturbance and maintain function*. Washington, DC: Island Press.

Walker, B., Pearson, L., Harris, M., et al. (2010). Incorporating resilience in the assessment of Inclusive Wealth: an example from South East Australia. *Environmental and Resource Economics*, **45**, 183-202.

Weintraub, E.R. (1979). *Microfoundations: the compatibility of microeconomics and macroeconomics*. Cambridge: Cambridge University Press.

CHAPTER FIFTEEN

Integrating biodiversity and ecosystem services into European agricultural policy: a challenge for the Common Agricultural Policy

ALLAN BUCKWELL
Institute for European Environmental Policy

Introduction: the nature of the challenge

Integrating environmental management into agricultural policy in the European Union is a significant challenge to all the actors directly concerned. Farmers and politicians have to be convinced of its necessity. Administrators have to devise ways of implementing the integration through practicable schemes which satisfy rules concerning market distortions and pass stringent audit and the public has to be satisfied that tangible environmental results are achieved at reasonable cost.

As farming is a dominant land-using economic sector it is bound to have a significant impact on the natural environment. In Europe, farming operates predominantly through highly fragmented, privately owned, geographically dispersed, individually managed family farms; therefore, this integration will not be accomplished unless the hearts and minds of farmers are engaged. If farmers are not convinced that their current activities are undermining their potential to continue into the foreseeable future, then self-interest may be insufficient to induce the level of environmental performance society wishes. The challenge then is to find the mix of incentives and sanctions to change this behaviour. While soil erosion, water shortages and perhaps even extreme climate events are gradually being understood by farmers as necessitating action on their part, it is less clear that they perceive the connection between biodiversity depletion and the sustainability of their activities.

The tension between maintaining production for the current generation and yet not undermining capacity to continue this for future generations was heightened when the century-old downward trend in global agricultural commodity prices was halted over the period 2005–08, and prices jolted significantly higher. This raised the opportunity cost of restrictions on farming justified by environmental concerns.

The challenge is further deepened by the technical complexity of the environmental management task itself. It is easy to summarise the goal as the protection of biodiversity, soil, water, climate and cultural landscape. However, each of these is complex, multi-faceted and dynamic and the elements are highly interactive. Notwithstanding the rapidly growing body of research in ecology and its interaction with human activity and natural conditions, there are big gaps in the understanding of the relationships between these five dimensions of the environment. Similar gaps exist in understanding the relationships between policies, induced agricultural management practices and consequential environmental outcomes. Thus even when there is willingness on the part of farmers consciously to engage in environmental land management and of politicians to fund such action, it has proved difficult to find the policy measures which bring tangible environmental results.

In the European debate on Common Agricultural Policy (CAP) reform to date and specifically the integration of environmental concerns into agricultural policy, there has been little or no explicit reference to the term resilience. One aspect of resilience has implicitly been at the core of the debate. This is the need to enhance the environmental sustainability of European Union (EU) farming and especially the more intensive sectors. This is partly premised on the notion that, *inter alia*, soil erosion and depletion of soil organic matter, exposure to more erratic weather and climate change-induced pest and disease attack all undermine agricultural resilience as well as being undesirable in their own right. Therefore, measures which improve sustainability may be consistent with strengthening resilience. However, such formal monitoring of the impacts of the CAP as exist make no specific reference to resilience.

It may also be that the lack of reference to resilience in European agricultural policy reflects the fact that farming seems to be a sector which survives whatever political, social, or economic disruption is thrown its way, or whichever natural conditions materialise: disasters, flood, drought, extremes of heat or cold. The very fact that generations of farmers have clung on under such volatility over centuries demonstrates a degree of resilience few other economic sectors have managed. Therefore, the EU debate seems rather to take this endurance for granted and focuses on alleviating some of the consequences such as rural poverty through rural development.

The core of this chapter is in two parts: first, a summary of the efforts to integrate environment into the CAP to date and second, a review of the current reform process due to be implemented over the period 2014–16 which will apply until the end of the current EU Multiannual Financial Framework (MFF) at the end of 2020.[1]

---

[1] See http://ec.europa.eu/budget/mff/index_en.cfm for explanation of how this EU budget works and details of the agreed MFF for 2014–20.

Integration of environmental service delivery into the CAP to date
The scale and nature of the environmental integration

Reform of the CAP is a slow incremental process. Because the EU has progressively expanded from six to 28 members, and the economic and political circumstances surrounding agriculture have changed greatly, the CAP has had to adapt and change. Indeed, it has undergone a significant development over the five decades since the 1960s.

The CAP has evolved from a set of policy measures exclusively regulating agricultural commodity markets to a two-pillar policy offering direct payments to farmers and residual market support in Pillar 1 and elaborate Rural Development Programmes in Pillar 2. The 1992 (MacSharry) reform introduced direct payments as a compensation for reducing market supports and the corresponding export subsidies (red and amber turn to blue). Then, from 2004 following the Fischler mid-term review of the CAP, and using the slogan of market orientation, direct payments have been increasingly decoupled from agricultural production (blue turns to green). Meanwhile, starting from tiny expenditures in the 1980s what have now become rural development measures have steadily grown to almost a quarter of EU CAP expenditure.[2] Throughout this reform process since the mid-1980s there has been an explicit and consistent process of adding environmental objectives and measures into the CAP.

Lowe and Whitby (1997) provide an account of the early history of the integration of environmental concerns into the CAP. Poláková et al. (2011) provide a more recent, highly detailed and comprehensive analysis of the contribution of CAP environmental measures. Their report focuses particularly on the effects for biodiversity and habitats, although it also summarises impacts for water, climate and landscape protection. The chronological steps in the integration process up to the 2012/13 reform process are listed in Table 15.1, which is adapted from Hart and Baldock (2010). The process started in 1985 and has been added to at each reform step since.

Several aspects of this process are worth noting. First, reflecting the inherent difficulty and the considerable political persuasion required, the integration of the environment into the CAP is an incremental, learning-by-doing process. Second, the process embraces both pillars of the CAP: Pillar 1 with its strongly agricultural production origins and its 100 percent EU-funded, EU-wide annual payments; and Pillar 2, which is structured around project-based, multi-annual, regionally defined and co-financed Rural Development Programmes. Third, it embraces a variety of measures which are considered below under the three broad categories:

---

[2] Given that most Pillar 2 measures have to be co-financed, this chart of EU expenditure understates the share of total public expenditure on rural development.

Table 15.1. *Elements of CAP reforms to integrate environmental objectives*

Year	Key reforms	Environmental measures
1985	Council Regulation EEC 797/85, Article 19	Allows Member States to introduce zonal schemes to protect farmland habitats and landscapes of Environmentally Sensitive Areas from the threat of agricultural intensification
1987	Council Regulation EEC 1760/87	Permitted Member States operating 'Article 19' schemes to claim up to 25 percent of the cost of payments from the CAP budget
1992	MacSharry Reforms – Council Regulation 2078/92	Agri-environment measure given greater prominence as one of three 'accompanying measures'. Compulsory for all Member States to introduce agri-environment schemes providing payments for managing agricultural land in ways that are 'compatible with protection and improvement of the environment, the countryside, the landscape, natural resources, the soil and genetic diversity' (Article 1c) Article 1e introduced provisions for the (voluntary) long-term set-aside of agricultural land for reasons connected with the environment
1999	Agenda 2000 – Introduction of Council Regulation 1259/1999	Range of environmental and socioeconomic measures brought together under the 'Rural Development Regulation' and given a more prominent role within the CAP as the 'Second Pillar' Fund switching from Pillar 1 to 2 (Modulation) introduced on a voluntary basis Introduction of standards of 'Good Farming Practice' to underpin LFA and agri-environment measures Support for Less-Favoured Area (LFA) payments on livestock farms moved from headage to area basis
2003	2003 CAP Mid-Term Review – Council Regulation 1782/2003	Decoupling of main support payments in Pillar 1; several variations and timescales permitted Cross-compliance Article 69 – up to 10 percent of decoupled payments able to be used for 'specific types of farming and quality production' Modulation becomes compulsory for EU-15 (3 percent in 2005 rising to 5 percent from 2007)
2005	Introduction of the European Agricultural Fund for Rural Development (EAFRD) – Council Regulation 1698/2005	Builds on the Rural Development Regulation, organising measures under three axes supporting competitiveness, the environment and quality of life. Member States required to spend a minimum of 25 percent of their budget on land management, including environmental measures, under Axis 2
2007	Council Regulation 378/2007	UK and PT permitted to raise voluntary modulation for a limited time to meet existing rural development commitments

Table 15.1. *(cont.)*

Year	Key reforms	Environmental measures
2008	CAP 'Health Check' – Council Regulation 73/2009 – replacing Council Regulation 1782/2003; Council Regulation 74/2009 – amending Council Regulation 1698/2005	Abolition of set-aside Further decoupling required for majority of sectors apart from suckler cows, sheep and goats Article 69 more flexible – becomes Article 68 'support for specific areas' Extension of cross-compliance requirements Compulsory modulation – higher standard rates (up to 10 percent in 2012), plus higher rates for larger farms. Extended to EU-10* for first time, and Romania/Bulgaria from 2012 Additional modulation receipts for biodiversity, water management, renewable energies and the dairy sector

Source: Adapted from Hart and Baldock (2010).
* European data often use the short-hand of EU15, to refer to the first 15 members; EU10 which refers to the 10 'new' Member States who joined in 2004; then EU27 after Romania and Bulgaria joined in 2007, and most recently EU28 including Croatia who joined on 1 July 2013.

(i) environmental conditionality,
(ii) voluntary environmental schemes and
(iii) regional, zonal and farm-type specific supports.

In addition to these measures which set out purposively to help farmers deliver environmental services, there are also CAP reform decisions which are taken primarily for economic reasons and yet may offer environmental benefit because, for example, they reduce environmentally harmful subsidies or offer positive environmental side effects. Two significant examples of such decisions are the dramatic reductions of price supports which commenced in the 1992 MacSharry reform and the introduction of supply control measures particularly milk quotas (in 1984) and arable land set-aside (in 1988). These decisions were taken to combat the international market distorting effects and the rapid escalation of the budget costs of the price support system which was the core of the CAP at that time. However, they were also associated with environmental benefits (Hart & Baldock, 2010, pp. 5–8).

Environmental conditionality

As the principal support instruments of the CAP changed from market price support to direct payments in the mid-1990s, policy had to interact directly with all individual farmers for the first time. This offered the possibility to attach conditions to the payments and *environmental conditionality* was born. Initially, in the 1992 reform the conditionality was voluntary at Member State level and few took up the option. The subsequent Agenda 2000 reform, which introduced the 'two Pillar' CAP, added maximum stocking rate conditions to payments to beef producers and also required farmers receiving agri-environment and less-favoured area payments to satisfy conditions of 'usual good farming practices'.

The next step was to make the environmental conditionality compulsory. This was achieved in the so-called Mid-Term Review in 2004. This reform was highly significant for environmental integration. It took the three major steps of: decoupling payments from production and signalling that market intervention was to be phased down and replaced by direct payments for all products,[3] introducing compulsory cross-compliance for the direct payments and requiring, not just allowing, funds to be switched from Pillar 1 to Pillar 2. The shift away from commodity market support was driven primarily by international trade considerations, to satisfy the conditions of the WTO Agreement on Agriculture.[4] However, there were potential environmental benefits from this policy shift which would reduce the incentives for higher cost, and environmentally damaging, production which was clearly beyond market demands in the EU. However, at the same time it could also reduce the rewards to those marginal parts of EU agriculture which are, at best, weakly viable economically and therefore at danger of land abandonment.

The cross-compliance framework comprised a series of Statutory Management Requirements (SMRs) and a series of Good Agricultural and Environmental Conditions (GAECs). The SMRs were existing EU regulations and directives, most of which were environmental including the Birds and Habitats Directives and the Nitrates Directive. The GAECs included good agricultural practices and a series of requirements based on regional and Member State regulations. These included measures for retaining landscape features, protecting permanent pastures, avoiding encroachment of unwanted vegetation on agricultural land, soil management to reduce erosion and retain soil organic matter and establishment of buffer strips – mostly for water protection.

The very existence of decoupled payments to farmers with cross-compliance conditions attached raises some controversy. This is because the legitimacy of the underlying payments is insecure because their fundamental purpose is not explicitly stated or agreed.[5] The payments were introduced in the mid-1990s as compensation for policy change, but it is rare that compensatory payments are made indefinitely. They will still largely be in place in

---

[3] This would bring the supports for dairy, sugar, wine, olive oil, cotton and tobacco into line with what had been done for arable crops (cereals, oilseeds and proteins) and the extensive livestock sectors (beef, sheep and goats). The compensatory payments for all products were consolidated into what was then called the Single Farm Payment.

[4] See Harvey (1997) and Swinbank and Daugbjerg (2006).

[5] See Buckwell (2008) for a discussion of five possible justifications for CAP direct payments and a more recent and fuller analysis by Matthews (2017).

2020 as ill-defined universal (that is non means-tested) income supports to farmers. Therefore, while they exist, it seems reasonable that they should be subject to conditions that farmers respect EU and National environmental (and other) regulations. Pragmatically this may be a practical way to achieve better compliance with regulations concerning diffuse pollution over a large proportion of European rural territory than otherwise possible given prohibitively high policing and enforcement costs. However, these payments can equally be described as no more than paying farmers for compliance with the law, especially as no agricultural production is required. This contradicts the polluter pays principle. A counter argument that they are environmental payments 'in waiting' carried some credibility in the light of the subsequent development of the CAP to shift the purpose of (some of) the direct payments towards environmental good provision.

Voluntary environmental schemes

Poláková et al. (2011) suggest that the 'agri-environment measure, now sitting within Pillar 2 of the CAP alongside other land management measures in Axis 2, is the oldest and the single most significant measure for pursuing environmental objectives across the farmed landscape, both in terms of the spatial coverage of schemes and the resources allocated to them'. While it is compulsory for all Member States to make use of this measure, there is considerable latitude for Member States to decide the objectives of the schemes and the detailed actions offered. This is necessary to cope with the range of natural conditions and farming systems around the EU. In total about 23 percent of the funds for Rural Development for the period 2007–2013 were committed to agri-environment, and the estimated 7 million agreements covered about the same proportion of the utilised agricultural area (Poláková et al., 2011, box 4). There is considerable variation in the scale of the use of the measure between the Member States.

The variety of ways these schemes can be used is indicated by the four situations analysed by Poláková et al. (2011), namely:

- maintaining and, in some cases, restoring, semi-natural agricultural habitats, mostly the extensively grazed high nature value farming systems;
- beneficial biodiversity management of un-farmed features as well as in-field options in arable farming and temporary grass, and occasionally focus on locally specific habitats and species;
- enabling continuation of traditionally managed permanent crop systems, olive groves, vineyards, fruit and nut orchards;
- conversion and maintenance of organic farming.

These agri-environment schemes are not simple to set up and run, either for the administrators or farmers. They work on a voluntary, multi-year contractual basis. They may offer annual payments for certain management actions or once-off capital grants for investments. The payment principles are established by regulation to be in accordance with World Trade Organisation (WTO) Green box rules for non-production and trade distorting support. This limits payments to the income forgone and direct costs associated with the specific environmental actions. These are not simple to calculate or understand and are not very sensitive to changing market conditions – features which make farmers cautious about signing up for periods of 5–10 years.

Regional, zonal or farm-type specific supports

A third broad category of CAP measures which can have environmental significance are those which attempt to deal with a large category of farms which have in common that they are economically highly marginal. The principal such measure is the payments made to farms in Less-Favoured Areas (LFAs). These payments are made to compensate for the natural disadvantages of farming in areas with unfavourable climate, poor soils, high altitude, steep slopes and often remoteness from markets with poor infrastructure. Of course economically this does not seem a particularly rational policy – in any other sector of the economy such businesses would be encouraged to relocate! It is this logic which perhaps indicates the real purpose of such payments. These areas are already subject to depopulation and certainly an outmigration of the young, therefore the LFA payments can be seen as contributing to keeping farming alive in such areas. The farming systems found are invariably (but not always) relatively low-intensity ruminant grazing. Past attempts to help such areas by trying to encourage them to intensify agricultural production have often doubly failed: there is little scope for such productivity improvement precisely because of the disadvantageous conditions and in the process the drainage, fencing, pasture 'improvement' and higher stocking rates have inflicted significant damage in lost biodiversity, water pollution and greenhouse gas emissions.

These were the reasons that the LFA payments were converted from headage to area basis in the Agenda 2000 reform. However, it is still the case that the payments are generally not directly linked to specific environmental management actions. They are seen as supplements to the (usually much larger) direct payments as a way of keeping the farms in business. They are therefore in some circumstances properly seen as complementary to the agri-environment measures. There are examples, e.g. in high nature value land, where Member States have made use of the scope to differentiate the payments according to the severity of the natural handicaps to support such farming systems.

Similar principles can also apply to both the specific supports available under Article 68 and to the use of coupled payments. These are sometimes used to restrict support to certain traditional forms of production or to differentiate payments according to the size or location of farms with environmental objectives in mind. While these zonal or farm type instruments can be used to useful environmental effect, there is little guarantee that they always will be, and they can in many circumstances be helping keep in business farms which are still practicing environmentally unhelpful stocking rates and practices.

This overview of the progress to date in integrating environmental measures into the CAP is by no means exhaustive; there are a number of other CAP measures in Pillar 2 which have, or could have, positive environmental effects (or indeed the opposite). Two are the development of skills and capacity and the provision of so-called non-productive investments. Other rural development measures which assist rural economic diversification, and the LEADER programme which helps strengthen local participation and communities can also directly and indirectly assist improved environmental land management.

What has been learned from this experience?

While there is currently no systematic process for evaluating and monitoring the efficacy of CAP Direct Payments, there is in place an elaborate apparatus for the Rural Development (RD) policy, called the Common Monitoring and Evaluation Framework (CMEF).[6] This sets procedures for monitoring the progress of Member States in implementing their Rural Development Programmes and for collecting information to evaluate the success or otherwise of these measures in achieving their objectives. The procedures involve *ex-ante*, mid-term and *ex-post* evaluations of the RD measures, and it is intended that this review process feeds information from one seven-year period of programming to the next and subsequent periods to inform policy choices and decisions.

Naturally, these processes are not simple; it takes considerable time to accumulate and collate the information from the member states. Unfortunately this process has not worked well to date. The mid-term evaluations are conducted too early before new Rural Development Programmes have been up and running long enough to show results. This information is then often all that is available when the regulations for the next period are being debated and the subsequently agreed Rural Development Regulations (RDR) have to be designed and implemented by the Member States. It will clearly take two or three such cycles to accumulate better information on which to base future policy.[7] Thus there are long time lags before the lessons can be learned and implemented in better policy.

---

[6] Article 80 of Council Regulation (EC) No. 1698/2005 of 20 September 2005 on support for rural development by the European Agricultural Fund for Rural Development (EAFRD) (OJ L 277, 21.10.2005, p. 1).

[7] Which indicates why a structure like the European Network for Rural Development (see http://enrd.ec.europa.eu/en/home-page_en.cfm), which provides an ongoing information exchange about implementation of rural development policy, is so important.

These problems are compounded by the multiplicity of goals for rural development policy (some of which may be in conflict – such as increasing agricultural competitiveness and productivity and reducing negative environmental impacts), the often vague statements of objectives, some of which are in any case extremely difficult to define (e.g. sustainability), the absence of established baselines for many of the indicators of interest and difficulties in establishing the counterfactual policy (i.e. what would have happened in the absence of the RDR).

It is therefore not so surprising that audits have consistently found weaknesses in these processes. Indeed, most of the European Court of Auditors' (ECA) reports have been critical. Thus the Court's 2008 report posed the question, 'Is Cross Compliance an effective policy?' (ECA, 2008). The answer was no. The objectives and scope were judged to be unclear, the legal framework too complex and they found that the Member States did not take their responsibility to implement effective control and sanctions. The 2011 report on agri-environment (ECA, 2011) criticised the lack of targeting and monitoring of environmental effects, along with insufficient clarification, collection of information, justification and reporting on outcomes of individual agri-environment schemes. The 2013 report on the operation of Article 68 (ECA, 2013a) said that Member States have failed to prove they are using Article 68 aid for the worthiest activities. They went on to say that governments had not adequately justified the payments and they identified weaknesses in administration and control. A further report on the CMEF (ECA, 2013b) concluded that the current monitoring arrangements failed to show that the seven-year €100 billion Rural Development budget and the additional €58 m in co-financed funds from the Member States were properly spent. The report pointed to shortcomings in the evaluation framework at national level and suggested that information collected during the programming period was unreliable, inconsistent and irrelevant to show the results achieved in relation to the objectives set.

This is not encouraging. These criticisms refer to some of the key measures introduced to help deliver ecosystem services. They introduce an unwelcome negativity into the project, especially as the Member States are all too aware of the high administrative costs of the RD measures. Yet if the judgments are fair they have to be addressed. Combined with farmers' understandable preference for the entitlement approach of more-or-less automatic, annual, direct payments in Pillar 1 these criticisms certainly do not make it any easier to advocate a long-run strategy of greatly expanding the use of CAP supports towards more measures directed at the environment.[8]

---

[8] However, it cannot be stressed enough that if schemes to promote environmental goals are found wanting by auditors, the direct payments measure which soaks up three times the budgetary resource of Pillar 2, with even less clear objectives, is unlikely to be judged more efficacious.

Despite the difficulties pointed out by the Court of Auditors, there is much evidence of environmental benefits achieved by some of the measures introduced into the CAP. The detailed analysis of Poláková et al. (2011) shows that for each of the four situations (above) there are many documented examples of improvements for biodiversity and ecosystem services.

For example, on agri-environment schemes they concluded that for semi-natural habitats:

There are a number of examples of agri-environment schemes that have been successful in supporting HNV (High Nature Value) farming, thereby maintaining semi-natural wooded pasture habitats (Sweden, Estonia), hay meadows and mountain pastures (Slovakia, Romania), the restoration of over-grazed pastures (Bulgaria), moorland grazing (UK) and traditional agro-forestry systems in Spain (dehesas).

And for other areas, their general conclusion was that:

Although the overall evidence is variable, it suggests that agri-environment measures have also proved successful in delivering benefits for widespread and common species in improved grasslands and intensive croplands. The benefits associated with agri-environment measures for intensive croplands are found mainly in instances where a combination of management options provide key ecological resources for vulnerable species, in particular breeding habitat and year-round food resources, as these tend to be reduced by agricultural intensification and specialisation. The main priority for most of the declining species of such habitats (especially birds), are measures that provide in-field resources (such as fallow patches or fields, over-wintered stubbles, diverse crops and crops with reduced pesticides). However, some species also benefit from field edge management measures, such as the planting of field margins with seed-rich or nectar-rich plants, or reductions in the use of pesticides in field headlands.

To this point it can be concluded that from the zero baseline of the mid 1980s, in the succeeding nearly three decades, there has been a considerable advance in the willingness to insert rural environmental goals into the CAP. A non-trivial part of support expenditure has been allocated for these goals. However, the complexity of the task is now evident and while acknowledging some real progress, the delivery on the ground has been disappointing. This is the background against which the most recent CAP reform debate has taken place and which reached agreement in the autumn of 2013. This is the subject of the next section.

The greening of the CAP in the 2014–2020 reform[9]
Politico-economic context and the language of the debate

With the new Commission in 2010, strong expectations were generated for a far-reaching reform to be put in place for 2014–2020. The previous 'Health

---

[9] The author acknowledges that this account may be a little coloured by his own experience, until December 2011, as a protagonist in the debates on the 2014 reform, although it attempts to be a fair and balanced account.

Check' reform had been conducted while the enlargement to embrace the new Member States from Central and Eastern Europe was still being finalised, so the next reform would be the first undertaken by the EU27. Also the coincidence of the CAP reform discussion taking place alongside the negotiations for Multiannual Financial Framework was likely to mean that the budget resources for the CAP would be a focal concern. In addition, the 2014 reform was the first to be decided by the ordinary legislative procedure which gives full co-decision powers on CAP reform to the Agriculture Council and the European Parliament.

Criticism of the CAP generally came from three directions: those who saw it as an extravagant policy which provides very poor value for public money and should therefore be sharply curtailed; those who were principally concerned with the environmental damage caused by intensive agriculture; and those whose concerns came from the perspective of international development and poverty reduction in the poorest rural areas of the world.

The British Government, backed to some extent by the so-called Stockholm group,[10] were in the lead of the first group of critics. Their early warning shot for the new reform negotiations was the 'Vision for the Common Agricultural Policy', published jointly by the Treasury and Defra in December 2005. This argued that the market support and direct payments first pillar should shrink and thus the CAP should comprise essentially the co-financed second pillar rural development measures dealing with market failures. While this appeared to cultivate few friends who would overtly support it, there is no doubt that it was noticed and affected the conduct of the debate by putting direct payment supporters on the defensive.

The second group comprised environmental NGOs, with the significant part of society, at least in north-west Europe, which supports green goals. It also included support in governmental environment ministries and agencies, and corresponding and influential parts of the European Commission. These interests made a well-researched and powerful case for what became the adopted jargon of 'greening the CAP'. The context through the early years of the 2000s until 2007–2008 was favourable to this approach. Climate change and the necessary policies to reduce greenhouse gases and assist carbon sequestration had come up the political agenda, particularly following the 2006 Stern Report.[11] The political cause of biodiversity conservation and restoration was also rising.

The language of the environmental case for further integration of environment in the CAP coalesced around the concept of public goods (see below for

---

[10] The core Stockholm group comprised UK, Sweden, Denmark and Netherlands with Germany 'attending'. A public-facing expression of particularly strong criticisms of the CAP is http://farmsubsidy.openspending.org

[11] Stern Review of the economic effects of climate change www.hm-treasury.gov.uk/independent_reviews/stern_review_economics_climate_change/sternreview_index.cfm

explanation). This long-established economic concept was often abused as much as it was correctly used. However, it was recognised by all as signalling the broad message that there are pervasive market failures surrounding the major land-using activities (farming and forestry) which will only be put right by collective actions. Indeed, the more the CAP shifts to a market-oriented position (by decoupling supports from agricultural production) then the less attention farmers might pay to the environmental services they could produce but for which there is no market reward.

This message in itself represented a significant change in farmer attitudes. There had undoubtedly been some shift from the confrontational mode of accusation and denial between environmentalists and farmers about the environmental damage caused by modern farming practices. There were beginnings of signs that each side appreciated they had to work together to deal with the interactions between the environment and farming. Farmers' organisations were themselves acknowledging that improved environmental land management was part of their role. In the UK the National Farmers Union (NFU) adopted the somewhat muted 'produce more, impact less' slogan, which they say recognises that 'protecting and enhancing the environment is an important side-effect of a sustainable agricultural policy' (NFU, 2010). Earlier, the Country Land and Business Association (CLA) was more explicit in recognising the interrelationship between agriculture and the environment in Europe – each depending on the other. They argued that the CAP should become an integrated policy for *Food and Environmental Security* (Buckwell, 2007).

The idea of explaining the under-supply of habitats and species on farms as well as diffuse pollution caused by agriculture as market failures, in particular externalities, is not new. These are natural concepts for economists and are at the heart of environmental economics. Indeed, in an early contribution to this field, Coase (1960) used the externalities between peripatetic ranchers and static farmers to explain the nature of the problem, how to arrive at an efficient solution and the relationship with property rights. The concept slowly moved out of the academic journals and explicitly into the CAP debate in the EU. It was helped to no small extent by studies commissioned by the Agricultural and Rural Development Directorate of the European Commission. Influential examples of two such studies are European Economy (1997) and the report by Cooper et al. (2009) entitled 'Provision of public goods through agriculture in the EU'. This latter paper explains the concept of public goods and documents in great detail and the nature of the (mostly) environmental market failures surrounding EU agriculture. It looks at the degree to which they exhibit the defining criteria for public goods, their non-rivalness in consumption and non-excludability. It then examines in detail the role which the instruments of the CAP could play in providing these public goods. A parallel report published by the RISE Foundation

(2009) picked up the public goods theme and also integrated it with the language of Ecosystem Services which has become part of the discourse of the CAP debate.

It is thus suggested that in the years running up to the opening of the latest reform discussions the ground had been well prepared to signal that further integration of environmental concerns into the CAP would be one of the main planks of the 2014 reform. How did it work out?

The reform consultation and the 2010 communication

The new Commissioner launched a wide consultation in 2010 to gather the views on the priorities for the reform culminating in a large conference. An independent concluding report on this conference by Matthews (2010) focused on public goods and on rural development – and in doing so indicated the challenges in defining and delivering environmental public goods, hinting that there was not a consensus that they should be delivered through Pillar 2.

It is important to note also that as the Commission was preparing its communication and the subsequent proposals for reform the market situation for agriculture had markedly changed. The credit market and financial crises which erupted in late 2007 were accompanied by a surge in energy prices and successive spikes in agricultural commodity prices. By common consent the century-old downward trend in international grain and oilseed prices had halted. Commodity prices have been significantly higher since 2007 than the previous decade and there are no predictions that the twentieth-century downward trend will resume in coming decades. This market turn-round improved the economic situation of arable farmers, but precipitated a crisis in the livestock sector, especially dairying. These developments stimulated a resurgence of concerns about food security and this played strongly into the CAP reform debate. Although higher market prices (admittedly moderated by higher energy and fertiliser costs) might be expected to diminish the case for farmer income support, farmers' organisations were skilled at projecting the argument that concerns about food insecurity strengthened the case for supporting agriculture. Combined with the emerging austerity programmes to deal with the public finance crisis, these arguments pushed the environment down the list of policy priorities.

The Communication on 'The CAP towards 2020' (Commission of the European Communities, 2010) identified the challenges facing the policy as: food security, environment and climate change and territorial balance. It talked explicitly about public goods and also about the need to make CAP support more equitable and balanced between Member States and farmers by reducing disparities. The three objectives of the new CAP were therefore to be: viable food production, sustainable management of natural resources and climate action and balanced territorial development.

The Commission's view of the principal aspects of the proposed reform were clearly signalled in the Communication. A key was the restructuring of the direct payments system to redistribute support, introduce more targeting towards farmers in areas with specific natural constraints and small farmers and to introduce 'a mandatory greening component of direct payments by supporting environmental measures across the whole EU territory'. It also signalled that the existing Water Framework Directive could be brought into cross-compliance. The importance of Rural Development was stressed and guiding themes for this pillar were stated as 'environment, climate change and innovation'.

The Communication set the three main themes which persisted through the debate, the negotiations and the reform decisions which emerged in the autumn of 2013. These themes were *redistribution*, more *targeting* of direct payments and a further significant *greening* of the CAP.

The big choice on the table was whether the greening should be done through Pillar 1 by attaching requirements to direct payments, or by expanding agri-environment and climate measures in Rural Development Programmes. The principal arguments for greening Pillar 1 were as follows. First, at the high political and financial level there were few Member States who wanted to expand Pillar 2. Even the UK never suggested a significant absolute expansion of resources for Pillar 2; they wanted it to grow relative to a big cut in Pillar 1. Several Member States already had trouble utilising the Pillar 2 funds at their disposal and were inclined to switch resources the other way – back to Pillar 1.[12] A key factor is the co-financing needs for Pillar 2. Every €1 m of funds switched from Pillar 1 to 2 involves additional national public expenditure to co-finance the Pillar 2 measures. This funding was unthinkable in the age of austerity. In addition, both administrators and farmers were very aware of the much greater bureaucracy involved in setting up and running RD measures. Farmers also, naturally, preferred the automatic, entitlement approach of annual payments in Pillar 1 to the uncertainty and additional effort required to apply for competitive funds though Pillar 2 measures. These negative views of expanding Pillar 2 predominated.

It is interesting that no organisations actively argued that greening Pillar 1 would be the most effective way to deliver environmental public goods. However, throughout the debates there was an abiding concern that generalised measures attached to annual direct payments, effectively an enlargement of the scope of cross compliance, were not the first-best way to deliver environmental public goods. An interesting attempt to square this circle had been suggested earlier by the European Landowners Organisation and BirdLife International (2010), who proposed that the greening was achieved by

---

[12] The new Member States were permitted to do this, and several have done so.

financing through Pillar 1 a wide-application, base-level environmental scheme which had the characteristics of Pillar 2 (i.e. a multi-annual, regionally specific, menu-driven approach). Even though the distinctions of purpose and types of measures in the two pillars was never clear-cut,[13] this mixing of 'the best features of both pillars' was considered a step too far.

The legislative proposals

Legislative proposals were published in October 2011 (Commission of the European Communities, 2011).[14] As expected, the main proposals were redistribution of support and more targeting of direct payments. A key part of the political narrative was that the CAP's call on a significant part of the EU budget was to be justified by the delivery of more environmental public goods. The explanatory memorandum of the proposals proclaimed:

> This reform accelerates the process of integration of environmental requirements. It introduces a strong greening component into the first pillar of the CAP for the first time thus ensuring that all EU farmers in receipt of support go beyond the requirements of cross compliance and deliver environmental and climate benefits as part of their everyday activities.

The greening component was proposed to take 30 percent of the funds for the direct payments.

The real negotiations took much time to gather momentum because there was some reluctance to get deep into details until the budget for the CAP had been decided. This did not emerge until the European Council of February 2013.[15] The decision was to cut the CAP budget by 13 percent for 2014–2020 (compared to the commitments for 2007–2013), with a somewhat higher cut for Pillar 2 than Pillar 1. Given that farmers had feared cuts up to 30 percent in the CAP budget, this was greeted by them with relief. Apparently, once the decision on the CAP budget had been made the steam

---

[13] The key differences between the two CAP pillars are that Pillar 1 contains (generally) obligatory, annual 100 percent EU-financed measures, and Pillar 2 voluntary, multi-annual, programmed and cofinanced measures. However, there is no clear distinction of purpose between the pillars; both contain measures for income support, environmental protection, risk management and improving productivity.

[14] Four regulations were proposed for Direct Payments (COM(2011) 625 final/2), Rural Development (COM(2011) 627 final/2), the Common Organisation of Markets (COM(2011) 626 final/2) and Financing, managing and monitoring (COM(2011) 628/2 final).

[15] Even then there was considerable uncertainty about the status of the outcome of this Council because it contained many statements on detailed aspects of the CAP, particularly payment capping, the switching of funds between pillars and even some aspects of greening.
The European Parliament was unhappy that decisions which should be made through the ordinary legislative process (i.e. co-decision between the agriculture Council and the Parliament) had been expropriated by the European Council. Resolving this dispute extended the negotiations.

went out of real meaningful greening. If greening was simply a rhetoric to defend the budget, then this had been largely achieved for another seven years!

The centrepiece of the CAP greening, spelled out in Articles 29 to 33, was to be achieved by allocating 30 percent of the funds for mandatory 'Payment for agricultural practices beneficial for the climate and the environment'. Three such practices were defined: crop diversification, maintenance of permanent pasture at farm level and ecological focus areas. Preamble 26 explained that these should be 'addressing as a priority both climate and environmental policy goals'. The practices should take the form of 'simple, generalized, non-contractual and annual actions that go beyond cross compliance'.

The three agricultural practices were seen by the Commission as just one component of the apparatus in the CAP for delivering environmental public goods. Cross-compliance was also to be adjusted to improve biodiversity and water protection, and these Pillar 1 measures would then be complemented by revising agri-environment schemes to build on the higher base level of environmental performance achieved through the Pillar 1 greening. In addition, there was great emphasis in the proposals on stimulating innovation and providing advisory services, information and skills enhancement to be devoted both to improving agricultural productivity and environmental land management. Also, the LEADER programme was seen, through its community-based, bottom-up approach, as a further way of bringing about a good balance of economic, environmental and social actions through rural development. Furthermore, the proposed restructuring of direct payments offered Member States three further ways in which they could, if they chose, direct more support to farms with potential for higher environmental service delivery.

First was the requirement that all Member States move away from historic-based payment to full regionalisation, i.e. flat-rate payments per hectare.[16] This offers the possibility to define regions and allocate funds to these regions embracing environmental considerations. Second, the proposals offer the option of a top-up for farmers in areas of natural constraints. The purpose suggested is 'to promote the sustainable development of agriculture in these areas'. Third, the Commission proposed to extend considerably the list of commodities for which coupled payments could be used to help 'specific regions facing particular situations where specific types of farming or specific agricultural sectors are particularly important for economic, environmental and/or social reasons'.

These three options illustrate one of the paradoxes of using the CAP to deliver environmental services and the conceptual model that is employed

---

[16] For the EU15 this had been an optional way of administering direct payments since 2004, but only Germany and England had made use of it. The simplified area payments system adopted by the new Member States was already paid on this regionalised, flat-rate basis.

to judge such policy. Each of these options could well be used to provide invaluable support which keeps in business farmers who are doing an excellent job of environmental management. However, a critical problem is that there is nothing in the measures themselves which ensures that the payments only go to the farmers who are doing this. Even for the farmers who are, the payments only enable continuation of existing good provision, there is no additionality. The audit results of such a policy can easily be imagined.

Summarising, on the face of it the Commission could reasonably claim that their proposals constituted a considerable step forward in integrating environmental objectives and actions into the CAP. Thirty percent of Pillar 1 payments devoted to environmental goals sounds radical; it is inconceivable that there could have been a proposal to move 30 percent of Pillar 1 funds to environmental measures in Pillar 2. Progress in integrating environmental concerns into agricultural practice on the ground was then up to stakeholder reactions, Council and Parliament decision-making, Member State implementation and finally farmer behaviour.

The negotiations

The tortuous negotiations occupied two years.[17] They had a different character to previous reform negotiations because of the additional dimension of the involvement of the European Parliament and its farmer-interest dominated Committee on Agriculture.[18] While the proposals on greening were highly controversial and provoked a good deal of debate, many Member States were more concerned with the redistribution and targeting of support.

Farming interests fought hard to reduce the restrictive effects of the Pillar 1 greening measures. They claimed that the measures imperiled food security and efficient farming, complicated CAP administration[19] and delivered very little for the environment. It was difficult for supporters of CAP greening to help support the case as the Commission's impact assessment offered no strong evidence of the likely environmental outcomes of the proposed measures. Environmental interests tried focusing their efforts on the measure which had greatest chance of delivering positive benefits, the Ecological Focus Areas. They also worked to improve the provisions for the protection of the most important biodiverse semi-natural permanent grass and to ensure that farmers could not be paid twice for

---

[17] See the postings of Matthews (2013a) for independent, in-depth and penetrating analysis of the reform as it took place. See Swinnen (2015) for a thorough analysis of the political economy of the 2013 reform.

[18] The influence of COMAGRI and the Parliament is now the subject of intense scrutiny by political economy experts and will be reported upon by a study commissioned by the Parliament Research Department, European Parliament (2013).

[19] Simplification of the CAP was a constantly aired aim of the reform, although amendments proposed by the Council and Parliament invariably complicated the Commission's proposals.

the same actions (in Pillar 1 greening and again for the same actions, e.g. buffer strips, in Pillar 2 environment schemes). The outcome of the negotiation was to exempt many more farmers from having to implement greening and to dilute the additional efforts required of farmers for their green payments, in both cases risking reduction in environmental service delivery. Table 15.2 lists the principal changes between the proposals and what was agreed.

At the time of writing, further important details of these greening practices were still to be agreed in the delegated acts and implementing regulations. Even then, the environmental impacts depend on how the measures are put into place by the Member States and then, critically, on the attitude and reaction of farmers.

Apart from the dilution of greening, the agreed reform included two other potential setbacks for the integration of the environment into the CAP. First is the reduction in core funds for Rural Development for the 2014–2020 period by 13.6 percent. This smaller budget was also slightly redistributed. The Commission's proposals had included the possibility to switch up to 10 percent of Pillar 1 funds to Rural Development, and 5 percent in the opposite direction. The agreement increased both amounts. Up to 15 percent could switch in either direction, and 12 Member States could if they wish switch up to 25 per ent back from Pillar 2 to 1. A second setback was that the proposals to introduce the provisions of the Water Framework Directive into cross-compliance were rejected.

During the negotiation there were strong fears that the principle that agri-environment and climate measures in Pillar 2 should require actions above and beyond the higher reference level in Pillar 1 established by greening practices was going to be dropped. In particular, some proposed that farmers could be paid in both pillars for the same environmental actions. Strong lobbying (see Hart, 2013) finally ensured that there should, in principle, be no such double-funding.

Two other aspects of the reform agreement concerned directions on how funds should be directed. The Multiannual Financial Framework included an agreement to 'mainstream' climate change obligations across a number of policy areas and for at least 20 percent of the EU budget to support climate change-related activities. This was certainly expected to apply to the CAP and Member States had to indicate how their implementation of greening and Rural Development Programmes satisfied this requirement. The new Rural Development Regulation no longer structures measures into axes with minimum spending requirements for each axis. However, it requires[20] that Member States should

---

[20] The purpose is spelled out in preamble 28, and detailed in Article 65(5a) of the Rural Development Regulation COM 2013/628 final/2.

Table 15.2. *Proposed and finally agreed greening of direct payments.*

Commission proposal	Final agreement
Greening is compulsory, non-compliance can mean forfeiting all direct payments	Effectively voluntary, non-compliance means loss of up to 30 percent of payments in 2015 rising to 37.5 percent after 2017
Three defined greening practices: crop diversification, maintenance of permanent grassland and Ecological Focus Area	The three practices or equivalent practices that yield equivalent or higher benefit for climate and environment
Organic farming is 'green by definition', no need to implement the three greening practices. Participants in the small farmer scheme are also exempt from greening	Greening equivalency added to green by definition: for farms in agri-environment schemes or in national or regional certification schemes
Crop diversification: for arable land > 3 ha, min. of three crops, smallest > 5 percent, largest < 70 percent	Two crops required for arable land between 10 and 30 ha. Three crops for arable land >30 ha, main crop max. 75 percent, two main crops max. 95 percent. Four cases defined where the thresholds do not apply, e.g. if 75 percent of farm is grass and arable area is <30 ha
Maintenance of permanent grassland: obligatory at holding level, max. of 5 percent of permanent grassland as of 2014 can be converted (by 2020)	Environmentally sensitive permanent grasslands are to be defined, and such grass shall not be converted or ploughed. The ratio of permanent grassland to agricultural area must not decrease by more than 5 percent compared to a reference ratio. This may be applied at national, regional or subregional level
Ecological Focus Area (EFA): at least 7 percent of eligible hectares excl. perm grass; defined as land left fallow, terraces, landscape features, buffer strips and certain afforested areas	EFA must be 5 percent of arable area > 15 ha. This figure may rise to 7 percent after review in 2017. EFA may be: land lying fallow, terraces, landscape features (including those not in eligible area), buffer strips (including those which are permanent pasture), supported agro-forestry, strips along forest edges, unfertilised short rotation coppice, certain afforested areas, catch crops, green cover, areas with nitrogen-fixing crops. Weighting factors may be used to convert some of these to equivalent ha of EFA. Collective implementation of up to half of the EFA may be permitted

maintain the level of efforts made during the 2007–2013 programming period and spend a minimum of 30 per cent of the total contribution from the EAFRD to each rural development programme for climate change mitigation and adaptation, biodiversity, resource efficiency and soil, water and land management, through the agri-environment-climate, organic farming and payments to areas facing natural or other specific constraints measures, forestry measures, payments for Natura 2000 areas and climate and environment related investment support contributing to environment and climate.

The Commission (2013) proclaimed the outcome of these negotiations as a 'new direction' with 'far-reaching changes' taking better account of society's expectations helping 'farming meet the challenges of soil and water quality biodiversity and climate change'. Can it now be concluded that biodiversity and ecosystem services are now fully integrated into the CAP?

Conclusions

Despite the Commission's proclamation, the questions remain on delivery. It is clearly the case that protection of biodiversity and delivery of ecosystem services have now been overtly recognised as part of the Common Agricultural Policy for over a quarter of a century. Their relative importance has undoubtedly grown over this period and with it the share of CAP financial resources. At the present time environmental objectives are explicitly part of: cross-compliance rules, the purpose of coupled payments and payments under Article 68. In Pillar 2 they are included in the objectives of agri-environment schemes, support for organic farming and less-favoured areas. They can also be part of the objectives of measures for farm advisory services, quality food production and rural diversification measures and LEADER.

The further greening of the CAP has been a headline reform objective for the policy for many years. The outcome was, on the face of it, a large step forward as 30 percent of the Pillar 1 direct payments until 2020 are defined as greening payments. This represents a public expenditure of around €85 billion over seven years and if likely expenditures on agri-environment and climate measures are added to this (with the corresponding national co-financing) there can easily be well over €100 billion of taxpayer contribution to the environmental stewardship of Europe's agricultural land over the seven years to 2020. This constitutes a fourfold increase in resources devoted to environmental management under the CAP compared to 2007–2013.

Yet the universal reaction to the reform from environmental organisations[21] has been disappointment and criticism. A general reaction has been that the proposals to deliver environmental services by greening Pillar 1 started with low ambition and were significantly weakened in the negotiation. Matthews (2013b) suggested that while the concept may survive, 'its practical environmental benefits will be negligible'. A Defra (2013) analysis estimated net environmental benefits of £1 billion from greening requirements, mostly from EFA. This is less than a quarter of the payments to farmers for the greening. If it transpires, it

---

[21] And from scientists, see Pe'er et al. (2014).

would represent abysmal value for public money. This pessimism suggests that the greening measures are not rigorous enough to guarantee significant, additional environmental benefits and there is no general spirit or determination among administrations and farmers to make this work. Indeed, a driving consideration, particularly for CAP administrators at Member State level, has been a preoccupation to avoid penalties and disallowance of payments.[22] The easier it is made for farmers to comply, the less the risk of such problems.

Turning to environmental delivery via Pillar 2 measures, there are concerns that some of the environmental capital built up in agri-environment schemes over the last decade or more may be lost if these measures are scaled back in the next period. Core funds for Pillar 2 have been reduced, and some Member States may choose to reduce them further by switching money back to Pillar 1. Fewer farmers may be inclined to join environment schemes partly because they feel they are doing enough through Pillar 1 greening and partly because the environment payments offered do not reflect higher opportunity costs.

This is not a happy situation. The reformed CAP for 2014–20 can be characterised as follows. It is smaller, its support redistributed and targeted a little more to social objectives (to smaller, younger, active farmers and those in disadvantaged regions), it will be greener in rhetoric but may be scarcely different in performance, it will be a less common policy, and considerably more complicated to administer. Few will be satisfied that this is a mature policy demonstrating the success of common European democratic political action.

If national administrations and farmers took the criticisms and concerns about feeble greening measures seriously, their joint actions in the years to come could prove the pessimists wrong. First evidence about this will be gleaned during 2014–2015 from the implementation choices made by Member States. Implementation of the new direct payments and greening does not come into effect until 2015. Contracts under the new agri-environment and climate measures are not likely to be in operation until 2016. It will then take one or two years of evidence gathering to test for changes in environmental outcomes based on the array of indicators collected (with all the usual problems of attributing cause and effect).

If the worst fears materialise that the €100 billion experiment on new greening produces only a fraction of this in environmental benefit, what

---

[22] When Commission checks and audit reveal that payments have been made to farmers in ways which do not fully comply with regulations the Member State is not reimbursed for such expenditures and therefore has to recover it from the farmers retrospectively or bear it as their own public expenditure. They can also be fined. Avoidance of such risks is now a high-level objective of CAP reform implementation. For example, it is one of the reasons the UK government chose not to implement greening through a more rigorous national certification scheme.

conclusions will be drawn? That the greening payments should be reduced or withdrawn? That the greening conditions should be significantly tightened? That environmental payments should be results-based rather than prescriptive? That the strategy of greening Pillar 1 was a mistake? That the institutional decision structure in the EU is inappropriate? That environmental contracts with individual farmers simply do not work so the process must operate through collectives of farmers at higher, landscape or river catchment scale? That a common European policy based on the CAP is the wrong instrument through which to operate? These questions set an agenda for the 2020–2027 CAP reform debate.

This chapter started by explaining the intrinsic difficulty of asking food producers in a tight market to devote more of their energies and land to environmental management. In the current reform of the CAP the Commission made a genuine attempt to suggest how this might be done mostly through Pillar 1 greening while trying to keep the proposals simple. If this doesn't work and yet the alternative of shifting resources to Pillar 2 is deemed too complicated and politically infeasible, then this seems to be closing off incentive payments through the CAP as the right policy framework. Discussions of the appropriate way to improve biodiversity protection and ecosystem service delivery over agricultural land will continue. Unfortunately in the interim there may be continued degradation of natural capital in Europe's farmed areas.

References

Buckwell, A.E. (2007). Land management in the 21st century. Speech given at the Centenary Conference of the Country Land and Business Association (CLA) Westminster Hall, 10 May 2007, available from the CLA, 16 Belgrave Square, London SW1X 8PQ.

Buckwell, A.E. (2008). *Analysis of the Health Check Proposals: the Reform of the Mechanism for Direct Support*. Policy Department B Structural and Cohesion Policies, IP/B/AGRI/IC/2008_056. Brussels: European Parliament.

Coase, R.H. (1960). The problem of social cost. *Journal of Law and Economics*, **3**, 1–44.

Commission of the European Communities. (2010). *The CAP Towards 2020: meeting the food, natural resources and territorial challenges of the future*. COM(20210) 672/5. Brussels: Communication from the Commission to the European Parliament, the Council, The European Social and Economic Committee and the Committee of the Regions.

Commission of the European Communities. (2011). *Proposal for a Regulation of the European Parliament and Council establishing Rules for Direct Payments to Farmers under Support Schemes within the Framework of the Common Agricultural Policy*. COM(2011) 625 final/2. Brussels: Commission of the European Communities.

Commission of the European Communities. (2013). Political agreement on a new direction for common agricultural policy. Press release. Brussels: Commission of the European Communities.

Cooper, T., Hart, K. & Baldock, D. (2009). *Provision of Public Goods through Agriculture in the EU*. London: Institute of European Environmental Policy.

Defra. (2013). *Implementation of CAP Reform in England, Evidence Paper*. London: Department for Environment, Food and Rural Affairs.

European Court of Auditors. (2008). *Is Cross Compliance an Effective Policy?* Special report No 8//2008. Luxembourg: European Court of Auditors.

European Court of Auditors. (2011). *Is Agri Environment Support Well Designed and Managed?* Special report No 7. Luxembourg: European Court of Auditors.

European Court of Auditors. (2013). *Can the Commission and Member States Show that the EU Budget Allocated to the Rural Development Policy is Well Spent?* Special Report No 12. Luxembourg: European Court of Auditors.

European Court of Auditors (2013a). *Common Agricultural Policy: is the specific support provided under Article 68 of Council regulation (EC) 73/2009 well designed and implemented?* Special Report No. 10. Luxembourg: European Court of Auditors.

European Economy. (1997). *Towards a Common Agricultural and Rural Policy for Europe*. Reports and Studies No 5. Brussels: European Commission, Directorate General for Economic and Financial Affairs.

European Parliament. (2013). *The First CAP Reform under the Ordinary Legislative Procedure: a political economy perspective*. Terms of reference for a study, IP/B/AGRI/IC/2013-156. Brussels: Directorate General for Internal Policies, Directorate B – Structural and Cohesion Policy.

European Landowners Organization and BirdLife International. (2010). *Proposals for the Future CAP: a joint position from the ELO and BLI*. Brussels.

Hart, K. (2013). *Principles of Double Funding*. London: Institute for European Environmental Policy.

Hart, K. & Baldock, D. (2010). *Impact of CAP reforms on the Environmental Performance of Agriculture: a report to the OECD from IEEP*. London: Institute of European Environmental Policy.

Harvey, D.R. (1997). The GATT, the WTO and the CAP. In: *The Common Agricultural Policy*, edited by C. Ritson & D.R. Harvey, 2nd edition. Wallingford: CAB International.

H.M. Treasury & Defra. (2005). *A Vision for the Common Agricultural Policy*. Norwich: HMSO.

Lowe, P. & Whitby, M.C. (1997). The CAP and the European environment. In: *The Common Agricultural Policy*, edited by C. Ritson & D.R. Harvey, 2nd edition. Wallingford: CAB International.

Matthews, A. (2010). *The CAP Post 2013 Conference on the Public Debate, Closing Report*. Brussels: European Commission, DG Agriculture and Rural Development.

Matthews, A. (2013a). CAP Reform.eu, Europe's common agricultural policy is broken – let's fix it! See the postings throughout 2012 and 2013 for a blow-by-blow account of the main steps in the 2014–2020 reform of the CAP found in http://capreform.eu/archives-2/, particularly his Short Bibliography on CAP Greening, 4 May 2013.

Matthews, A. (2013b). *Greening CAP Payments: a missed opportunity?* Dublin: Institute for International and European Affairs.

Matthews, A. (2017). Why Further Reform? Appendix 1 of RISE report CAP: Thinking outside the box. Further modernization of the CAP: why, what and how? RISE Foundation, Brussels. www.risefoundation.eu/images/files/2017/2017_RISE_CAP_APPENDIX_1.pdf

NFU. (2010). www.nfuonline.com/business/cap/cap-paper-completely-misses-the-mark/

Poláková, J., Tucker, G., Hart, K., Dwyer, J. & Rayment, M. (2011). Addressing biodiversity and habitat preservation through measures applied under the Common Agricultural Policy. Report Prepared for DG Agriculture and Rural Development, Contract No. 30-CE-0388497/00-44. London: Institute for European Environmental Policy.

Pe'er, G., Dicks, L.V., Visconti, P., et al. (2014). EU agricultural reform fails on biodiversity. *Science*, **344**, 1090–1092.

RISE Task Force. (2009). *Public Goods from Private Land*. Brussels: RISE Foundation, Rural Investment Support for Europe.

Swinbank, A. & Daugbjerg, C. (2006). The 2003 CAP reform: accommodating WTO pressures. *Comparative European Politics*, **4**, 47–64.

Swinnen, J. (2015). The Political Economy of the 2014–2020 reform of the Common Agricultural Policy: an imperfect storm, CEPS, Brussels. www.ceps.eu/publications/political-economy-2014-2020-common-agricultural-policy-imperfect-storm

CHAPTER SIXTEEN

Using environmental metrics to promote sustainability and resilience in agriculture

JONATHAN R. B. FISHER
The Nature Conservancy
and
PETER KAREIVA
University of California, Los Angeles

Introduction

Under current projections of human population growth and improving standards of living, worldwide demand for food is expected to more than double by 2050 (Hunter et al., 2017). Already, cultivated cropland covers about a quarter of the Earth's land surface (Millennium Ecosystem Assessment, 2005). When land used for the grazing of livestock is added in, roughly 40 percent of the world's terrestrial area is dedicated to producing food for people (FAO, 2013). This makes agriculture the single largest land use on the planet (Foley et al., 2005).

Over the last 15 years agriculture has been intensifying (higher production per unit of land area). Total agricultural land area worldwide has not increased since 1998, with agricultural expansion in South America, Southeast Asia and much of Africa being offset by a decline in agricultural lands elsewhere (Figures 16.1 and 16.2). At the same time, food produced per capita has increased (Figure 16.3). However, increasing food prices have recently begun to accelerate land conversion for new farms and pasture (FAO, 2013, Figure 16.2). If current trends continue, roughly 1 billion hectares of new land will have to be cleared by 2050 to meet demand (Tilman et al., 2011). Faced with these trends in human population growth and the potential for an acceleration of land conversion, it is clear that one key to maintaining biodiversity and ecosystem function is guiding the future path that agricultural development takes.

The pressing question is: how can the world's lands and waters meet the needs of both food production and biodiversity protection? A truly sustainable agricultural system needs to be capable of meeting human food and fibre needs over long time periods, have minimal adverse effects on the environment and be practical (and profitable) for farmers to implement. It also must be resilient to ordinary stresses (such as dry weather), extreme weather (heatwaves or unusual drought), novel pests, natural disasters such as floods and

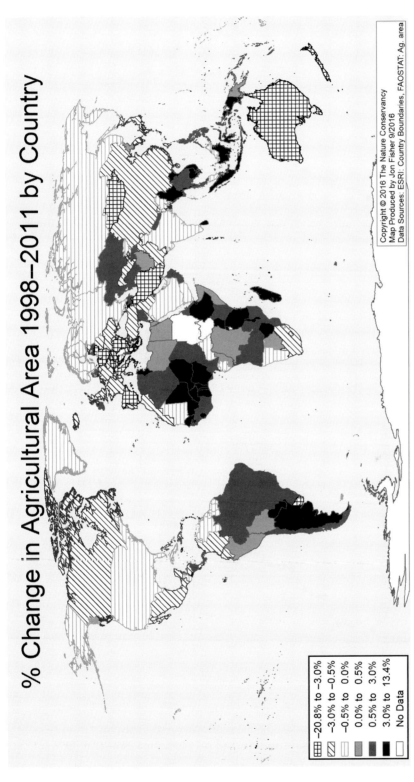

Figure 16.1 Percentage change in each country's agricultural land area (annual row crops, pasture/rangeland and permanent crops) between 1998 and 2011. Countries with darker grey colour indicate greater agricultural expansion, countries with more hatching indicate greater agricultural contraction. Data from FAO (2013).

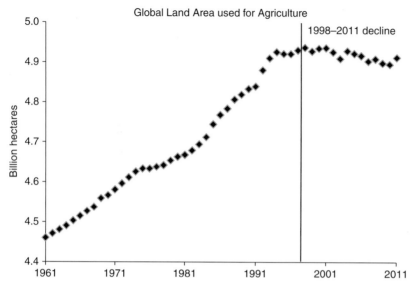

Figure 16.2 Global land area used for agriculture (annual row crops, pasture/rangeland and permanent crops) from 1961 to 2011, with the period of decline from 1998 to 2011 indicated. Note that the y-axis does not begin at 0 to show the variation more clearly. Data from FAO (2013).

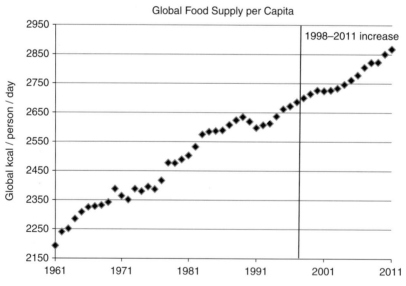

Figure 16.3 Global food supply per capita (in kilocalories per person per year) from 1961 to 2011, with the period of decline from 1998 to 2011 indicated. Note that the y-axis does not begin at 0 to show the variation more clearly. Data from FAO (2014).

climate change (Pretty, 2008). This requires considering the entire 'agroecosystem' (not only the crop, but all of the other living and abiotic components and their relationships) rather than a traditional definition of a farm (Conway, 1985; Kremen et al., 2012).

One path forward that has been advanced as a solution to the challenge of producing more food without degrading the environment is what has been labelled 'sustainable intensification'. Sustainable intensification is the process of 'increas[ing] food production from existing farmland in ways that place far less pressure on the environment and that do not undermine our capacity to continue producing food in the future' (Garnett et al., 2013). Given the projected increase in demand for food, it is critical to find ways to sustainably intensify production in order to avoid the conversion of native habitat to new agricultural lands. It is also important to consider that sustainability initiatives that offer a profit incentive for farmers (via increased yields and/or improved resilience) are much more likely to be adopted than practices with environmental benefits but no direct economic benefits to the farmer. There is evidence that yield and sustainability can be increased simultaneously; a review of 198 sustainable agriculture projects in the developing world reported a mean relative yield increase of 79 percent relative to yields prior to the sustainability improvements (Pretty et al., 2006).

However, while sustainable intensification is an admirable aspiration, intensification is sometimes achieved at the cost of decreased diversity (both crop genetic diversity and biodiversity more broadly in the surrounding ecosystems) and thus carries with it a risk of decreased system resilience (Matson et al., 1997; Thomas, 1999; Tscharntke et al., 2005). Complicating matters is the fact that ideas about what type of agriculture is good for the environment can be based more on opinions than evidence. Beliefs about food and food production systems can be strongly rooted in culture and political ideology (Harmon, 2014; McWilliams, 2009). For this reason, there is a need for objective, practical metrics that could be applied to any given agricultural plot or landscape in order to assess the degree to which different combinations of practices are indeed sustainable.

For metrics to be more than simply academic they need to be responsive on a timescale that is actionable. In other words, metrics have to respond fast enough that if a new agricultural practice is undermining sustainability (or simply failing to improve it), the negative impact can be detected and used as feedback to farmers, policymakers and businesses. Conversely, if a new practice is improving sustainability, the farmer or business implementing that new practice deserves to get credit for it and others should be able to find out about it and replicate their success. Ideally, credible sustainability metrics can be a lever for altering the behaviour of major agribusinesses, by holding the major links in food production systems (retailers such as Walmart, food distributors such as Unilever, agricultural producers/ traders such as Cargill and food processors such as General Mills) accountable. Consumers also have a role to play through the pressures they put on industry via market forces and choices. However, consumers need credible, transparent labels – and here again sustainability metrics can play a role in evaluating whether a 'green label' is well-deserved. Later in this chapter we elaborate on the potential for environmental metrics in shaping both agribusiness and consumer choice.

The challenge of assessing sustainability and resilience in agroecosystems

In considering how to measure sustainability, there are several choices. One could simply measure *resource inputs/costs* such as the money being spent on soil management or the amount of water withdrawn for irrigation. Alternatively, the measures could focus on *conservation actions* like the number of acres on which best management practices (BMPs) are implemented. Finally, actual outcomes could be measured like volume of water consumed or fish species richness in a stream. There are several organisations promoting measurement schemes for sustainable agriculture. In the USA the two largest are Field to Market and The Sustainability Consortium (TSC). Globally, the Sustainable Agriculture Network and Rainforest Alliance are the most widely used. While all of these organisations have likely helped to foster improvements in agriculture, they generally focus on agricultural practices rather than achievement of anything that could be rigorously called sustainable production. For example, one question TSC asks is 'How is your organisation engaging this product's supply chain to address energy consumption during fertiliser manufacturing?' and the top score is for buying fertilisers from companies that have an energy management plan (The Sustainability Consortium, 2013a). Such questions do not provide any information about conservation actions or resource inputs, much less about whether the production system is truly sustainable in the sense of durable incomes and productivity without environmental degradation (note that since this chapter was written they have started shifting towards more quantitative outcome metrics). Other groups ask questions that are more directly related to farming practices, such as 'How many pounds of nitrogen were added per ton of crop harvested?' (Stewardship Index for Specialty Crops, 2013); but even here the long-term consequences of specific nitrogen regimes, and the impact of other biophysical variables, go unmonitored on farms.

At its most essential level, sustainable agriculture should ensure that future generations can obtain food and the many other ecosystem services that nature provides. Measures of soil and water depletion and degradation and changes to habitat and biodiversity all capture some elements of sustainability, as do metrics of yield and profit. The more challenging questions concern resilience to shocks and the possibility of crossing some sort of threshold from which ecosystem recovery (or crop production recovery) is difficult. A good example of potentially reduced resilience is the loss of pollinator biodiversity and increasing dependence on the domesticated species *Apis mellifera* (the European honeybee). At least 87 crop types (70 percent of the major crop species, representing roughly one-third of global food supply) are dependent on pollinators (Klein et al., 2007). Declining diversity of pollinators and reliance almost exclusively on *Apis mellifera* could potentially lead to lower crop production (Allen-Wardell et al., 1998; Hoehn et al., 2008). Such yield losses are not yet detectable at a global scale (Aisen et al., 2008) and crop yields are still on the rise thanks to synthetic fertilisers, irrigation, chemical pesticides and better management in many places. However, the United Nations Environment Program (UNEP) cautions that widespread reports of honeybee colony mortality and an increase in the fraction of our food crops that require pollinators put the stability of our food production systems in peril due to pollinator scarcity (UNEP, 2010). Maintaining pollination services requires the conservation of sufficient resources for wild pollinators within agricultural landscapes, including both suitable habitats and sufficient floral resources for pollen and nectar (Kremen et al., 2007; Garibaldi et al., 2016; Williams et al., Chapter 6).

Soil biodiversity may also play an important role in providing ecosystem services to agricultural systems and enhancing their resilience. Garbeva et al. (2006) found the highest suppression of a soil-borne potato pathogen in plots with the highest soil microbial diversity and it is possible that such diversity-based suppression might be present in other systems as well. Similarly, mycorrhizal diversity contributes positively to nutrient and water-use efficiency and plant productivity (Brussaard et al., 2007; Maherali & Klironomos, 2007; van der Heijden et al., 1998). The more diverse the mycorrhizal community, the better able plants are to extract nutrients and water, and thus to tolerate adverse conditions. If seeking to optimise yield leads to a decline in the diversity of mycorrhizae and other soil biota, it is likely that the resulting system will have decreased resilience. In fact, there are a range of ecosystem services in agroecosystems that have been shown to increase under diversified farming systems (Kremen et al., 2012). Soil science is re-evaluating several long-held beliefs, and while there is not yet a clear and consistent relationship between soil diversity and other outcomes, the importance of microbial activity as drivers of key soil properties is increasingly evident (Lehmann & Kleber, 2015).

Other examples of possible thresholds and collapse of systems include overgrazing to the point of desertification, and excessive nutrient loading that causes algal blooms and anaerobic conditions or dead zones. In the Cerrado of Brazil, where cattle production is on the rise, 'sudden death' outbreaks due to poisonous plants associated with overgrazing may represent a threshold whereby intensified production yields a collapse of the entire production system (Merrill & Schuster, 1978).

In the face of relentlessly increasing greenhouse gas emissions, increased droughts and extreme weather events, sustainability requires resilience to a rapidly changing and uncertain climate. In a constant environment, it is challenging enough to identify appropriate measures of soil condition, water use and landscape change that indicate sustainability. In a changing environment, there are no fixed standards that can be counted on to guarantee sustainability over the long term, but there is a clear need for learning. Human communities and societies vary widely in their ability or willingness to adapt to climate change (Palmer & Smith, 2014). In the context of agriculture, adaptability to climate change will be essential, and this adaptability is likely to depend on networks of farmers that experiment with new cropping systems and practices (MacMillan & Benton, 2014).

It is doubtful that any of the sustainability metrics discussed in this paper could identify key thresholds and a heightened risk of ecosystem collapse. However, the metrics should be able to reveal accelerated change, which could be a harbinger of an approaching threshold (Scheffer et al., 2012). By implementing standardised sustainability metrics with extensive and meaningful reporting on a regular basis, it should be possible to test which of the proposed indicators of sustainability are linked to higher resilience to pest and pathogen outbreaks and a reduced likelihood of suffering productivity collapses. The ideal metrics should allow the comparison of different intensification strategies such as the planting of one genetically modified clone (which might reduce resilience due to a loss of genetic variety) versus precision agriculture which uses highly targeted micro-applications of fertiliser (which could improve resilience by delivering nutrients in response to the need of the plants as soil and climate conditions change, rather than using a fixed approach). The increased availability of remote sensing data coupled with computerised inventory tracking systems and advances in ecosystem modelling suggest we have an unprecedented opportunity to manage for improved sustainability and test hypotheses about what system attributes confer resilience.

The importance of outcome metrics

The problem with focusing on actions or practices is that they may not achieve the desired outcomes in spite of the best intentions. For example, Lemke et al. (2011) found that implementation of wider riparian buffers and conservation tillage had no impact on water quality in the watershed where these best practices were implemented. In the watershed where this study took place, the presence of tile drainage allowed nitrate dissolved in water to flow directly into the stream, bypassing the improvements on the soil surface.

Even promoting efficiency, which seems like an obvious component of sustainability, can backfire. With more efficient irrigation, more water is typically delivered to the root zone of the crops and transpired (consumed), which means less of it returns to streams and groundwater. So even though less water is 'wasted' (applied but not used by the crop), water withdrawals often remain nearly the same after efficiency improvements and as a result water *consumption* may go up, meaning less water is available for other downstream users (Ward & Pulido-Velazquez, 2008; Samani et al., 2012; Ward, 2014). For example, a data-rich economic model applied to 800,000 ha of irrigated mixed cropland in Southern Idaho predicts that improving irrigation efficiency from 60 percent to 80 percent would result in a reduction in water applied to the field by 15 percent, but still lead to a 3 percent increase in total water consumed (Contor & Taylor, 2013).

Proper water accounting is essential to understanding the role of both technology and policy in achieving desired outcomes (Foster & Perry, 2010). Richter et al. (2017) lay out a framework of how to actually reduce agricultural water consumption and highlight the importance of being able to transfer water savings to other users as a critical element along with water budgeting and changes in crop water use. The failure of increases in efficiency to reduce consumption (the 'Jevons paradox') has also been observed in relation to energy use (Polimeni et al., 2008). The key point is not that efficiency is an unwise objective, but rather that outcomes are more important than actions. Hence sustainability metrics need to focus much more on ecological outcomes (in this case, groundwater levels and stream flows) than on practices and actions than has been the case up until now (see also Buckwell, Chapter 15).

Can remote sensing become a foundation for sustainability metrics?

In agriculture, there are several scales at which sustainability metrics can be useful. At the broadest scale (i.e. global or national levels), sustainability metrics will generally be limited to highlighting areas of concern. Ecologically, the landscape or watershed level is perhaps the most relevant scale; outcomes such as water quality, landscape characteristics and biodiversity can be measured at this scale and compared with aggregated metrics of agricultural practices. Finally, certain outcomes would be best measured at the field or plot scale (e.g. soil organic carbon), either on the ground or with high-resolution imagery (e.g. up to 77 percent accuracy in detecting soil organic carbon levels on bare soil has been achieved (Chen et al., 2000)). The field scale is also probably best for engaging farmers in learning networks that test ideas about how to adapt to climate change and climate shocks (MacMillan & Benton, 2014). Fisher et al. (2014) provide a framework for sustainability

Table 16.1. *Proposed metrics for sustainable agriculture, adapted from Fisher et al. (2014).*

Category	Metric	How to measure
Soil	Soil erosion	Modelled based on soil/site, agricultural management practices and climate
	Soil organic carbon	Field samples; for large areas remote sensing may be useful in reducing samples needed
Water	Water consumption	Modelled evapotranspiration using climate data (adjusted for crop type and water availability)
	Water quality	N & P concentration (measured in stream, or modelled using land cover for landscapes)
Landscape ecology	Habitat conversion	Remotely sensed land use (percent of study area covered by natural habitat)
	Habitat composition	Remotely sensed land use (# land cover classes, indicating habitat diversity)
	Habitat connectivity	Remotely sensed land use (calculated connectivity score)
Biodiversity	Species richness	Field samples (richness relative to reference natural landscape)
	Species abundance	Field samples (abundance relative to reference natural landscape, i.e. the most undisturbed nearby similar habitat)
Agronomy	Yield/area	Harvested mass of crop per unit area over a 5-year rolling average, perhaps using geometric mean to penalise variation in yield

measures that is outcome-based and can be applied from the field to the global scale, along with potential data sets available to measure each characteristic at the various scales (Table 16.1).

While there are many other potential metrics available, these were chosen for the combination of impact/importance, relevance at multiple scales and practicality. For example, greenhouse gas emissions are certainly important at a global scale, but they cannot practically be independently measured (instead requiring information from companies and farmers), and are less useful when considering local impacts. This does not mean they should not be measured, but it requires a different approach. Similarly, Fisher et al. (2014) recommend focusing on birds and amphibians in part because of the widespread availability of data on these taxa, but other taxa (e.g. aquatic macroinvertebrates) may be more appropriate for a specific area or as local data availability varies. The intent is to provide a simple framework that can be applied almost everywhere, rather than to provide a complete framework.

Finally, we have added a simple agronomic metric (geometric mean of yield per area, measured over 5 years to capture some information on resilience) to the original suite of environmental performance metrics. We recognise that properly integrating economic and social metrics with environmental metrics is not a trivial endeavour, but would argue a minimum first step is some measure of yield that also penalises variation from year to year (because even a single low-yield year could put farmers and food security at risk in spite of a generally high average yield). Further research is needed where ecologists, agronomists and economists collaborate to determine the most appropriate metrics.

The metrics above can all be normalised to a 0 to 1 scale, where a score of 1 can be considered fully sustainable. Details are available in Fisher et al. (2014), but for example a soil erosion score of 1 would mean no soil loss (or even soil accretion) and a water quality score of 1 would mean nutrient run-off is low enough to not significantly impact streams (e.g. below the Total Maximum Daily Load in the USA). Additional contextual data will typically be necessary to identify what 'fully sustainable' would entail. It is more appropriate to quantify relative sustainability, rather than simply pronounce any system as either sustainable or not.

An obstacle to outcome-based measurement schemes for agriculture is a lack of data (or perceived lack of data). Two concerns often cited are physical or budget difficulties in obtaining measurements and an unwillingness of farmers to collect or share such data. Some characteristics like soil organic carbon and water quality generally rely on in-person sampling, which tends to be time-intensive and expensive. Farmers often express concern about being 'graded' or compared to their neighbours if they share data from their farm. Addressing these concerns will require a dialogue to discover what safeguards or conditions on use of the data (or other incentives) could encourage more farmers to collect and share data. Another opportunity is for large corporations with significant buying power to start requiring their suppliers to share data, or at least to reward those who do share data with longer contracts.

Given the expense of collecting data on the ground, and the large areas devoted to agriculture, remote sensing will play an increasingly critical role in tracking sustainability. For example, there are many ways to measure water consumption, but typically evapotranspiration (ET) is used (Sinclair et al., 1984). ET is a good proxy for total consumptive water use, as < 1 percent of water consumed is used to create plant biomass (Condon et al., 2004). However, directly measuring ET (e.g. via a soil lysimeter) is complicated, so modelling it with remotely sensed climate data (air temperature, wind speed and relative humidity) is generally more practical (Allen et al., 1998; Rana & Katerji, 2000; Farahani et al., 2007), at least until networked soil probes advocated by companies like Climate Corp become more common. Remote sensing methods are also being developed to measure soil organic carbon in conjunction with field samples (Gomez et al., 2008), as well as water quality parameters like turbidity (Olmanson et al., 2013). As the resolution of satellite imagery improves, and the use of unmanned aerial vehicles (UAVs or drones) for remote sensing increases, the accuracy and utility of remote sensing should continue to increase. Already UAVs are being evaluated for in-stream monitoring of water quality (Ayana, 2015) and other novel applications are likely in development. It may be that in 20 years most tracking of environmental outcomes for agriculture can be done via remote sensing.

Measurement of outcomes could allow targeting of sustainability practices where they will do the most good

An agricultural innovation that is receiving a great deal of attention and investment is the use of precision agriculture, which promotes the use of varying practices within a field such as altering irrigation and fertilisation according to soil and terrain variability (Cassman, 1999; Gebbers & Adamchuk, 2010).

The idea of spatially varied agricultural practices that is the foundation of precision agriculture could also be a foundation for sustainable agriculture more broadly. Advocates of sustainability often tend to promote the same practices not only uniformly within a field, but uniformly across a landscape. Recently some promoters of sustainable agriculture have begun to borrow from the precision agriculture concept and asked where one can get the greatest environmental benefits from site-specific sustainability practices, as opposed to insisting on implementing the same sustainability practices everywhere. For example, the Wisconsin Buffer Initiative conducted an analysis to determine which fields had the potential to make the biggest improvements in stream and lake quality (specifically reduced phosphorus and sediment), as well as aquatic biological community health, at the lowest cost (University of Wisconsin, 2005). The focus was on low-cost practices that result in improved water quality as well as healthier fish and invertebrate communities (all of which are being monitored); picking the best practice for each site rather than taking a uniform approach. This approach is being tested with a paired watershed study (comparing water quality in a treatment watershed to a control watershed), and so far has resulted in 50 percent less phosphorus during storms (when most run-off enters the stream; Carvin et al., 2018). This is one of the first cases where changing agricultural practices has been demonstrated to improve water quality at a landscape scale. This outcome is encouraging as it was achieved with voluntary participation, unlike the mandatory framework that has effectively improved agricultural water quality in the Everglades (Daroub et al., 2011). The general idea of spatially targeting sustainability practices where they will do the most good makes ecological sense and could help government incentive programmes achieve better outcomes for less money (PCAST, 2011).

Can corporate sustainability reporting be a force for improved agricultural practices?

Increasingly, agroecosystems are shaped by the decisions of major global agribusiness operations. For that reason, it is worth asking whether these businesses themselves could be levers for promoting sustainable agriculture. There is a revolution going on with global corporations – as a result of stakeholder pressure, reputational risk and competition for talent, corporations around the world are taking seriously their social and environmental impacts. The Government & Accountability Institute reports the number of S&P 500 companies releasing sustainability reports more than doubled from 2010 to 2011, from 20 percent to 53 percent (Clark & Master, 2012). CorporateRegister.com (CR), the largest directory of non-financial reporting, is now adding roughly 1000 new reporting institutions every year. Interest from mainstream investors in Environmental, Social and Governance (ESG) integration is also growing so that an increasing number of investors are considering environmental and sustainability information when making investment decisions. A 2012 survey of 4000 business leaders, including 2600 executives, found that 60 percent felt that pursuing sustainability is necessary to be competitive (Kiron et al., 2013). The top benefits these business leaders saw from pursuing sustainability were better brand reputation and improved innovation (independent of the actual environmental benefits, which may be primarily positive externalities from the business standpoint). Another survey of senior business executives found that 83 percent identified spending on sustainability (defined here as a company's effort to drive profitable growth while achieving a positive economic, social and environmental impact) as an investment rather than a cost, and 92 percent said that sustainability is either critical or very important (Accenture, 2012). Admittedly, many business leaders do not think of sustainability in the same way ecologists might, but several multinational corporations are working closely with scientists in order to gain a deeper and more scientific understanding of sustainability and the potential private benefits it may include. Examples include the Dow Chemical Company with The Nature Conservancy (Kroeger et al., 2014; Reddy et al., 2015) and Unilever with the Natural Capital Project (Chaplin-Kramer et al., 2017). Manufacturers and retailers are responding by marketing an increasing number of products as 'green'; TerraChoice found a 73 percent increase in product offerings claiming to be green from 2009 and 2010, following 79 percent growth from 2008 to 2009 (TerraChoice, 2010).

While most sustainability reporting and actions focus on energy and greenhouse gas emissions, there is also a growing trend toward whole-system approaches that include dimensions such as land conversion and water quality. Notably, the athletic wear company PUMA issues an annual environmental profit and loss report that documents its material debt to the planet and tracks impacts such as land-use conversion stratified by habitat type (tropical forest, grassland, temperate forest, etc.) (PUMA, 2011). Remarkably, an increasing percentage of the average corporation's value can be attributed to intangible assets (reputation, management, innovation, employee quality, retention of employees, etc. and other non-physical, non-monetary assets). The Ocean Tomo Intangible Asset Market Value study indicates this intangible value has increased to 80 percent in 2010 from a mere 17 percent in 1975 (Ocean Tomo, 2010). This reflects the fact that brand, talent and reputation (along with intellectual capital and innovation) are perhaps more important than physical and financial assets in today's world. The relevance of this to conservation is that reputation provides a pressure point with which to influence corporations to take conservation seriously.

In the agricultural sector, several of the biggest players are already committed to sustainability reporting. The Sustainability Consortium (which was founded in 2009 with support from the Walmart Foundation) has over 100 members, including BASF, Cargill, Dow, Monsanto, Unilever and more (The Sustainability Consortium, 2013b). Field to Market is another initiative that has attracted considerable interest from large agricultural companies; while this programme is more farmer-focused than some other initiatives, its focus on quantitative measures for commodity crops in the USA is noteworthy (Field to Market, 2014). The challenge for scientists is to link the corporate desire to improve sustainability to metrics that are ecologically meaningful and that aptly capture impacts on the ground and in the water, rather than simply highlighting sustainability successes for marketing purposes. The key to making this ecologically credible is obtaining spatially explicit information on soil, water, habitat and farming practices, as well as traceability of food products to their source. While traceability might seem impractical, in fact it is the exact type of information one needs to exercise effective inventory control in addition to being critical to assess environmental impact. We hypothesise that improved traceability can actually help agribusinesses manage their inventories and even save money under the right conditions (Attaran, 2012; see discussion below).

Food labels and sustainability

There is some evidence that agricultural practices can be driven by consumer choices as well as corporate sustainability concerns (Ottman et al., 2006). For example, US sales of organic food and beverages have risen from $1 billion in 1990 to $26.7 billion in 2010 (Organic Trade Association, 2011), representing a startling increase in market share (USDA ERS, 2014). Globally the land area used for organic farming more than doubled from 14.9 million hectares (0.4 percent of all agricultural area) in 2000 to 37.5 million hectares (0.9 percent of all agricultural area) in 2012 (FiBL, 2014). While this is still relatively small, considering the rate of change and the fact that it takes several years for a farm to be certified as organic it is an encouraging start to demonstrating consumer interest in sustainability.

The success of the dolphin-safe label in persuading customers to buy labelled tuna (in conjunction with strong legislative support for the definition of the label) demonstrates the potential for green labels to shape consumer decisions in a way that is relevant to conservation (Ramach, 1996; Teisl et al., 2002). A recent survey of American consumers found that 35 percent said they would pay more for 'environmentally friendly' products (Mintel, 2010). The proposed metrics above might be too complex for consumers, but there is evidence that consumers are willing to pay more for better performance (Basu & Hicks, 2008), meaning that information beyond a simple binary label (e.g. quantitative data on water quality rather than 'organic') could lead to an increase in willingness to pay. The new requirements in France for certain products to have a label that includes carbon, water use and biodiversity impacts (under Grenelle 2) provides an opportunity to study how these types of more complex green labels affect consumer choices.

In addition to changing purchases by consumers, label standards can also help drive manufacturers to make improvements (Caswell & Padberg, 1992). For example, several businesses are shifting their seafood procurement in response to rating systems like the Monterey Bay Aquarium's Seafood Watch and the Marine Stewardship Council's certification (Aramark, 2008; Compass Group North America, 2009; Whole Foods, 2014).

But while there are several eco-label success stories (Thøgersen, 2000; Kemmerly & Macfarlane, 2009), there is no guarantee that sustainable labelling will be an effective force for change. Information overload and time pressure while grocery shopping limit the willingness of consumers to read labels (Caswell & Padberg, 1992) and sustainability metrics on their own may not be sufficient to sway consumers (Hallstein & Villas-Boas, 2009). Even when consumers want to be sustainable, obstacles include the perception that sustainable products are difficult to find and that the claims on the labels may not be justified (Tanner & Kast, 2003; Vermeir & Verbeke, 2006). Surprisingly, income or monetary barriers are not significantly related to green purchasing behaviour (Huffman et al., 2003; Tanner & Wölfing Kast, 2003), but product quality apart from sustainability remains a primary factor – it appears that consumers must perceive a product to be of high quality in order for sustainability to command a premium (Loureiro & Hine, 2002; McCluskey & Loureiro, 2003). In addition to the actual *content* of a label (the metrics chosen, design and actual performance of a product), credibility is one of the most important factors in willingness to pay a premium for a more sustainable product (Hicks, 2012).

More data on the influence of labels on consumer choice are needed. The use of eye-tracking technology is yielding insights into the degree to which customers read labels and how label design affects reading the label – which is the first step in altering choice (Jones & Richardson, 2007). One promising proposal for sustainability labelling arranges scores for several aspects of sustainability in a simple diagram where each aspect is scored on the same traffic-light colour scheme (red, green and yellow: see Sustain, 2007). Ultimately, the truthfulness and correlation of labels to sustainability outcomes is perhaps the biggest hurdle to overcome. Most sustainability labels are at least partially misleading (although they are getting more accurate overall over time); in 2010, 95 percent of 'green' labels were found to have at least some form of greenwashing (disinformation that presents a sustainable image) (TerraChoice, 2010). One solution is to combine standardised and transparent metrics with sustainability reporting through institutions such as The Sustainability Consortium. Key to informative labelling is spatially explicit traceability of foods. While retailers and producers have argued that traceability is too expensive, in fact there are examples of it saving money because it requires better inventory tracking and analysis (Seuring & Müller, 2008; Roh et al., 2009; Attaran, 2012). For example, cost savings from adopting radio frequency identification (RFID) tracking can come from reduced theft, decreased labour costs from faster scanning and the ability to reduce bottlenecks and low inventory levels (Roh et al., 2009). Just as sustainable practices can spur innovation in production systems, sustainable labels can spur innovation in inventory management and flow.

Sustainable certification is another potential solution to combat greenwashing; while companies may benefit from overstating their sustainability (Laufer, 2003; Genç, 2013), the organisations who manage certifications have an incentive for it to be meaningful. The International Standards Organisation (ISO) laid out guidelines for eco-labels based on certification in 1999 (ISO, 1999) with the goal of promoting rigour and consistency. Products certified by a programme complying with these standards have been found to be six times more likely to be free of 'greenwashing' (TerraChoice, 2010). Certification can also enhance consumer willingness to pay for sustainable products, and/or give sustainable producers access to additional markets. For example, certification by the Forest Stewardship Council is increasingly providing a price premium (NEPCon, 2008; Germain & Penfield, 2010; Shoji et al., 2014) in addition to providing some assurance of sustainability, as does USDA organic certification (Lin et al., 2008). However, even certifications that represent a meaningful difference may still need improvement. For example, although stocks for certified seafood are 3–4 times more likely to be managed sustainably than non-certified stocks, 19–31 percent of certified fish stocks were still found to be overfished and subject to ongoing overfishing (Froese & Proelß,

2012). In the end, a combination of strong meaningful certification programmes, additional eco-labels that are simpler (but meaningful) and educating consumers on how to identify truly sustainable products will likely be needed. The increasing prevalence of online shopping presents an opportunity to test and iterate different approaches in presenting sustainability information to consumers and examining the impact it has on purchasing behaviour.

The answer is yes – environmental metrics are part of the sustainability solution

The concept of sustainability in general, and as applied to agriculture in particular, has been criticised as being vague, circular and unhelpful (Hansen, 1996; Glavič & Lukman, 2007). While it is clear there are many different uses of the word 'sustainability', as soon as specific quantitative metrics for it are proposed, it becomes actionable and potentially useful. It is in this spirit that we argue that properly designed sustainability metrics can promote sustainable food in several ways. First, they allow us to determine whether we have been successful in actually *achieving* tangible outcomes, as opposed to simply *working* towards sustainability as measured by implementation of best management practices. Where we are falling short, we can change our approach until we find what works. Metrics also allow comparisons among supply chains and agricultural products or businesses, thereby enabling competition and innovation. Good metrics may mean the difference between meaningful corporate sustainability and greenwashing. Finally, the right metrics will also help us to test hypotheses about system attributes that confer resilience. It is only through such testing that we can promote more resilient systems.

Better incorporation of social and economic metrics with environmental metrics will make sustainability more appealing and practical to a broader audience. Just as farmers have an incentive to care about factors like soil retention and water efficiency in scarce environments (as they are essential to their economic future), environmentalists are starting to understand the importance of yield increases (to reduce the pressure to clear new farms) and profitability (to keep farmers interested in conservation practices without external payments). Better integration of these disciplines should help to identify 'win–win' scenarios, such as the potential of fertiliser optimisation to reduce water pollution, carbon footprint, and costs to the farmer. In addition, corporate sustainability reporting is catching on, but the key is helping corporations set the right goals, which requires the right metrics. Finally, meaningful, credible labels can help consumers make more sustainable choices and drive improvement on the manufacturing end. Further research is needed on what label design will best motivate consumers, and what aspects of sustainability they are most willing to act on.

Ultimately, the path to useful and actionable sustainability metrics is affordable and accurate remote sensing technology. Fortunately, such technology is at hand, although it cannot be the sole solution to data capture. Increasingly, satellite images allow detection of surface soil types and soil moisture, sediment loads in lakes, agricultural management practices, crop yields, vegetation types and changes in any of these variables. By combining this technology with creative scientific hypotheses concerning sustainability and resilience, and pragmatic metrics that draw on the data, a great opportunity for pursuing sustainable food systems lies before us. The limiting factor now is simply ideas about what metrics to use, what hypotheses to test and finally implementing programmes that collect the data. Given the importance of food systems to both humans and the planet, sustainability labelling and reporting should be a priority for agricultural scientists. The fastest progress will be made by working with corporations, consumer groups and NGOs to make food sustainability part of everyday life as opposed to an academic discussion topic.

References

Accenture. (2012). Long-term growth, short-term differentiation and profits from sustainable products and services: a global survey of business executives. www.accenture.com/SiteCollectionDocuments/PDF/Accenture-Long-Term-Growth-Short-Term-Differentiation-and-Profits-from-Sustainable-Products-and-Services.pdf (accessed 24 January 2014).

Aisen, M.A., Garibaldi, L.A., Cunningham, S.A. & Klein, A.M. (2008). Long-term global trends in crop yield and production reveal no current pollination shortage but increasing pollinator dependency. *Current Biology*, **18**(20), 1572–1575.

Allen, R.G., Pereira, L.S., Raes, D. & Smith, M. (1998). Crop evapotranspiration – guidelines for computing crop water requirements. FAO Irrigation and Drainage paper 56. Rome: FAO.

Allen-Wardell, G., Bernhardt, P., Bitner, R., et al. (1998). The potential consequences of pollinator declines on the conservation of biodiversity and stability of food crop yields. *Conservation Biology*, **12**(1), 8–17.

Aramark. (2008). Monterey Bay Aquarium partners with ARAMARK to develop sustainable seafood practices to protect the world's oceans. www.aramark.com/PressRoom/PressReleases/2008/Monterey-Bay-Aquarium.aspx (accessed 24 January 2014).

Attaran, M. (2012). Critical success factors and challenges of implementing RFID in supply chain management. *Journal of Supply Chain and Operations Management*, **10**(1), 144–167.

Ayana, E. (2015). Field test: can we use drones to monitor water quality? https://blog.nature.org/science/2015/11/05/drones-in-the-field/ (accessed 15 May 2017).

Basu, A.K. & Hicks, R.L. (2008). Label performance and the willingness to pay for Fair Trade coffee: a cross-national perspective. *International Journal of Consumer Studies*, **32**(5), 470–478.

Brussaard, L., De Ruiter, P.C. & Brown, G.G. (2007). Soil biodiversity for agricultural sustainability. *Agriculture, Ecosystems & Environment*, **121**(3), 233–244.

Carvin, G., Good, L.W., Fitzpatrick, F., et al. (2018). Testing a two-scale focussed conservation strategy for reducing phosphrous and sediment loads from agricultural watersheds. *Journal of Soil and Water Conservation*, **73**(2), 298–309.

Cassman, K.G. (1999). Ecological intensification of cereal production systems: yield potential, soil quality, and precision agriculture. *Proceedings of the National Academy of Sciences*, **96**(11), 5952–5959.

Caswell, J.A. & Padberg, D.I. (1992). Toward a more comprehensive theory of food labels. *American Journal of Agricultural Economics*, **74**(2), 460–468.

Chaplin-Kramer, R., Sim, S., Hamel, P., et al. (2017). Life cycle assessment needs predictive spatial modelling for biodiversity and ecosystem services. *Nature Communications*, **8**, 15065.

Chen F., Kissel, D.E., West, L.T., & Adkins, W. (2000). Field-scale mapping of surface soil organic carbon using remotely sensed imagery. *Soil Science Society of America Journal*, **64**, 746–753.

Clark, L. & Master, D. (2012). *Corporate ESG/Sustainability/Responsibility Reporting – Does It Matter*. New York, NY: Governance & Accountability Institute.

Compass Group North America. (2009). Compass Group announces results of landmark policy to purchase sustainable seafood. http://compass-usa.com/pages/News-Release-Viewer.aspx?ReleaseID=60 (accessed 24 January 2014).

Condon, A.G., Richards, R.A., Rebetzke, G.J. & Farquhar, G.D. (2004). Breeding for high water-use efficiency. *Journal of Experimental Botany*, **55**(407), 2447–2460.

Contor, B.A. & Taylor, R.G. (2013). Why improving irrigation efficiency increases total volume of consumptive use. *Irrigation and Drainage*, **62**(3), 273–280.

Conway, G.R. (1985). Agroecosystem analysis. *Agricultural Administration*, **20**(1), 31–55.

Daroub, S.H., Van Horn, S., Lang, T.A., & Diaz, O.A. (2011). Best management practices and long-term water quality trends in the Everglades agricultural area. *Critical Reviews in Environmental Science and Technology*, **41**(6), 608–632.

FAO (Food and Agriculture Organisation). (2013). FAOSTAT. www.fao.org/faostat/en/#data/EL (accessed 6 December 2013).

FAO (Food and Agriculture Organisation). (2014). FAOSTAT. www.fao.org/faostat/en/#data/FBS (accessed 2 July 2014).

Farahani, H.J., Howell, T.A., Shuttleworth, W.J. & Bausch, W.C. (2007). Evapotranspiration: progress in measurement and modeling in agriculture. *Transactions of the ASABE*, **50**(5), 1627–1638.

FiBL (Research Institute of Organic Agriculture). (2014). Statistics on organic agriculture worldwide: organic agricultural land and share of total agricultural land 2000–2012; fully converted and in-conversion areas. www.organic-world.net/fileadmin/documents/statistics/data-tables/world-statistics/TABLE-03-WORLD-organic-agricultural-land.xlsx (accessed 2 July 2014).

Field to Market. (2014). Members. www.fieldtomarket.org/members/ (accessed 28 January 2014).

Fisher, J.R.B. (2017). Global agricultural expansion – the sky isn't falling (yet). In: *Uncomfortable Questions and Confirmation Bias in Conservation*, edited by P. Kareiva, B. Silliman & M. Marvier. Oxford: Oxford University Press

Fisher, J.R.B., Boucher, T.M., Attwood, S.K. & Kareiva, P. (2014). How do we know an agricultural system is sustainable? www.researchgate.net/publication/264002578_How_Do_We_Know_an_Agricultural_System_is_Sustainable (accessed 29 January 2014).

Foley, J.A., DeFries, R., Asner, G.P., et al. (2005). Global consequences of land use. *Science*, **309**(5734), 570–574.

Foster, S.S.D. & Perry, C.J. (2010). Improving groundwater resource accounting in irrigated areas: a prerequisite for promoting sustainable use. *Hydrogeology Journal*, **18**(2), 291–294.

Froese, R. & Proelß, A. (2012). Evaluation and legal assessment of certified seafood. *Marine Policy*, **36**(6), 1284–1289.

Garibaldi, L.A., Carvalheiro, L.G., Vaissière, B.E., et al. (2016). Mutually beneficial pollinator diversity and crop yield outcomes in small and large farms. *Science*, **351**(6271), 388–391.

Garnett, T., Appleby, M.C., Balmford, A., et al. (2013). Sustainable intensification in agriculture: premises and policies. *Science*, **341**(6141), 33–34.

Garbeva, P., Postma, J., Van Veen, J.A. & Van Elsas, J.D. (2006). Effect of above-ground plant species on soil microbial community structure and its impact on suppression of *Rhisoctonia solani* AG3. *Environmental Microbiology*, **8**(2), 233–246.

Gebbers, R. & Adamchuk, V.I. (2010). Precision agriculture and food security. *Science*, **327**(5967), 828–831.

Genç, E. (2013). An analytical approach to greenwashing: certification versus non-certification. *Journal of Management & Economics*, **20**(2), 151–175.

Germain, R.H. & Penfield, P.C. (2010). The potential certified wood supply chain bottleneck and its impact on leadership in energy and environmental design construction projects in New York State. *Forest Products Journal*, **60**(2), 114–118.

Glavič, P. & Lukman, R. (2007). Review of sustainability terms and their definitions. *Journal of Cleaner Production*, **15**(18), 1875–1885.

Gomez, C., Viscarra Rossel, R.A. & McBratney, A.B. (2008). Soil organic carbon prediction by hyperspectral remote sensing and field vis-NIR spectroscopy: an Australian case study. *Geoderma*, **146**(3), 403–411.

Hallstein, E. & Villas-Boas, S.B. (2009). Are consumers color blind? An empirical investigation of a traffic light advisory for sustainable seafood. Department of Agricultural & Resource Economics, UC Berkeley, Working Paper Series.

Hansen, J.W. (1996). Is agricultural sustainability a useful concept? *Agricultural Systems*, **50**(2), 117–143.

Harmon, A. (2014). A lonely quest for facts on genetically modified crops. *The New York Times*. www.nytimes.com/2014/01/05/us/on-hawaii-a-lonely-quest-for-facts-about-gmos.html (accessed 21 January 2014).

Hicks, R.L. (2012). Product labelling, consumer willingness to pay, and the supply chain. In: *Sustainable Supply Chains*. New York, NY: Springer, pp. 165–174.

Hoehn, P., Tscharntke, T., Tylianakis, J.M. & Steffan-Dewenter, I. (2008). Functional group diversity of bee pollinators increases crop yield. *Proceedings of the Royal Society B: Biological Sciences*, **275**(1648), 2283–2291.

Huffman, W.E., Shogren, J.F., Rousu, M. & Tegene, A. (2003). Consumer willingness to pay for genetically modified food labels in a market with diverse information: evidence from experimental auctions. *Journal of Agricultural and Resource Economics*, **28**(3), 481–502.

Hunter, M.C., Smith, R.G., Schipanski, M.E., Atwood, L.W. & Mortensen, D.A. (2017). Agriculture in 2050: recalibrating targets for sustainable intensification. *Bioscience*, **67**(4), 386–391.

ISO (International Standards Organisation). (1999). Environmental labels and declarations: type I environmental labeling – principles and procedures. ISO 14024:1999.

Jones, G. & Richardson, M. (2007). An objective examination of consumer perception of nutrition information based on healthiness ratings and eye movements. *Public Health Nutrition*, **10**(3), 238–244.

Kemmerly, J.D. & Macfarlane, V. (2009). The elements of a consumer-based initiative in contributing to positive environmental change: Monterey Bay Aquarium's Seafood Watch program. *Zoo Biology*, **28**(5), 398–411.

Kiron, D., Kruschwitz, N., Reeves, M. & Goh, E. (2013). The benefits of sustainability-driven innovation. *MIT Sloan Management Review*, **54**(2), 69–73.

Klein, A.M., Vaissiere, B.E., Cane, J.H., et al. (2007). Importance of pollinators in changing landscapes for world crops.

Proceedings of the Royal Society B: Biological Sciences, **274**(1608), 303–313.

Kremen, C., Williams, N.M., Aisen, M.A., et al. (2007). Pollination and other ecosystem services produced by mobile organisms: a conceptual framework for the effects of land-use change. *Ecology Letters*, **10**(4), 299–314.

Kremen, C., Iles, A. & Bacon, C. (2012). Diversified farming systems: an agroecological, systems-based alternative to modern industrial agriculture. *Ecology and Society*, **17**(4), 44.

Kroeger, T., Escobedo, F.J., Hernandez, J.L., et al. (2014). Reforestation as a novel abatement and compliance measure for ground-level ozone. *Proceedings of the National Academy of Sciences*, **111**(40), E4204–E4213.

Laufer, W.S. (2003). Social accountability and corporate greenwashing. *Journal of Business Ethics*, **43**(3), 253–261.

Lehmann, J. & Kleber, M. (2015). The contentious nature of soil organic matter. *Nature*, **528**(7580), 60–68.

Lemke, A.M., Kirkham, K.G., Lindenbaum, T.T., et al. (2011). Evaluating agricultural best management practices in tile-drained subwatersheds of the Mackinaw River, Illinois. *Journal of Environmental Quality*, **40**(4), 1215–1228.

Lin, B.H., Smith, T.A. & Huang, C.L. (2008). Organic premiums of US fresh produce. *Renewable Agriculture and Food Systems*, **23**(3), 208–216.

Loureiro, M.L. & Hine, S. (2002). Discovering niche markets: a comparison of consumer willingness to pay for local (Colorado grown), organic, and GMO-free products. *Journal of Agricultural and Applied Economics*, **34**(3), 477–488.

Macmillan, T. & Benton, T.G. (2014). Agriculture: engage farmers in research. *Nature*, **509**(7498), 25–27.

Maherali, H. & Klironomos, J.N. (2007). Influence of phylogeny on fungal community assembly and ecosystem functioning. *Science*, **316**(5832), 1746–1748.

Matson, P.A., Parton, W.J., Power, A.G. & Swift, M.J. (1997). Agricultural intensification and ecosystem properties. *Science*, **277**(5325), 504–509.

McCluskey, J.J. & Loureiro, M.L. (2003). Consumer preferences and willingness to pay for food labeling: a discussion of empirical studies. *Journal of Food Distribution Research*, **34**(03), 95–102.

McWilliams, J.E. (2009). *Just Food: Where locavores get it wrong and how we can truly eat responsibly*. New York, NY: Hachette Digital, Inc.

Merrill, L.B. & Schuster, J.L. (1978). Grazing management practices affect livestock losses from poisonous plants. *Journal of Range Management*, **31**(5), 351–354.

Millennium Ecosystem Assessment. (2005). *Ecosystems and Human Well-being: synthesis*. Washington, DC: Island Press.

Mintel. (2010). Are Americans willing to pay more green to get more green? www.mintel.com/press-centre/press-releases/514/are-americans-willing-to-pay-more-green-to-get-more-green (accessed 2 July 2014).

NEPCon. (2008). Premium on FSC wood sparks interest among Swedish forest owners. www.nepcon.net/2112/English/HOME/News_2008/June/FSC_price_premium/ (accessed 2 July 2014).

Ocean Tomo. (2010). Intangible asset market value. www.oceantomo.com/productsandservices/investments/intangible-market-value (accessed 28 January 2018).

Olmanson, L.G., Brezonik, P.L. & Bauer, M.E. (2013). Airborne hyperspectral remote sensing to assess spatial distribution of water quality characteristics in large rivers: the Mississippi River and its tributaries in Minnesota. *Remote Sensing of Environment*, **130**, 254–265.

Organic Trade Association. (2011). Industry statistics and projected growth. www.ota.com/organic/mt/business.html (accessed 24 January 2014).

Ottman, J.A., Stafford, E.R. & Hartman, C.L. (2006). Avoiding green marketing myopia: ways to improve consumer appeal for environmentally preferable products.

Environment, Science and Policy for Sustainable Development, **48**(5), 22–36.

Palmer, P.I. & Smith, M.J. (2014). Earth systems: model human adaptation to climate change. *Nature*, **512**(7515), 365–366.

PCAST (President's Council of Advisors on Science and Technology). (2011). Sustaining Environmental Capital: Protecting Society and the Economy. Executive Office of the President of the United States. www.whitehouse.gov/sites/default/files/microsites/ostp/pcast_sustaining_environmental_capital_report.pdf (accessed 28 January 2014).

Polimeni, J.M., Mayumi, K., Giampietro, M. & Alcott, B. (2008). *The Jevons Paradox and the Myth of Resource Efficiency Improvements*. London: Earthscan.

Pretty, J. (2008). Agricultural sustainability: concepts, principles and evidence. *Philosophical Transactions of the Royal Society B: Biological Sciences*, **363**(1491), 447–465.

Pretty, J., Noble, A.D., Bossio, D., et al. (2006). Resource-conserving agriculture increases yields in developing countries. *Environmental Science and Technology*, **40**(4), 1114–1119.

PUMA. (2011). PUMA's Environmental Profit and Loss Account for the year ended 31 December 2010. http://about.puma.com/wp-content/themes/aboutPUMA_theme/financial-report/pdf/EPL080212final.pdf (accessed 29 January 2014).

Ramach, J. (1996). Dolphin-safe tuna labeling: are the dolphins finally safe. *Virginia Environmental Law Journal*, **15**(4), 743–784.

Rana, G. & Katerji, N. (2000). Measurement and estimation of actual evapotranspiration in the field under Mediterranean climate: a review. *European Journal of Agronomy*, **13**(2), 125–153.

Reddy, S.M.W., McDonald, R.I., Maas, A.S., et al. (2015). Finding solutions to water scarcity: incorporating ecosystem service values into business planning at The Dow Chemical Company's Freeport, TX facility. *Ecosystem Services*, **12**, 94–107.

Richter, B.D., Brown, J.D., DiBenedetto, R., et al. (2017). Water policy opportunities for saving and reallocating agricultural water to alleviate water scarcity. *Water Policy*, **19**(3), 886–907.

Roh, J.J., Kunnathur, A. & Tarafdar, M. (2009). Classification of RFID adoption: an expected benefits approach. *Information & Management*, **46**(6), 357–363.

Samani, Z., Skaggs, R., Bawazier, A., et al. (2012). Remote sensing of agricultural water use in New Mexico from theory to practice. *New Mexico Journal of Science*, **46**, 1–16.

Scheffer, M., Carpenter, S.R., Lenton, T.M., et al. (2012). Anticipating critical transitions. *Science*, **338**(6105), 344–348.

Seuring, S. & Müller, M. (2008). From a literature review to a conceptual framework for sustainable supply chain management. *Journal of Cleaner Production*, **16**(15), 1699–1710.

Shoji, Y., Nakao, N., Ueda, Y., Kakisawa, H., & Hirai, T. (2014). Preferences for certified forest products in Japan: a case study on interior materials. *Forest Policy and Economics*, **43**, 1–9.

Sinclair, T.R., Tanner, C.B., & Bennett, J.M. (1984). Water-use efficiency in crop production. *BioScience*, **34**(1), 36–40.

Stewardship Index for Specialty Crops. (2013). Nitrogen Use Working Metric Version 1.0. www.stewardshipindex.org/amass/documents/document/12/SISC_Metric_NitrogenUse_2013-07.pdf (accessed 16 January 2014).

Sustain. (2007). Pictorial representations for sustainability scoring. www.sustainweb.org/publications/?id=228 (accessed 24 January 2014).

Tanner, C. & Wölfing Kast, S. (2003). Promoting sustainable consumption: determinants of green purchases by Swiss consumers. *Psychology & Marketing*, **20**(10), 883–902.

Teisl, M.F., Roe, B. & Hicks, R.L. (2002). Can eco-labels tune a market? Evidence from dolphin-safe labeling. *Journal of Environmental Economics and Management*, **43**(3), 339–359.

TerraChoice. (2010). The sins of greenwashing: home and family edition. http://sinsof greenwashing.org/findings/greenwashing-report-2010/index.html (accessed 24 January 2014).

The Sustainability Consortium. (2013a). Key performance indicator category: beans, lentils and peas. www.sustainabilityconsortium.org/what-we-offer/measurement-reporting-system (accessed 21 Janaury 2014).

The Sustainability Consortium. (2013b). Members. www.sustainabilityconsortium.org/members/ (accessed 28 January 2014).

Thøgersen, J. (2000). Psychological determinants of paying attention to eco-labels in purchase decisions: model development and multinational validation. *Journal of Consumer Policy*, **23**(3), 285–313.

Thomas, M.B. (1999). Ecological approaches and the development of "truly integrated" pest management. *Proceedings of the National Academy of Sciences*, **96**(11), 5944–5951.

Tilman, D., Balzer, C., Hill, J. & Befort, B.L. (2011). Global food demand and the sustainable intensification of agriculture. *Proceedings of the National Academy of Sciences*, **108**(50), 20260–20264.

Tscharntke, T., Klein, A.M., Kruess, A., Steffan-Dewenter, I. & Thies, C. (2005). Landscape perspectives on agricultural intensification and biodiversity–ecosystem service management. *Ecology Letters*, **8**(8), 857–874.

UNEP (United Nations Environment Programme). (2010). UNEP Emerging Issues: global honey bee colony disorder and other threats to insect pollinators. www.unep.org/dewa/Portals/67/pdf/Global_Bee_Colony_Disorder_and_Threats_insect_pollinators.pdf (accessed 6 September 2014).

University of Wisconsin–Madison, College of Agricultural, Life Sciences, & Wisconsin. Natural Resources Board. (2005). The Wisconsin Buffer Initiative: A Report of the Natural Resources Board of the Wisconsin Dept. of Natural Resources by the University of Wisconsin-Madison, College of Agriculture and Life Sciences. Office of the Dean, College of Agriculture and Life Sciences.

USDA ERS (Unites States Service). (2014). Table 1 – Food and alcoholic beverages: total expenditures. www.ers.usda.gov/datafiles/Food_Expenditures/Food_Expenditures/table1.xls (accessed 24 January 2014).

Van der Heijden, M.G., Klironomos, J.N., Ursic, M., et al. (1998). Mycorrhizal fungal diversity determines plant biodiversity, ecosystem variability and productivity. *Nature*, **396**(6706), 69–72.

Vermeir, I. & Verbeke, W. (2006). Sustainable food consumption: exploring the consumer 'attitude–behavioral intention' gap. *Journal of Agricultural and Environmental Ethics*, **19**(2), 169–194.

Ward, F.A. (2014). Economic impacts on irrigated agriculture of water conservation programs in drought. *Journal of Hydrology*, **508**, 114–127.

Ward, F.A. & Pulido-Velazquez, M. (2008). Water conservation in irrigation can increase water use. *Proceedings of the National Academy of Sciences*, **105**(47), 18215–18220.

Whole Foods. (2014). Seafood sustainability FAQ. www.wholefoodsmarket.com/mission-values/seafood-sustainability/seafood-sustainability-faq (accessed 24 Janaury 2014).

CHAPTER SEVENTEEN

Conclusions on agricultural resilience

SARAH M. GARDNER
GardnerLoboAssociates
STEPHEN J. RAMSDEN
University of Nottingham
and
ROSEMARY S. HAILS
Centre for Ecology & Hydrology

Introduction
Resilience is the study of the ability of a system to cope with change. The pathway that a system adopts in developing resilience may include measures to absorb change, to enable a system to adapt to change, or to cause it to transform in response to change. Transformation can include measures that enable a system to function via a different pathway or to build additional pathways that support and diversify a system's original function. The strategies available for coping with change take many forms and it is evident from the studies reported in this book that there are many pathways to resilience. In the sections that follow, we explore four themes that emerge from these studies.

The first theme considers the characteristics needed to enable resilience to emerge in agricultural systems. The second recognises that ecosystem services that flow from natural capital are fundamental to any form of agricultural production. Capturing the economic and non-economic values of these services and the capital generated on agricultural land is one mechanism by which the resilience of agricultural systems can be enhanced. The third theme focuses on the need to balance efficiency and resilience in agricultural systems. Efficiency facilitates system growth and resilience is needed for system persistence, but achieving a balance between the two, while maintaining profitable agricultural livelihoods and avoiding system collapse, is a topic requiring further research. The final theme notes that as an emergent property of complex adaptive systems, resilience is not a state that can be prescribed for. Instead, a more effective strategy is to adopt measures that foster the characteristics and conditions that enable resilience to emerge rather than create policies that attempt to prescribe for it directly.

Theme 1: the characteristics that enable resilience in agricultural systems
Ecological and socio–ecological systems such as agriculture are complex adaptive systems: they are dynamic and self-organising, with system properties

such as resilience emerging from the interactions and behaviours of numerous, autonomous and heterogeneous agents operating at the lower levels of the system (Ladyman et al., 2013; Holland, 2014). It is the occurrence and interactions of these agents that enable complex adaptive systems to absorb, adapt and transform in response to change and shocks. Loss of numerosity, autonomy and heterogeneity among the lower-level agents of a complex adaptive system causes the system to lose its adaptive ability and increases the chance of system collapse.

Within ecological systems, heterogeneity of individual agents is the resource on which natural selection acts to generate a diversity of ecological forms and functions (Levin, 1999). Different ecological forms may deliver the same or similar functions (functional redundancy) or may undertake different functions (functional diversity). Diversity of form and function enables ecological systems to adapt to changing conditions, while redundancy provides some insurance against loss of individual agent types. Both categories therefore contribute to resilience in different ways.

Agriculture as a social–ecological system can be viewed as being nested within a hierarchy of systems that comprise the modern food supply system. The latter includes all the elements that contribute to the production, processing, retail and supply of food. Within the food system, agriculture occurs as the first link of the chain and is the principal source of unprocessed food commodities. In considering agriculture as a complex adaptive system, Gardner (Chapter 2) suggests that agricultural resilience is generated by both the biological agents that form agroecosystems and by agricultural producers that develop and manage farm production systems. Maintaining heterogeneity, autonomy and numerosity among both of these sets of agents is difficult. For the biological agents of agroecosystems, the aforementioned characteristics are constrained by agricultural production systems which, for commercial purposes, focus on a relatively narrow range of food crops and, in intensive agriculture, adopt many practices that reduce variability within the growing environment in order to increase yields and production efficiency. Similarly, the autonomy and heterogeneity of producers is unintentionally eroded by the structure and functioning of modern food supply chains (Gardner, Chapter 2). In particular, the scale of commercial food processor and retailer operations and the volume of goods handled, together with the undifferentiated nature of agricultural commodities, means that farmers are 'price-takers' and not 'price-makers'. This price-taking position, together with the efficiency practices used by processors and retailers such as minimum purchase volumes, constrains the autonomy of producers, the range of goods they can supply and their ability to experiment with new products.

For complex adaptive systems, heterogeneity among agents is one of the characteristics needed to enable higher-order properties such as resilience to emerge. Tilman (Chapter 3), Williams et al. (Chapter 6) and Meyer (Chapter 8) highlight the role of heterogeneity in facilitating adaptation and persistence of plant communities, pollination services and genotypes for pest resistance. Diversity in plant communities enhances a community's capacity to cope with limiting resources, species invasion and environmental stresses such as drought; it also confers benefits in terms of productivity (plant biomass and yield) and soil quality (Tilman, Chapter 3). Similarly, Williams et al. (Chapter 6) highlight response diversity – a form of functional redundancy – among pollinators as a critical element for maintaining the resilience of pollination services to climate variation and other stressors. Different pollinators exploit plant resources in different ways and under different conditions, thus enabling the ecosystem service of pollination to persist under conditions of climate change, for example. The behaviour of pollinators highlights not only their heterogeneous form but also their autonomous nature – each species/individual adopting its own foraging strategy to exploit the nectar and pollen resources present in the available plant community. In the context of plant breeding and modern biotechnology, Meyer (Chapter 8) notes that a 'lack of diversity renders crops and cropping systems vulnerable to pest and disease attacks'. While diversity is built into approaches such as marker-assisted selection which combines 'several genes together into a single genotype in order to develop durable, broad-spectrum resistance' (Meyer, Chapter 8), historically, plant breeding and modern biotechnology have tended to favour the development of only a few traits in a few crop species, thereby exacerbating the lack of diversity in cropping systems. Moreover, if a trait (e.g. pest resistance in *Bt* maize) proves very effective, the uptake of that crop becomes widespread, further increasing the loss of diversity in cropping systems.

Within agriculture, the introduction of heterogeneity, for example as heterogeneous cropping systems, mixed farming or even areas for conserving farm biodiversity, complicates agricultural production and can give rise to opportunity or transaction costs that adversely affect the already tight margins of many farm production systems. However, in a world of increasing uncertainty – economic, environmental and social – focusing on efficiency alone leads to fragility. Under conditions of uncertainty, maintaining diversity in both ecological and economic resources is important for adaptability. An important question is thus: how to farm commercially without further reducing diversity (one of the goals of 'Sustainable Intensification'), or alternatively, how to integrate more diversity into commercial farms to enhance their resilience while maintaining, or perhaps increasing, agricultural productivity. Tilman (Chapter 3) has a number of practical suggestions to increase biodiversity including increasing genetic and species diversity, introducing more diverse planting and improving soil quality.

'Biodiversity in its broadest sense (from within species to variation in landscapes) is thought to be important in enhancing the resilience of ecosystem functions and services' (Hails et al., Chapter 4). Viewed from the perspective of complex adaptive systems, biodiversity can be considered as the set of freely interacting autonomous, heterogeneous agents which generate natural capital and ecosystem services used in agriculture and for other human benefits. For example, the diversity of soil organisms which generate soil fertility, structure and composition, the diversity of pollinators which maintain pollination services, the diversity of plants which regulate water, carbon and climate resources and the diversity of plant communities within a landscape which provide resources for recreation, cultural identity and aesthetic appreciation. In the context of agriculture, biodiversity provides the source of heterogeneity that generates both the fundamental services on which agriculture depends and the adaptive capacity that enables these services to persist under conditions of uncertainty. In this way, biodiversity is essential for facilitating the resilience of agroecosystems which underpin agricultural production.

The tensions between maintaining biodiversity and increasing agricultural productivity are well-illustrated in the agricultural systems of Africa and Asia. Ewbank in Chapter 7 points out that on these continents, small-scale cropping systems traditionally use a diverse mix of cropped species and livestock, with organic waste being recycled as input into the cropping regime. Attempts to increase the productivity of these systems have, until recently, focused on introducing external inputs, such as fertilisers and pesticides and more intensive production methods. The consequent reduction in the variety of crops and animal breeds, combined with, in some cases, poorly informed use of external inputs by farmers, soil and water degradation and climate change has compromised the resilience of agriculture on these continents. Ewbank argues that the tension between these different approaches and the persistence of small-scale farms is being undermined by external interests that favour intensive production methods: the use of external inputs to enhance crop productivity rather than coupling cropping systems with changes in natural capital as practised in more traditional approaches to agriculture.

A similar scenario is described by Homewood et al. in Chapter 9, for mobile pastoralists who move their livestock through the open rangelands of sub-Saharan Africa. This movement and consequent spatial heterogeneity in the use of land by livestock has helped maintain these arid and semi-arid lands over thousands of years, despite their ecological fragility and unpredictable rainfall. Policies of restricting grazing rights and closing off land to pastoralists for National Parks and for settled, agricultural cultivation have undermined the pastoralist lifestyle and are not compatible with the resource-limited nature of these lands. Pastoralists have responded by diversifying their livelihood portfolio in order to maintain their resilience. In this respect, pastoralists are behaving in a similar way to producers in more intensive production systems who wish to build resilience by diversifying their farming approach.

The response of pastoralists to constraints on traditional grazing practices highlights another approach available to producers for building resilience into their production systems: the diversification of income streams. Examples include contracting out agricultural land or services, developing non-agricultural and/or agri-environment goods and services, adding value to existing products by local processing and selling directly to customers (see examples in Chapter 2). Such activities help restore autonomy and heterogeneity among producers and offer flexibility for producers working under conditions of uncertainty.

In complex adaptive systems, heterogeneity at one level helps to support the heterogeneity of agents in the levels surrounding it (Sengupta, 2006). Gardner (Chapter 2) notes that in the modern food supply chain, the occurrence of a relatively small number of processors and retailers handling large volumes of goods has accelerated the drive towards uniformity in production practice and in the types of goods produced. Such uniformity undermines producer resilience and also impacts on agroecosystems and ecosystem services. In a similar way, Williams et al. (Chapter 6) observe for pollinators that 'Regional specialisation in agricultural production … creates a temporal homogenisation (of flowering) with truncated periods of bloom followed by periods of relative dearth'. Such specialisation in cropping increases the difficulty of 'achieving temporal and spatial synchrony between pollinators and receptive flowers', which is necessary for successful pollination and for the persistence of pollination services into the future. Once again, loss of heterogeneity on one level of the system (the plants) impacts upon the diversity in another level (the pollinators).

In Chapter 3, Tilman reminds us that resilience at the system level does not necessarily confer stability at the individual agent level. In his study of plant communities, Tilman observed that 'plots containing high plant species numbers (heterogeneity) were about five times more resistant to drought than plots that were very species-poor. This greater ecosystem stability did not carry over to the stability of individual species'. This observation suggests that for complex adaptive systems in general, heterogeneity among agents while facilitating system resilience, does not 'guarantee' the persistence of individual agents themselves. This is important in the context of agricultural systems as it suggests that even when heterogeneity among producers exists, individual producers may still go out of business as conditions change, although the agricultural system itself can persist as other producers develop new business pathways. Indeed, the creation of an environment that fosters innovation is one way of ensuring that new businesses will emerge (see theme 4). In the absence of heterogeneity, however, both producers and the system itself are more likely to collapse as the system will lack adaptive ability.

Intensive, input-dependent agriculture is a relatively recent phenomenon. Already we are seeing substantial environmental effects, for example, in the form of increased levels of greenhouse gases, soil erosion and through land-use change, loss of biodiversity. How might a focus on resilience in agricultural production systems lessen these environmental impacts? As we argue in the following sections, capturing and accounting for change in ecosystem stocks and services and ensuring that these costs are included in assessments of the cost-effectiveness of agricultural systems will help ensure that the effects of agriculture on natural capital are not overlooked. Moreover, as authors in this book argue, greater resilience to shocks can be achieved by building up natural capital and associated ecosystem services by reducing demand on external inputs and encouraging recycling in agricultural systems, diversifying agricultural livelihoods and recognising that agriculture provides not only food, but also goods for human health, well-being, recreation and culture.

Theme 2: capturing the economic and non-economic value of ecosystem services as a mechanism for enhancing the resilience of agricultural systems

Natural capital and ecosystem services are fundamental for agricultural production 'because they generate and maintain the resources needed for food production and ultimately human health and well-being' (Hails et al., Chapter 4). However, agriculture also interacts with natural capital to deliver cultural services, e.g. characteristic agricultural landscapes that are valued by human society.

At the present time, the contribution of natural capital and ecosystem services to agriculture is not fully accounted for. Similarly, the contribution of agriculture in delivering public goods for society (e.g. in the form of landscapes, carbon, soil and water management, biodiversity conservation) is not fully rewarded and the externalities of agriculture – e.g. nitrate pollution of water – are not fully accounted for in its production costs. This lack of accountability, both with respect to the contribution of natural capital and ecosystem services to agriculture and of agriculture to the maintenance (or loss) of natural capital and ecosystem services for society, means that these assets have been often overlooked, dismissed or ignored in decisions concerning the development, management and use of land for agriculture.

For these natural capital and ecosystem service assets to be included in decisions about agricultural land management, their contribution to society has to be valued, not just in economic terms but also with respect to aspects such as human physical and mental health and shared social values

(Hails et al., Chapter 4). Valuing natural capital and ecosystem service assets is difficult due to the variety of assets and the different methods used to assess them. A particular challenge is to choose 'metrics' that are appropriate for determining trade-offs between different assets: carbon management as opposed to food production for example; or for the same asset used for different purposes – for example, a landscape with food production, energy and cultural values. Economic value is the common metric used for comparing different assets (ecosystem services or otherwise), but other metrics such as impact on human health and well-being will help to capture the benefits of ecosystem services to people more fully. Hodge (Chapter 10), for example, suggests that 'demands for agricultural land to generate higher levels of ecosystem services that are valued by other sectors' could increase the value of ecosystem services.

Hails et al. (Chapter 4) suggest that 'An alternative approach (for valuation) is to constrain the set of possible decisions to include only those which do not reduce the biophysical metric in question (e.g. biodiversity indices)'. This approach, exemplified by Bateman et al. (2013), is similar in thought to that of Martin (Chapter 13), who uses Viability Analysis (a form of Set-Valued Analysis – Aubin & Frankowska, 2009) to determine whether different sets of values (pertaining, for example, to agricultural policies and management practices) maintain or return an agricultural system within a pre-determined 'desired' set of values. The level at which to establish the 'desired' set of values of Martin (Chapter 13) or the biophysical metrics of Hails et al. (Chapter 4) is raised by Hodge (Chapter 10). He concludes that the reference level on which these values or metrics were based would need to be a social judgement on what constitutes 'good agricultural practice' with respect to the integration of agriculture and ecosystem services. In particular, Hodge (Chapter 10) reminds us that it is important 'to be explicit about the counterfactual' used to determine the level of 'negative and positive contributions to ecosystem services' that arise from agriculture. Hodge gives the example of conservation tillage as a practice for mitigating greenhouse gas emissions. This practice is cited by some as 'an ecosystem service from agriculture' when considered against a counterfactual of conventional agriculture, and by others as having a smaller negative effect than conventional agriculture when considered against a counterfactual of 'the unfarmed environment'.

Providing information on and a better understanding of the benefits of natural capital and ecosystem services to agricultural producers and other interest groups is an important output of the valuation of these assets, particularly where conflict arises over the management of agricultural land between private and common interests and between the production of private and public goods. An example of such a conflict is given in Homewood et al. (Chapter 9), where private interests undervalue fragile savannah land held in common and used by pastoralists and wildlife, and where these interests also exploit conflicts between pastoralists and wildlife to undermine both.

To help tackle conflicts between different interest groups over the management of agricultural land, Hails et al. (Chapter 4) advocate the design and establishment of multi-functional agricultural landscapes 'that co-produce multiple ecosystem services rather than (landscapes) that maximise a single service'. The authors argue that a multi-functional agricultural landscape approach can help enhance the resilience of both agroecosystems and agricultural producers for example by linking the delivery of ecosystem service benefits from agriculture to financial incentives or rewards – so-called 'payments for ecosystem services'. Such payments provide additional income streams that can help build producer resilience. Agri-environment stewardship schemes such as those linked to Pillar 2 of the EU's Common Agricultural Policy (Buckwell, Chapter 15) can be viewed as such an incentivising scheme, although these payments are linked to 'ongoing management practices rather than the delivery of ecosystem services' (Hails et al., Chapter 4). However, payment schemes such as those suggested by Drechsler and Wätzold (Chapter 12), for example, can be a useful tool for promoting both agricultural and ecosystem service benefits that help build agricultural resilience.

The idea of multi-functionality acknowledges the multiple roles that producers fulfil in managing land for production and in stewarding its resources for society and for future generations. A multi-functional approach might also offer agricultural producers an opportunity for developing a broader range of income streams with which to diversify their production system. This diversification is important for maintaining producer autonomy and heterogeneity and for giving value to the non-food goods and services that agricultural producers provide.

Theme 3: assessing resilience in agricultural systems: the problem of balancing efficiency and resilience

A challenge for agriculture is how to integrate resilience with efficiency in agricultural systems: efficiency facilitates system growth while resilience facilitates system persistence. A tension arises between these two parameters because efficiency tends to drive systems towards greater uniformity among lower-level agents, while resilience requires heterogeneity among such agents. This tension lies at the heart of agriculture and today's food system, as a focus on productivity has emphasised the importance of efficiency at the expense of resilience.

Hodge (Chapter 10) highlights this tension by pointing out that 'a focus on resilience . . . challenges assumptions about (the use of) optimisation' as a tool for determining the 'best' outcome for agricultural production. He notes that 'a resilient system will . . . include greater diversity than a system that seeks to maximise expected net present value on the basis of current knowledge and projections'. To be resilient, a system is likely to include redundant elements that are of value only at a time of shock and change (see discussion of pollinators in Williams et al., Chapter 6). Hodge notes that 'modern agricultural systems ... tend ... to emphasise control and simplification', a strategy that has been adopted largely 'at the expense of externalising costs' such as 'water and air pollution and reductions in the quality of the landscape'. This exclusion of environmental costs, combined with policy support for output prices in developed countries (see Buckwell, Chapter 15), has encouraged producers to increase and intensify production and so exacerbate these external costs. The consequent degradation of the agroecosystem then undermines the resilience and ultimately, the efficiency of agricultural production.

In considering the resilience of intensively managed agricultural systems, Ramsden and Gibbons, following others, suggest in Chapter 5 that resource degradation will reduce both the amount of current production and the choice – the 'option' – over how to produce in the future. The degree to which this degradation matters is partly dependent on whether new technologies can substitute for lower resilience in the natural environment. The issue of substitution raises significant debate among ecologists and economists (see discussions in Gardner & Ramsden, Chapter 1 and Hails et al., Chapter 4). Ramsden and Gibbons (Chapter 5) suggest that while technology can substitute for some aspects of the environment, substituting for core functions such as elements of the carbon or nitrogen cycles, or for biodiversity, is considerably harder. Meyer, in Chapter 8, considers whether modern biotechnologies for plant breeding can be used to sustainably intensify agricultural production in different farm systems across Europe and concludes that such 'biotechnologies are most suitable for intensive and specialised cropping systems with high external inputs'. As we have seen, intensive cropping systems tend to operate to tight financial margins and emphasise efficiency to maintain their viability. Meyer (Chapter 8) points out that introducing GM crop varieties with 'monogenic' traits (linked to herbicide tolerance or insect resistance for example) into intensive cropping systems tends to reinforce current cropping practices and may hinder the development of sustainable intensification and resilience in these systems. Moreover, varieties with polygenic traits (e.g. biotic and abiotic stress resistance and yield potential) which have the potential to be used over a wider range of agricultural systems are not easily transferable to different crops and locations. Consequently, Meyer concludes that at present the potential for plant breeding biotechnologies to substitute for natural capital and ecosystem services and contribute to sustainable intensification and resilience in agriculture is limited and that agro-ecological and other management-based approaches such as Integrated Pest Management will remain important.

In the face of continuing economic and environmental uncertainty and ongoing resource degradation (e.g. soil erosion, water shortage) in many areas, integrating resilience into current agricultural systems while retaining or enhancing productivity remains a challenge. A standard approach for studying agricultural systems is 'bio-economic modelling' where features and processes of natural biological and physical systems are integrated into the microeconomy of the farm. The resultant model is used to explore the outcomes of different management practices or interventions and when combined with a constrained optimisation framework, to identify the approach best suited to meeting a stated economic or environmental objective. Ramsden and Gibbons (Chapter 5) show how these models have been used to address adaptations that can make a farm more resilient to market variability – such as commodity price volatility – although much less has been done to assess ecological resilience. They argue that the constrained optimisation format enables these models to evaluate trade-offs of the type discussed earlier: for example, the trade-off between efficiency of resource use and management of risk. Furthermore, the optimisation framework helps to explain – to some degree – why producers adopt production systems that are considered fragile by ecologists: the cost of building resilience is too great or the economic benefits are too small when measures to enhance resilience are adopted within a policy environment of price support such as that highlighted by Hodge above and discussed by Buckwell in Chapter 15. Based on their review of bio-economic models, Ramsden and Gibbons (Chapter 5) conclude that greater economic resilience will leave farmers less concerned about ecological resilience, other things being equal. As an example, short-term market stabilisation measures (e.g. use of futures markets by farmers to reduce exposure to price risk) discourages the exploration of measures, e.g. diversification, that enhance the resilience of the farm system over the longer term. Thus, the adoption of 'economic' stabilisation measures tends to compromise the resilience of long-term natural capital, such as soil, which is slow to recover from degradation. These observations highlight a need for these bio-economic models to capture more fully the interactions between the economic management of the agricultural production system and the ecological functioning of the agroecosystem within which production takes place and in particular, to take account of the differences in timescale over which economic and ecological resources are built.

In Chapter 14, Harris et al. consider the different strengths that economics and resilience thinking bring to the challenges of natural resource management in social–ecological systems. They remind us that an important element in integrating these two approaches will be identifying and resolving differences over the concepts and language used by the two disciplines. For example, they suggest that efficiency as used in economics has no analogue in resilience thinking. Indeed, Harris et al. observe that the concept of efficiency as 'achieving a given goal at lower cost' and as a driver of system change (seeking ever-greater efficiency) is likely to lead to ever-decreasing resilience, an observation supported by the work of Ulanowicz et al. (2009). Harris et al. suggest that the discussion over efficiency is not so much a matter of disagreement between the two disciplines but rather a tendency to place too great an emphasis on efficiency and insufficient consideration of externalities in other parts of the system. They argue that part of this problem might be overcome if the concept of efficiency was broadened from a focus on cost-effectiveness to include greater consideration of risk and stochasticity or of the effects of externalities on other parts of the system. In this way the goals of efficiency and resilience can start to align, each with its associated cost and benefits.

In Chapter 12, Drechsler and Wätzold use an integrated ecological–economic approach to design a cost-effective agri-environment scheme for conserving farmland biodiversity. They emphasise the importance of researchers developing a clear understanding of each discipline to ensure that feedbacks between the different elements, the dynamics and costs of the ecological and economic systems, are captured fully. In considering the differences in approach of the two disciplines to modelling biodiversity conservation, Drechsler and Wätzold observe that economic models tend to be formulated and solved analytically, that they are often static, ignoring uncertainty and with spatial and temporal functions (if included) modelled on an abstract scale. On the other hand, ecological models are often specific to particular species and/or regions, use rule-based approaches to simulate ecosystem dynamics and uncertainties and adopt specific spatial and temporal scales. These differences reinforce the points highlighted by Ramsden and Gibbons and Harris et al. (Chapters 5 and 15, respectively), that current approaches to modelling agricultural systems need to be broadened to capture the full range of dynamics in and interactions between the economic and ecological elements of agricultural production systems. It is pertinent to ask what elements might need to be included. Ramsden and Gibbons suggest: feedback loops between management interventions and the biological part of the farm system, thresholds and environmental tipping points and poorly understood 'uncertainty' rather than 'reasonably well-known variability'.

The issue of poorly understood uncertainty is also raised by Hodge (Chapter 10) and Harris et al. (Chapter 14) and is a key argument for retaining heterogeneity within a system – to facilitate system buffering and adaptation in the face of unforeseen change. As noted earlier, resilience emerges from the interactions of many heterogeneous agents and it is the diversity of agents' responses that provides the mechanism by which a specific goal can be achieved by multiple potential pathways. Incorporating the diversity of agents' responses within system models is difficult because this diversity cannot be easily captured by adopting a single representative response or even a set of responses to represent agent behaviour at a population or macro-scale. As Harris et al. note in Chapter 14, 'aggregation from micro to macro is difficult ... and ... strong assumptions are required for behaviour in the large to mimic behaviour in the small'. Studies assessing the resilience of agricultural systems therefore need to capture the heterogeneity of responses among the underlying agents and understand the complexities of scaling up from individual elements to the system as a whole, since the retention of multiple pathways is important for systems facing unknown sources of uncertainty.

In Chapter 13, Martin uses viability analysis to discern the set of management options that all meet a prespecified policy goal, while maintaining or returning a system within a predetermined set of system limits. The starting point for this approach is for users to first decide on the set of limits beyond which system change is unacceptable. For agriculture, this set of limits would include the environmental, economic and social boundaries beyond which change is undesirable. As highlighted by Hodge in Chapter 10, determining this set of limits is a societal judgement. Once the system model and set of limits have been established, viability analysis is used to define the state space within which changes in management will not result in undesirable system change. This zone is referred to as the viability kernel and takes account of both the system's dynamics and of the limits specified by the user. The viability kernel enables the user to distinguish between the set of input options that enable the system to remain within the user-specified set of limits and the set of options that cause the system to fall outside of the user-specified set of limits. In this way, viability analysis can be used to identify the set of options that can all achieve a particular policy target. It is also possible to use viability analysis to identify the set of policy options that can return a 'damaged' system back into the viability kernel and the management changes that might be needed to achieve this. In this way, viability analysis can be used to identify multiple pathways by which an agricultural system might achieve a production goal (both food and the production of other ecosystem services). For heterogeneous agents, operating within differing production contexts, this approach presents more pathways to achieving a particular policy goal and offers flexibility and a means by which agents can persist in meeting the goal even if production conditions change.

In Chapter 11, Abson et al. compare the resilience of conventional and holistic farming approaches to managing livestock in Australia. They consider these two approaches within the conceptual framework of land-sparing and land-sharing, applying these terms to conventional and holistic farming approaches, respectively. The land-sparing/-sharing framework presents two alternative land-use options for managing the trade-offs between food and biodiversity production: (i) land-sparing focusing on intensive, specialised production on existing agricultural land and sparing non-production land for biodiversity; (ii) land-sharing adopting an agro-ecological approach to land management, with more extensive agricultural production and lower yields and promoting 'wildlife-friendly' more multifunctional landscapes. Abson et al. compare conventional/land-sparing and holistic/land-sharing livestock management approaches with respect to general and specific resilience, the latter linked to uncertainties associated with fluctuations in climate and agricultural input prices.

The two types of producers studied by Abson et al. use different pathways to cope with uncertainty and to maintain a profitable livelihood. Conventional/land-sparing producers rely on external inputs to reduce variation in grazing resources and financial capital as a buffer against uncertainty. The latter provides these producers with better short-term resilience compared to their land-sharing colleagues because financial capital is more easily converted into resources to deal with unexpected shocks. Conventional/land-sparing producers also enjoy better institutional support in the form of advisory services and social networks. The authors note, however, that this production system lacks natural capital protection in the farmed areas which can then become vulnerable ecologically to shocks. Moreover, there is a potential risk of damage from farmed areas spilling-over to 'spared' areas. By comparison, holistic/land-sharing producers rely on dynamic adaptive management to cope with disturbances, coupling farm management and natural capital. This approach generates greater general resilience in the agroecosystem, but the practice of varying stocking levels can leave land-sharing producers exposed to fluctuations in market price and increase the vulnerability of their production system because of a lack of financial capital.

The authors observe that both systems have strengths and weaknesses with respect to resilience, suggesting that a mixture of the two would offer the most resilience for livestock management overall. They reach a similar conclusion with respect to biodiversity conservation; thus, while specialist species may be more readily conserved in the protected patches provided under land-sparing, in the face of unforeseen disturbances, there is greater connectivity for species in the land-sharing option.

In concluding this section, it is evident that the integration of resilience and efficiency into assessments of agricultural production systems is still in its infancy. As noted in Theme 2 and specified as a requirement in the work of Martin (Chapter 13), there is a need for clear thinking on the reference point or set of desired limits within which agriculture needs to practice. In particular, further interdisciplinary thinking is needed to consider how the effects of long- and short-run variables, environmental tipping points, unknown uncertainties and heterogeneity of agent behaviours might be included in future assessments of the performance of different agricultural systems. A resilience approach recognises that there are multiple pathways for achieving agricultural production, each suited to a particular context. Under conditions of uncertainty an important question for agricultural producers and policymakers is to consider how many paths (income paths or production approaches) are needed to ensure system survival without undermining producer livelihoods and the supply of goods to society. In addition, research might also explore the costs and benefits of pursuing production paths that are suboptimal in production output but provide more options for enhancing the adaptive capacity of agriculture to shocks and unknown uncertainty.

Theme 4: fostering resilience in agricultural systems

As a property that emerges from the interactions of many, heterogeneous agents, it is evident that there is no standardised path or methodology for building resilience. Indeed, as studies of agricultural producers highlighted in this book indicate (see Chapters 2, 9, 10 and 11), producers adopt many pathways for developing resilience. This observation suggests that policymakers and land managers with responsibility for building resilience should focus on fostering the conditions that give rise to it – namely, on maintaining the characteristics of autonomy, heterogeneity and numerosity among the underlying agents. These characteristics enable these agents to self-organise and adapt to unexpected change.

Resilient systems can deploy several strategies – buffering, adapting or transforming in response to change. As noted by Abson et al. in Chapter 11, not all of these forms lead to the long-term persistence of the system: thus the financial capital generated by the conventional 'land-sparing' approach in their study provided a robust short-term buffer for the production system, but the management approach undermined the long-term resilience of the farmed land. Conversely, the holistic land-sharing approach provided long-term resilience of the farmed land, but left producers vulnerable to short-term fluctuations in market price. While these outcomes might be seen as an inevitable consequence of short- and long-run trade-offs, studies of resilience among US and Austrian agricultural producers undertaken by Lengnick (2015) and Darnhofer (2010), respectively (and highlighted in Chapters 2 and 10 of this volume) emphasise another important requirement for building resilience, namely, the need to maintain significant reserves of *all* types of capital – human, natural, social, financial and physical. Such reserves help to buffer existing systems against short-term shocks and facilitate adaptation and transformation in response to unexpected and prolonged periods of change.

Maintaining a balanced portfolio of capital reserves is not easy, as we have seen in the study of Abson and colleagues. The choice of production path strongly influences the type of capital that is accumulated. This point is also noted by Harris et al. (Chapter 14), who suggest that conventional frameworks for assessing the cost-effectiveness of agricultural policy options should be extended to include the lock-in costs of each option. The lock-in costs represent the 'additional costs that path dependencies triggered by the (policy) option would create for subsequent efforts to adapt or transform the option'. An imbalance between capital types can, as Ramsden and Gibbons (Chapter 5) observe, lead to a trade-off between the economic and ecological assets of agricultural production systems. This trade-off reflects an overemphasis on developing one part of the production system to the detriment of the rest. In the land-sparing example of Abson et al., a focus on maintaining productivity over the stock of natural capital of farmed land; in the land-sharing example, of maintaining natural capital at a possible cost of production income. As noted in the previous section, a balance needs to be maintained between growth (efficiency) and persistence (resilience) in maintaining agricultural production systems. This may require limits being set on the growth of one type of capital to enable another to also thrive, or to permit some diversification in the production system to encourage the development of different types of capital. A starting point for achieving this balance is recognising the need to include both long-term and short-term costs and benefits of production within the evaluation of policy options and management practices for different production systems.

While much of this discussion has focused on the components needed to build resilience in agriculture, there is also a need for practical methods to determine the extent to which a system is resilient. In Chapter 16, Fisher and Kareiva present a set of environmental metrics that can be used to assess the persistence and sustainability of agroecosystems under different production management regimes. The authors emphasise the importance of adopting metrics that are responsive over a measurable timescale, credible in showing the delivery of some improvement and creditable so that producers can be rewarded for improvements or held accountable for losses and for successful practices to be passed onto others. Fisher and Kareiva favour the use of outcome-based metrics that evaluate system performance, arguing that these provide a measure of whether objectives have been achieved, allow comparisons across businesses and supply chains and facilitate testing of hypotheses about what confers system resilience. The authors acknowledge that to appeal to the broader audience of producers and food corporates, measures for assessing the persistence and sustainability of agriculture will need to include social and economic metrics as well as environmental ones. A similar conclusion is reached by Cabell and Oeloefse (2012), who include a set of social criteria, alongside environmental criteria, that enable the contributions of human and social capital to production resilience to be considered alongside those of natural, financial and physical capitals.

Rebuilding resilience into agricultural systems is possible, but it requires a production environment that encourages self-organisation among producers and enables them to diversify, experiment and innovate. Such an environment is created by fostering heterogeneity, autonomy and numerosity among producers and by their maintaining capital reserves of all types. These reserves buffer producers during the early stages of system adaptation and in the development and testing of new ideas.

Agriculture is facing a time of significant change and uncertainty. To date, the most common response to change has been to become more efficient and more productive, but a focus on being more adaptive would offer greater resilience. In agriculture, the drivers of change are usually external, often global and outside of the producer's control. For a producer, facing conditions of global uncertainty, a production path that delivers a viable livelihood and maintains a capital reserve in anticipation of unexpected change, might be more resilient than an expansionist path that requires specialisation, scaling-up and high productivity. For researchers and policymakers concerned with the sustainability and resilience of agriculture, providing tools and information that help producers shift from one path to another is important. Equally critical is for producers to maintain all their capital assets and seize opportunities to make that shift.

Concluding remarks

Of the many topics discussed in this book, the four themes highlight lessons that are fundamental for the rebuilding of resilience in agriculture and offer opportunities for ecologists and economists, alongside mathematicians, social scientists, policymakers and others to work together. Recognising that agriculture (including the agroecosystems on which it is founded and the production systems that deliver food commodities) operates as a complex adaptive system is important in considering how agricultural systems might be managed and how beneficial properties such as resilience and ecosystem services arise. Resilience is an emergent property of complex adaptive systems and its occurrence relies on maintaining particular characteristics among the sets of agents that underpin the system. In agriculture, the underpinning agents are the biological species that build agroecosystems and the producers (farmers) that build agricultural production systems. For agriculture to be resilient, the characteristics of heterogeneity, autonomy and numerosity need to be present in both sets of agents. The question for researchers is how much of these characteristics is needed among biological species and producers to build resilience within agroecosystems and agricultural production systems. Resilience is not the only goal for agriculture, because as Ulanowicz et al. (2009) note, too much resilience can lead to system stagnation while too little leads to system fragility. Finding the balance that facilitates system adaptation while maintaining system growth under conditions of uncertainty presents an interesting challenge for research.

It is clear that both the production practices of intensive agriculture and the structure and purchasing practices of the food processing and retailing industry tend to diminish heterogeneity and autonomy among producers and the production environment. This effect is a consequence of the adoption of certain practices and a focus on short-term efficiency rather than a direct intention, but nevertheless the outcome is a loss of resilience and increased fragility in the agricultural production system. An important goal for policymakers then is to devise policies that enable and encourage producers to diversify their production both with respect to their business models and to the persistence of the agroecosystems on which they are founded.

An important part of this diversification is recognising and reinstating the relationship between agriculture and ecosystem services. The agroecosystem and the ecosystem services that emerge from it are the foundations on which agriculture is built. Change in these natural capital assets needs to be properly accounted for in the evaluation of agricultural systems. Considerable research is being undertaken at present on approaches for valuing ecosystem services, but greater focus needs to be placed on determining how changes in these natural capital assets can be included and evaluated in models and decision tools used to assess the performance of agricultural systems. The recognition of agriculture as a multi-functional system producing food, ecosystem and societal goods and services would do much to ensure that the non-food production elements of agriculture are properly valued and rewarded. There remains plenty of scope for economists and ecologists to work with producers to explore ways of rebuilding natural capital that both benefits their business and delivers benefits for society. An important gap, noted by Abson et al. (Chapter 11), is the provision of technical advice and follow-up support on integrating natural capital into the farm business and on how to move production successfully towards greater resilience while remaining economically viable.

A dominant mantra in agricultural production is the need to focus on efficiency in order to maintain productivity and cost-effectiveness. Efficiency is needed, but it is not aligned with adaptability for resilience. As the production environment becomes increasingly uncertain, greater attention has to be given to considering how well each producer is placed with respect to absorbing, adapting or 'transforming' to change. Metrics such as those proposed by Fisher and Kareiva (Chapter 16) are certainly needed to assess the capacity of the agroecosystem and farm production system to withstand change. Similarly, as indicated by several authors in this book, reserves of capital of all types are needed, as well as an understanding of how these different capital reserves interact with each other over different timescales. A policy approach that focuses on evaluating the capacity of agricultural systems to withstand ongoing change (e.g. climate change) and unexpected shocks, and offers measures to enhance adaptation capacity, will be more appropriate for fostering resilience than an approach that advocates particular management practices as a means for delivering resilience.

As we argued at the start of this chapter, the pathways to resilience are many and it is by embracing this diversity that we will gain agricultural resilience. Agriculture is a social–ecological system that operates across many disciplines and touches every person's life. To build agricultural systems that are resilient to change and able to persist within the limits of our planet's natural resources, it is essential that those engaged in designing future systems understand the ecological, economic and social context within which agriculture operates. This context varies both within and between nations. This variation is the source of future adaptive ability and the resource from which resilience emerges. Now is the time to foster it.

References

Aubin, J.P. & Frankowska, H. (2009). *Set-Valued Analysis*. Boston, MA: Birkhäuser.

Bateman, I., Harwood, A., Mace, G., et al. (2013). Bringing ecosystem services into economic decision-making: land use in the United Kingdom. *Science*, **341**, 45–50.

Cabell, J.F. & Oelofse, M. (2012). An indicator framework for assessing agroecosystem resilience. *Ecology and Society*, **17**(1), 18. http://dx.doi.org/10.5751/ES-04666-170118

Darnhofer, I. (2010). Strategies of family farms to strengthen their resilience. *Environmental Policy and Governance*, **20**, 212–222.

Holland, J. (2014). *Complexity: a very short introduction*. Oxford: Oxford University Press.

Ladyman, J., Lambert, J. & Wiesner, K. (2013). What is a complex system. *European Journal for Philosophy of Science*, **3**(1), 33–67.

Lengnick, L. (2015). *Resilient Agriculture: cultivating food systems for a changing climate*. Gabriola Island: New Society Publishers.

Levin, S.A. (1999). *Fragile Dominion: complexity and the commons*. Cambridge, MA: Perseus Publishing.

Sengupta, A. (2006). Chaos, nonlinearity, complexity: a unified perspective. In: *Chaos, Nonlinearity, Complexity: the dynamical paradigm of nature*, edited by A. Sengupta. Berlin: Springer-Verlag, pp. 270–352.

Ulanowicz, R.E., Goerner, S.J., Lietaer, B. & Gomez, R. (2009). Quantifying sustainability: resilience, efficiency and the return of information theory. *Ecological Complexity*, 6, 27–36.

Index

abiotic stress, 170
Abrams Strogatz model for language competition, 286
adaptive capability, 33–34, 159, 165, 218, 375, *See also* producers
adaptive cycles, 4, 6, 105, 296, 301
African Postharvest Losses Information System (APHLIS), 148
African Stockpiles Program, 145
Agricultural and Rural Development Directorate (European Commisson), 327
agricultural depression, 220–222, 223, 224, 229
agricultural intensification, 253, 254–255, 340, 363, 367
   effects on pollinator habitats, 112, 115
   efficiency in, 27
   mitigation of the effects of, 117–120, 325
   production practices, 378
   reduction in floral forage plants, 114
agricultural systems. *See also* good agricultural practice; public goods; sustainability; sustainable intensification
   as complex adaptive systems, 19, 27
   bioeconomic modelling and, 90–96
   characteristics enabling resilience of, 362–367
   designing policies for resilience, 278–292
   in multifunctional landscapes, 82, 369, 374
   pollinator vulnerability and, 110–112
   value of ecosystems services and, 74–79
Agriculture Council, 326
agri-environment schemes, 83, 212, 253–254, 325
   as a policy instrument in Europe, 254–255
   cost effectiveness, 255–256
   development of decision support tools, 269
   ecological–economic modelling for cost-effectiveness, 256–267
   importance of social learning and adaptive governance, 255–256
   integrating protection of ecosystem services and biodiversity in design of, 268–269
agri-tourism, 77

agrochemicals, 137, 142, 143–145, 151, 234, 268
agroforestry, 143, 148, 174
aldrin, 144
Almar Farm and Orchards (Michigan), 28, 30, 31, 32
almond orchards, 110, 113
amphibians, 348
APES (Agricultural Production Externalities Simulator), 101
Apis mellifera (European honeybee), 109, 111, 345
Arab Spring, 23
arthropods, 48
autarchy, 216
autonomy. *See also* complex adaptive systems; producers
   role in self-organisation and resilience, 15–16, 19, 34, 362–363

Bacillus thuringiensis. *See* Bt crops
banana xanthomonas wilt, 146
BASF, 352
bats, 235
bees, 100
   colony mortality in honeybees, 345
   habitat management, 117–120
   measures to protect, 67
   sensitivity to change, 61
   threats to wild populations, 117–119
   vulnerability of honey bees, 110–112
benevolence, 216
best management practices (BMPs), 344
biodiversity, 39, 54–55, 364–365, *See also* grassland biodiversity; soil biodiversity
   agricultural production and, 67
   agri-environment schemes and, 83, 212, 253–255, 268–269, 372
   control of crop disease, 50–51
   control of ecosystem functioning, 50
   ecosystem functioning and, 39–41, 340
   ecosystem services and, 61, 315–316, 321, 325, 335
   effects on crop yields, 51–53
   food production and, 236, 340, 374
   foodweb effects, 48–49

invasive species and, 47–48
land-sharing and, 239, 245, 374
land-sparing and, 238, 243–245, 374
loss of and effects on small-scale farmers, 145–146
loss through intensive farming, 367
nutrient abatement by buffers, 53–54
pollination and, 116–117, 345
rangeland, 181–182, 183, 202
relevance to agriculture, 50
soil fertility and carbon sequestration, 48
stability and, 41–43
value, 74, 78
bio-economic models, 90–92, 94, 95–96, 371
future directions in, 103–106
studies of resilience in, 96–103
biofuels, 135, 149
biological resilience, 127
biophysical metrics, 78, 368
biotechnologies for plant breeding, 81, 160, 162–163, 175–177, 370
drought tolerance and abiotic stress, 168–171
herbicide resistance and weed management, 164–165
in extensive farming in less-favoured areas, 174
in extensive small-scale, semi-subsistence farming, 173–174
in intensive, larger-scale crop farming, 175
in large-scale corporate farming, 175
in medium intensive, mixed-farming systems, 174–175
increased yield potential and reduced yield gaps, 171–173
insect resistance, 165–168
intellectual property rights and seed industry concentration, 173
marker-assisted selection (MAS), 146, 163–164
birds, 3, 15, 16, 120, 235, 269, 348
benefit of fertilisers, 268
decline in farmland, 253
habitats, 241
measures to protect, 67, 70, 83, 245
sensitivity to change, 61
Birds and Habitats Directives, 320
blackgrass, 93
bovine spongiform encephalopathy (BSE), 24
Brexit, 223
British Government, 326
Bt (*Bacillus thuringiensis*) crops, 165–168
cotton, 166
maize, 162, 168
buffer capability, 33, 159, 165, 168, 218, 223, 224, 375, 377
buffer strips, 53–54, 64, 70, 71, 212, 238, 333, 346
burning, 93, 181
butterflies, 269
effects of mowing regimes, 264–268

C_3 plants, 172
capital reserves, 35–36, *See also* financial capital; human capital; manufactured capital; natural capital; social capital
maintenance of, 375–376
carbon, 50, 54, 61, 150
carbon cycle, 14, 93, 370
carbon dioxide (CO_2), 48, 50, 95, 287
effect on crop production, 141
carbon sequestration, 71–72, 76, 326
biodiversity and, 48
carbon storage, 49, 65, 72
Cargill, 344, 352
CGIAR, 152
climate change, 13, 103, 135, 136–137, 142, 154, 169, 223, 365
adaptation to, 98, 100, 113, 115, 153, 187, 346, 347
agri-environment schemes and, 267–268
effect on production, 26, 33, 80, 99
effects on pests and diseases, 167, 316
effects on small-scale farmers, 140–141
EU policy, 326, 328, 333–335
mitigation, 72, 76
plant breeding and, 146
rainfall intensity and, 143
weather extremes induced by, 5
Climate Corp, 349
climate extremes, 33
climate management, 65–66
coffee production, 149
Common Agricultural Policy (CAP), 83, 91, 102, 177, 223, 225, 226, 316, 369
environmental conditionality, 319–321
legislative proposals, 330–332
MacSharry reform (1992), 319
negotiations, 332–335
politico-economic context and language of reform debate, 325–328
reactions to reforms, 335–337
reform consultation and 2010 communication, 328–330
regional, zonal or farm type specific supports, 322–323
scale and nature of environmental integration, 317–319
voluntary environmental schemes, 321–322
Common Monitoring and Evaluation Framework (CMEF), 323, 324
community-based natural resource management (CBNRM), 185, 187, 195
complex adaptive systems, 3, 4, 27, 295–296, 302, 362–363, 377
agriculture as a, 18–19
behaviour of, 16–18
biodiversity, 365
characteristics, 15–16
heterogeneity, 17, 363, 366
resilience in, 14–15
study and analysis of, 296–298

complexity, 304–305
complexity theory, 27
conventional managment. *See* land-sparing
corporate farming, 161, 175
corporate sustainability, 351–352, 355
CorporateRegister.com (CR), 351
Costa Rican model system, 67
Country Land and Business Association (CLA), 327
cover crops, 25, 70, 91, 165
creative destruction, 105, 225
crop diseases
  climate change and, 146
  genetic diversification and, 50–51
crop rotation, 25, 100, 104, 165, 166, 215
cross-scale resilience, 115
cultural services, 66–67
  valuation of, 76–78
cyclones, 137, 140, 151, 298

dairy farming, 328
  change in production pathway, 28–30
  slurry disposal, 99
Darwin, Charles, 39, 43
DDT, 144
decision-making costs, 255
deforestation, 65, 142
Defra, 335
density compensation, 114
dieldrin, 144
direct marketing, 32, 33
disease, 25, 364
  bee populations, 109
  biodiversity and resistance to, 40, 41, 48, 61
  chemical control of, 39, 143, 145
  climate change and, 146, 167, 316
  competition from, 211
  effects and coping mechanisms, 222–223
  genetic diversification and control of, 50–51, 81
  global spread of, 146
  marker-assisted selection and resistance to, 166–167, 174, 175
  meat saftey regulation and, 24
  organic production and prevention of, 30
  spread of, 16, 23
  wild and domesticated herbivores, 181
disturbances, 5
  anticipated, 279, 281, 288, 292
  assessment of the impact of, 289–291
  crisis management, 285
  viability-based measure of, 291
diversification, 369
Dow Chemical Company, 351, 352
drought, 137, 346
  biotechnological solutions for, 168–171
  land-sharing adaptations to, 246
  resilience of rangeland to, 279, 281–284
DroughtGard corn, 169
drought-resistant species, 33
DSS-Ecopay, 269
dynamic programming models, 90, 98, 103

Dynamic Stochastic General Equilibrium (DSGE) model, 299, 302

E. coli 157:H7, 24
ecological models. *See* ecological–economic models
ecological resilience, 3–4, 61, 67, 106, 201, 245, 274, 371
ecological–economic modelling, 254, 256–258, 372
  case study of butterfly abundance, 264–267
  development of decision support tools, 269
  ecological models, 254, 258–261, 269
  economic models, 254, 261–263, 269
  optimisation methods, 254, 265
econometric models, 80, 261, 262, 263
economic metrics, 349, 355, 368, 376
economic resilience, 127
economic theory, 298, 300, 311
economic valuation, 72
economics
  ceteris paribus assumptions, 80, 298, 301
  complex adaptive systems and, 297
  complexity and resilience in, 305–306
  equilibrium, 295, 297, 298, 299–300, 301–302, 304
  macroeconomics, 298–302
  microeconomics, 298, 301
  natural resource management and resilience thinking, 306–307, 308–311, 371–372
  reductionism and aggregation, 301–302
  resilience thinking and, 295–296
  risk, stochasticity and uncertainty, 302–304
ecosystem services, 18, 212–214, 345
  adaptations for future agricultural resilience, 81–83
  agriculture and, 60–62
  agri-environment schemes and, 253–254, 267, 268–269
  cultural services, 66–67
  financial schemes and subsidies, 83–84
  from natural capital, 62–64, 362
  management of, 71–72
  policy and, 324–325, 335
  regulating services, 65–66
  supporting services, 64–65
  valuation of, 79–81, 216, 367–369, 378
efficiency. *See also* agricultural intensification; producers
  biodiversity, fertilisers and nutrient use, 54, 345
  gains of integrated research, 270
  heterogeneity and production, 34–35
  in economics, 306–307
  input use by farmers, 104, 160
  Jevons paradox and, 347
  resilience and, 26–27, 219, 369–375
  small-scale farmers and, 136
  sustainability and, 218
Efficient Markets Hypothesis, 303

Elton, Charles, 40
endosulfan, 144
endrin, 144
engineering approach to farm management, 211, 218
engineering resilience, 2
Environmental, Social and Governance (ESG) integration, 351
equilibrium. *See also* economics
  in ecological systems, 2–3
  in resilient farming systems, 219
  measuring resilience and, 273–275
Ernst & Young, 20
European Agricultural Fund for Rural Development (EAFRD), 318, 334
European Council, 330
European Court of Auditors (ECA), 324
European Parliament, 326
  Committee on Agriculture, 332
European Union (EU), 316, *See also* Common Agricultural Policy
  environmental policy, 212
  farm holdings and supply chains, 19–21
  integration of environmental management, 315–316
  pesticide regulation, 168
  regulations, 255
eutrophication, 94, 273, 291
evapotranspiration (ET), 349
extreme climate events, 28, 103, 140, 315, 346

faba bean, 52
Fama, Eugene, 303
farm-level resilience, 223–225
Farmer First, 152
farming systems
  comparison of general resilience between, 243–246
  comparison of specified resilience between, 246–247
  diversified, 215, 345
  ecosystem approach, 212–214
  environmental impacts, 211–212
  European, 161–162
    biotechnology and, 173–175, 176–177
  feedback, 3, 15, 16, 21, 27–28, 224, 233, 248, 285, 296, 343
  non-linear, 16–17
Ferme de l'Or Blanc cheesemakers (Cheniers), 31, 32
fertilisers, 81, 101, 136, 147, 211, 254
  benefits of, 268, 345, 346, 355, 365
  conversion of, 13
  costs, 22, 30, 265
  discontinued use of, 30
  effects of, 25, 53, 64, 79, 136, 142–143
  greenhouse gas emissions from, 65
  intensive use of, 150
  reduced need for, 151
  reduced use of, 268
  regulation of, 152
  sustainable manufacture, 344

Field to Market, 344, 352
financial capital, 18, 31, 36, 82, 216, 233, 243, 246–247, 248, 374, 375
flower strips, 33, 119, 125, 268
Food and Agriculture Organisation, 147
food chains, 15, 20, 28, 161, 222
food labels, 352–355
food system, multi-layered, 19, 34–36
  connectivity effects, 22–23
  food supply chain, 19–21
  regionalisation, 35
  scale effects, 21–22
  standards and controls, 24
foodwebs, effects of biodiversity, 48–49
Foot and Mouth Disease, 23, 222–223
Forest Stewardship Council, 354
fossil fuels, 136
fruit production, 29, 30
FSSIM model, 101
functional redundancy vs. functional diversity, 17
fungal diseases, 48, 51

Gabra pastoralists, 184
General Mills, 344
general resilience, 232–234, 248, 374
  effect of land-sharing and land-sparing, 243–246
genetic diversity, 43, 145–146, 232–234, 343
  control of crop disease, 50–51
genetically modified crops, 162–163, 370
  disease resistance, 50–51
  herbicide tolerance and weed management, 164–165
  insect resistance, 165–168
Global Financial Crisis (GFC, 2008), 299
glyphosate, 164
GM crops. *See* genetically modified crops
Good Agricultural and Environmental Conditions (GAECs), 320
good agricultural practice, 320, *See also* best management practices
  reference level for, 214, 368
Government & Accountability Institute, 351
grassland biodiversity
  effects of grazing and haying, 53
  foodweb effects, 48–49
  invasive species and, 47
  productivity and, 43–47
  stability and, 41–43
grazing practices, 235–236, 291
  change from feed to pasture-based, 30, 31
  effects on biodiversity, 53
  intensive, 30
  Lachlan river area (Australia), 237
  land-sharing approach, 239, 244, 246, 248, 373–374
  land-sparing approach, 237–239, 241, 244, 245, 247–248, 373–374
  low intensity, 322
  pastoralist, 181, 195–200, 365–366
  reduced intensity, 212

grazing practices (cont.)
    scrubland system, 3
grazing pressure, 16, 275, 277–284
greenhouse gases (GHGs), 54, 95, 135, 287, 346, 348, 351, 367
    agricultural generation of, 7, 65, 136, 142
    mitigation of, 54, 65, 96, 150, 213–214, 326, 368
greenwashing, 354, 355
gross margin calculations, 261, 263

habitat loss, 253
Happy Cow Creamery (South Carolina), 28–30, 31, 34
Hawaii, management of ecosystem services, 71–72
haying effects on biodiversity, 53
hedgerow management, 67, 212
herbicides, 136,
    GM crop tolerance, 162, 164–165, 174
herbivores, 15, 41, 48, 50, 181
heterogeneity. *See also* complex adaptive systems; producers
    modelling and spatial heterogeneity, 259
    of land-sharing landscapes, 236
    role in self-organisation and resilience, 19, 34, 363
heterogeneous cropping systems, 364
High Nature Value areas, 226, 322
    maintenance of, 212, 321, 325
holistic management. *See* land-sharing
human capital, 6, 18, 19, 29, 31, 33, 35, 61, 82, 216, 233, 376
    substitute for natural capital, 94
Hurricane Irma, 22

insecticides. *See* pesticides
insects. *See also* bees
    diversity, 48
    hedgerow management and, 67
    herbivorous, 41, 48
    life stages, 259
    parasitoid, 39, 41
    predatory, 39, 41
integer programming models, 90, 98, 99, 264
integrated catchment management (ICM), 308
integrated pest management (IPM), 30, 151, 152–153, 370
    genetically modifed crops and, 167–168
intensive, larger-scale crop farming, 161, 175
intercropping, 52–53, 54, 100, 167
International Assessment of Agricultural Knowledge, Science and Technology (AKST) for Development (IAASTD, 2009), 146, 152
International Standards Organisation (ISO), 354
invasive species, 41, 50

biodiversity and, 47–48
Investment Framework for Environmental Resources (INFFER), 308–311
irrigation, 65, 81, 98–99, 142, 148, 344, 345, 347, 350

Jones, James Holland, 305

Keynesian macroeconomics, 302
Knightian uncertainty, 302–303

Lachlan river area (Australia), 237
    environmental setting, 241
    protection initiatives and stewardship schemes, 242
    socioeconomic setting, 241–242
lake management, 94
land degradation, 7, 137, 142–143, 307
Land Matrix, 149
land-sharing, 232, 239, 247–249, 373–374
    effect on general resilience, 243–246
    effects on specified resilience, 246–247
    social difficulties, 239, 242
    vs. land-sparing, 234–235
land-sparing, 232, 237–239, 247–249, 373–374, 375
    effect on general resilience, 243–246
    effects on specified resilience, 246–247
    vs. land-sharing, 234–235
land tenure security, 148–150
landslides, 171
language competition, 286, 291
Latin America, management of ecosystem services, 69–70
LEADER programme, 323, 331, 335
LEAF certification scheme, 19
Lepidoptera, 67
less-favoured areas
    extensive farming in, 161, 174
    financial support for, 202, 320, 322, 335
lindane, 144
linear programming models, 90, 98, 261–262, 263, 264
livestock. *See also* grazing practices
    as status symbol, 7, 195–198, 199–200
    disease, 23, 24, 222–223
    emissions from, 7, 65
    farm models, 95, 103–104
local and regional food, 35
Lotka–Volterra competition model, 43
Lucas Critique, 300, 303
Lucas, Robert, 300
Lupinus perennis, 46

Maasai household, 191
Maasai Mara National Reserve (MMNR), 183, 192–195
maize, 171
    effects of climate change, 140, 141
    genetically modified, 162, 166, 168
    intercropping, 52

yield improvements, 150
manufactured capital, 6, 18, 29, 31, 33, 61, 216
manure, 25, 147
Marine Stewardship Council's certification, 353
marker-assisted selection (MAS), 146, 163–164, 166–167, 172, 173, 174, 175, 176
　for drought tolerance, 169–170
　for increased yield potential, 172
meat safety regulation, 24
mechanisation, 26
mechanistic models, 258–261
methane, 65, 95, 136
　from rice production, 136
milk production, 27, 29–30, 32, 103, 222
milk quotas, 319
Millennium Ecosystem Assessment (2003), 60, 80
millet, 141, 150
mineral fertiliser, 101, 142
Ministry of Food and Agriculture, 144
Minnesota biodiversity experiment, 41–43, 44, 46
　effect of invasive species, 47
　effects on soil fertility and carbon sequestration, 48
　foodweb effect, 48–49
Minsky moment, 301
mixed farming, 161, 174–175, 215, 364
Monsanto, 164, 169, 352
Monte Carlo analysis, 103
Monterey Bay Aquarium's Seafood Watch, 353
Multiannual Financial Framework, 333
multifunctionality. *See also* producers
　in agricultural landscapes, 82, 369
　land-sharing and, 234
　value of of agriculture and, 378
multiple-objective programming models, 90
multi-species coexistence, 40–41
mycorrhizal fungi, 48

National Action Plan for Salinity and Water Quality (Australia), 308
National Farmers Union (NFU), 327
National Landcare Program, 307
Natura 2000 areas, 335
natural capital, 18, 25, 31, 33, 35, 60–62, 68, 74, 104, 233, 365, 367–368
　agricultural impacts on, 216–217, 367
　decline in, 22, 80, 247, 248, 337
　economic stabilisation measures and, 371
　ecosystem services from and agriculture, 62–64, 362, 367–369, 378
　increased use of, 29
　protection of, 216–217, 224, 228, 236, 243, 248–249, 374, 376
　role of, 64–67
　substitutes for, 82, 95, 370
　weak and strong sustainability and, 6

Natural Capital Project, 351
Natural Capital Protocol (2016), 68
Natural Heritage Trust (Australia), 308
natural resource management
　economics and resilience thinking, 306–307, 371–372
　in Australia, 307–308
　integration of economics and resilience thinking, 308–311
Nature Conservancy, The, 351
neonicotinoid chemicals, 145
Nitrate Vulnerable Zone regulations, 99, 214
Nitrates Directive, 320
nitrogen, 48, 50, 52, 54, 61
　loss of, 90, 95, 141, 143, 214
　regimes, 344
　storage, 91
nitrogen cycle, 93, 136, 370
nitrogen fertilisers, 65, 95, 136, 142, 168
nitrous oxide, 65, 95, 136, 142
Northern Rangeland Trust (Kenya), 199
no-till farming, 25, 65

Ocean Tomo Intangible Asset Market Value study, 352
optimisation
　ecosytem approach and, 211
　in bio-economic modelling, 97–103
　in ecological–economic modelling, 256
　methods of, 264
　path dependency and, 307
　shocks and, 217
organic farming, 19, 161, 321, 334, 335,
　benefits to pollination, 119, 121
　conversion to, 29–30, 150
　economic benefits of, 151–152
　global land area use, 352
　plant breeding, 174–175, 176
organic produce, 29–30
　labelling of, 352–353
　USDA certification, 354
outcome metrics, 346–347, 350, 376
overgrazing, 212, 245, 345
overyielding, 43–45, 52

parasitoids, 50
　arthropods, 48
　insects, 39, 41
Park Grass Experiment (Rothamsted, UK), 53
pastoralism and conservation, 180–184, 198–202, 365–366, 369
pastoralist household, 190–191, 198–202
　ecological and economic approaches to analysing drivers and outcomes of decisions, 192–198
　factors shaping decisions, 188–190
pastoralist social–ecological systems, 202
　resilience thinking and, 182–183, 184–188
pasture
　agri-environment schemes, 325
　benefits of biodiversity, 53
　conventional management of, 246

pasture (cont.)
    conversion to native forest, 71
    holistic management of, 239, 246
    invasion by exotic species, 48
    plant diversity, 39, 43, 54
    productivity, 238
    semi-natural, 268
payments for environmental services (PES), 185, 187, 195, 199
payments for wildlife conservation (PWC), 185, 187, 195, 199
perennial crops, 65, 100, 149
persistence, 216-217
pest control, 61, See also pesticides
    biodiversity and, 39, 54, 61, 215
    biological, 148, 150
    genetically modified crops and, 165-168, 175
    habitat management and, 65-66
    integrated pest management (IPM), 30, 151, 152-153, 167-168, 370
    natural, 79, 81-82
    valuation of, 76
    wildflower plantings and, 33, 268
pesticides, 25, 76, 79, 136, 150, 167, 172, 211, 260, 268, 345, 365
    detrimental effects of, 143-145
    effects on pollinators, 110, 117
    expenditure, 151,
    genetically modified crops and reduced use of, 166-168, 175
    move from intensive use of, 151
    regulation of, 152, 168
    removal of residues from water, 74
    resistance to, 25
    toxicity, 98
pests, 25, 364
    competition from, 211
    crop species, 141, 146
    livestock species, 137
    novel, 340
Petalostemon purpureum, 46
phosphorus, 16, 52, 274, 275
photosynthesis, 171-172
plant breeding, 93, 94, 145-146, See also biotechnologies for plant breeding
    Participatory Plant Breeding (PPB), 170, 176
pollinators/pollination, 39, 61, 65-66, 67, 74, 109-110, 127-128, 364, 365, 366
    diversity, buffering and resilience, 112-117
    habitat enhancement, 33, 66, 79, 81, 117-120, 268
        case study, 121-127
    loss of, 245, 345
    threats to wild bee populations, 117-119
    valuation of, 76
    vulnerability of, 110-112
Porcine Epidemic Diarrhoea Virus, 23
Positive Mathematical Programming (PMP), 98
potato production systems, 97-99
precision agriculture, 160, 175, 176, 346, 350

predators, 15, 50
    arthropod, 48
    density, 260
    habitat enhancement, 79
    insect, 39, 41
primary producers, 15
primary productivity, 40, 41, 44, 50
producers, 34-36
    adaptive capability, 33-34, 366-367
    autonomy, 36
    balancing resilience and efficiency, 26-27
    buffering capability, 33, 377
    connectivity effects, 22-23
    development of more resilient pathways, 28-32
    factors effecting resilience of, 14
    management of uncertainty, 25-26
    multifunctionality, 368-369
    production pathways, 27-28
    role in the food supply chain, 19-21
    scale effects, 21-22
    standards and controls, 24
    transformative capability, 34
    within a complex adaptive system, 19
production costs, 8, 255, 261, 268, 367
provisioning services, 61, 74-75
Prunus dulcis, 110
public goods, 8, 64, 76, 212, 326, 328-330, 331, 367, 368
    provision of private vs., 79
    valuation of, 71-72
puma, 351
pyrethroid insecticides, 144

quantitative trait loci (QTLs), 169-170

Rachel's Dairy (Wales), 28-29
rainfall, 16, 90, 103, 137, 141, 143, 169, 246, 273, 291, 365
rangeland
    cost function in management, 279
    impact of conservation interventions, 185-190, 198-202
    pastoralist/wildlife synergies, 181-184
    resilience evaluation, 282
    resilience to drought, 279, 281-284
rational choice theory, 296
Red Tractor mark, 19
reductionism, 301-302
redundancy. See also functional redundancy
    in farm resilience, 228
    in holistic management/land-sharing, 243, 244, 248
    optimal level of, 138
    principles of resilience, 14, 217-218
    resilience of ecological systems and, 363
reference state, 2-3
remote sensing, 347-350
    data, 346
resilience. See also adaptive capability; buffer capability; complex adaptive systems; ecological resilience; engineering

resilience; producers; social–ecological
resilience; transformative resilience
  definitions, 2–5, 135, 159, 273–277, 297
  general vs. specified, 232–234
  importance of, 5–7
  in agriculture, 7–9
  principles for building, 14–15, 217
response diversity, 113–115
rewilding, 225
rice production
  effects of climate change, 140, 141
  genetic diversification and disease resistance, 51
  loss of diversity, 145
  marker-assisted selection, 172
  methane emissions, 7, 65, 136
  System of Rice Intensification (SRI), 151, 153
  women's labour in, 147
riparian buffers, 64, 71, 238, 346
riparian fencing, 238, 242
RISE Foundation, 327
risk modelling, 105–106
Royal Society, 216
Rural Development Programmes, 317, 323, 329, 333–335
Rural Development Regulations (RDR), 323
rust diseases, 50, 146

savannah, 273, 368
  biodiversity, 202
SCaMP (Sustainable Catchment Management Programme), 84
Schelling, Thomas, *Micromotives and Macrobehaviour* (1978), 94
scrubland system, 3
seed banks, 50
seed industry, 173
selection effect, 114
Shiller, Robert, 303
small-scale farmers, 135–137, 161
  effects of agrobiodiversity loss, 145–146
  effects of land, soil and water degradation, 142–143
  exclusion from the research and extension agenda, 146–148
  involvement with plant breeding technologies, 173–174
  land tenure insecurity, 148–150
  productivity and sustainable agriculture, 150–152
  risk perception, 137–140
  support and incentives for, 35, 152–154
  threat from agrochemicals, 143–145
  vulnerability to climate variation and change, 140–141
social capital, 18, 19, 31, 33, 35, 216, 224, 228, 233, 247, 299, 376
  livestock ownership and, 189
social metrics, 349, 376
social response capacity, 33

social–ecological resilience, 92, 104, 232–234, 235, 248, 249
social–ecological systems
  agricultural resilience and, 159, 218
  analysis of, 305
  building resilience in, 15
  effect of fast and slow variables on, 16
  pastoralist, 182
soil acidity, 308
soil biodiversity, 25, 75, 151, 345
soil degradation, 142–143, 365
soil erosion, 6, 22, 30, 70, 98, 171, 212, 308, 315, 316, 349, 367, 371
  rainfall intensity and, 141
  terracing to reduce, 148, 150
soil fertility, 21, 50, 64, 75, 171, 215, 261, 365
  biodiversity and, 48, 53, 54
  loss of, 69
  management of, 79, 81, 150, 160, 173
  water scarcity and, 169
soil formation, 6, 61, 62
soil management, 25, 33, 100, 246, 320, 344
  organic, 148
soil organic carbon, 48, 75, 94, 96, 101, 102, 104, 347, 349
  management of, 76
soil organic matter, 14, 75, 93–94, 316
soil productivity, 64, 256
soil protection measures, 143
soil, value of, 75
Solow, Robert, 299
Sonnenschein–Mantel–Debreu (SMD) Theorem (Anything Goes Theorem), 301–302
sorghum, 141, 150
species diversity, 254, 364
  ecosystem stability and, 40, 43
  effect on carbon storage, 48
  effects of land-sharing and land-sparing, 243
  loss of, 253
specified resilience, 232–234
  effects of land-sharing and land-sparing, 246–247
spiritual connection to nature, 61, 66, 76–77
statistical models, 258–261
Statutory Management Requirements (SMRs), 320
Stern Report (2006), 326
stochastic programming models, 98, 100, 102–103
Stockholm group, 326
Stockholm Resilience Centre, 217
straw burning, 93
sustainability, 27, 82
  at a farm level, 218–223
  biodiversity and, 39, 50
  capacity for at farm scale, 225–226
  challenges in assessing, 343
  corporate involvement in, 351–352
  economic, 68
  food labelling and certification, 352–355

sustainability (cont.)
  governance arrangements, 226–227
  indicators, 102
  interpretations of, 5–7, 287
  measurement. *See* sustainability metrics
  of agricultural systems, 216–218
  pollination and, 113, 117, 119
  productivity and, 136, 150–152
  research and technology for, 152–153
  vs. optimisation issues, 96
Sustainability Consortium, The (TSC), 344, 352
sustainability metrics, 344, 346, 376
  remote sensing and, 347–350
  success of, 355–356
Sustainable Agriculture Network and Rainforest Alliance, 344
Sustainable Catchment Management Programme (SCaMP), 84
sustainable development, 5, 287
sustainable intensification, 79, 159, 211, 214–216, 224–225, 226, 343
  approaches for, 160–161
  strategies, 346
sustainable livelihoods framework, 188

Tanzania Land Act (1999), 149
Tanzanian Wildlife Management Areas (WMAs), 192, 199, 201
technological developments, 7, 13, 153
temperatures, effect of increasing, 103, 140, 141
TerraChoice, 351
terracing, 81, 100, 143, 148, 150
tillage
  changes in (England/Wales), 220–221
  conservation, 213–214, 346, 368
  low/reduced regimes, 25, 70, 71, 143, 164, 171
  no-till regimes, 164, 171
Timeless Seeds (Montana), 30, 31
tipping points, 96, 102, 103, 106
Tolerable Windows approach, 287
Tonnemaker Hill farm (Washington), 29, 31, 32
tourism
  agri-tourism, 77
  declines in, 192
  environmental quality and, 292
  farm diversification, 106
  impact of disease on, 23, 222
  impact of farming systems on, 98
  income from, 181, 199
trade-offs (in land management), 78–79
  bio-economic modelling and, 92
  in pastoralist decision-making, 188–190
  InVEST and, 72
  land-sharing/land-sparing, 234
  loss of ecological resilience and, 104
  valuation and, 67

transformative capability, 33, 34, 159, 177, 218, 375

UK National Ecosystem Assessment, 80
Ulanowicz, R.E., 27, 377
UN Special Rapporteur on the Right to Food, 152
uncertainty, 372–373
Unilever, 344, 351, 352
United Nations Environment Program (UNEP), 345
unmanned aerial vehicles (UAVs or drones), 350

valuation of public goods, 71–72
value, estimation of, 72–74, *See also* ecosystem services
value-added processing, 33
viability analysis, 259, 277, 291, 292, 368, 373
viability theory, 276
  aim and scope, 284
  available numerical tools, 286
  crisis management, 285–286
  exit time, 286
  implementation to the measure of resilience, 288–291
  intertemporal optimality, 285
  main concepts, 284–285
  scope and application to measure of resilience, 291
  survey of applications, 286–287
  viability kernel, 280, 286, 287, 288, 289–290, 291
viability-based measure of resilience, 274–277, 279–284, 287–288
  implementation in the measure of mathematical viability theory framework, 288–291
  in models of socio-ecological systems, 291–292
  scope and applications, 291
Village Land Act (1999), 149

Walmart, 343
Walmart Foundation, 352
water availability, 25, 65, 75, 81, 168
water consumption, 347, 348
  measurement of, 349
water cycle, 14, 61
water degradation, 142–143, 365
Water Framework Directive, 333
water pollution, 7, 54, 308, 322, 355
water quality, 64–65, 69, 75, 76, 79, 346, 347, 348, 351
  biodiversity and, 53
  carbon storage and, 72
  enhanced pollinator habitat and, 66
  improvement schemes, 84, 350
  measurement of, 349–350
  water funds and, 68, 70